Lecture Notes in Computer Science 15459

Founding Editors

Gerhard Goos
Juris Hartmanis

Editorial Board Members

Elisa Bertino, *Purdue University, West Lafayette, IN, USA*
Wen Gao, *Peking University, Beijing, China*
Bernhard Steffen ⓘ, *TU Dortmund University, Dortmund, Germany*
Moti Yung ⓘ, *Columbia University, New York, NY, USA*

The series Lecture Notes in Computer Science (LNCS), including its subseries Lecture Notes in Artificial Intelligence (LNAI) and Lecture Notes in Bioinformatics (LNBI), has established itself as a medium for the publication of new developments in computer science and information technology research, teaching, and education.

LNCS enjoys close cooperation with the computer science R & D community, the series counts many renowned academics among its volume editors and paper authors, and collaborates with prestigious societies. Its mission is to serve this international community by providing an invaluable service, mainly focused on the publication of conference and workshop proceedings and postproceedings. LNCS commenced publication in 1973.

Sanju Tiwari · Boris Villazón-Terrazas ·
Fernando Ortiz-Rodríguez · Soror Sahri
Editors

Knowledge Graphs and Semantic Web

6th International Conference, KGSWC 2024
Paris, France, December 11–13, 2024
Proceedings

Editors
Sanju Tiwari
Sharda University
Greater Noida, Uttar Pradesh, India

Fernando Ortiz-Rodríguez
Universidad Autónoma de Tamaulipas
Ciudad Victoria, Tamaulipas, Mexico

Boris Villazón-Terrazas
Universidad Autónoma de Tamaulipas
Ciudad Victoria, Tamaulipas, Mexico

Soror Sahri
Université Paris Cité
Paris, France

ISSN 0302-9743 ISSN 1611-3349 (electronic)
Lecture Notes in Computer Science
ISBN 978-3-031-81220-0 ISBN 978-3-031-81221-7 (eBook)
https://doi.org/10.1007/978-3-031-81221-7

© The Editor(s) (if applicable) and The Author(s), under exclusive license
to Springer Nature Switzerland AG 2025

This work is subject to copyright. All rights are solely and exclusively licensed by the Publisher, whether the whole or part of the material is concerned, specifically the rights of translation, reprinting, reuse of illustrations, recitation, broadcasting, reproduction on microfilms or in any other physical way, and transmission or information storage and retrieval, electronic adaptation, computer software, or by similar or dissimilar methodology now known or hereafter developed.
The use of general descriptive names, registered names, trademarks, service marks, etc. in this publication does not imply, even in the absence of a specific statement, that such names are exempt from the relevant protective laws and regulations and therefore free for general use.
The publisher, the authors and the editors are safe to assume that the advice and information in this book are believed to be true and accurate at the date of publication. Neither the publisher nor the authors or the editors give a warranty, expressed or implied, with respect to the material contained herein or for any errors or omissions that may have been made. The publisher remains neutral with regard to jurisdictional claims in published maps and institutional affiliations.

This Springer imprint is published by the registered company Springer Nature Switzerland AG
The registered company address is: Gewerbestrasse 11, 6330 Cham, Switzerland

If disposing of this product, please recycle the paper.

Preface

This volume contains the main proceedings of the Sixth Knowledge Graph and Semantic Web Conference (KGSWC 2024), held during December 11-13, 2024, at Université Paris Cité in Paris, France. KGSWC is established as a yearly venue for discussing the latest scientific results and technology innovations related to Knowledge Graphs and the Semantic Web. At KGSWC, international scientists, industry specialists, and practitioners meet to discuss knowledge representation, natural language processing/text mining, and machine/deep learning and large language model research. The conference's goals are (a) to provide a forum for the AI community, bringing together researchers and practitioners in the industry to share ideas about innovative projects, and (b) to increase the adoption of AI technologies in these regions.

KGSWC 2024 followed on from successful past events in 2019, 2020, 2021, 2022, and 2023. It was also a venue for broadening the focus of the Semantic Web community to span other relevant research areas in which semantics and web technology play an important role and for experimenting with innovative practices and topics that deliver extra value to the community.

The main scientific program of the conference comprised 23 full research papers, selected out of 58 reviewed submissions, which corresponds to an acceptance rate of 40%. The program was completed with 2 workshops sessions, and a Winter School where researchers could present their latest results and advances and learn from experts. The program also included five high-profile experts as invited keynote speakers: **Nathalie Aussenac-Gilles**, IRIT (CNRS, Université de Toulouse), France; **Sven Groppe**, University of Lübeck, Germany; **Fabien Gandon**, Université Côte d'Azur, Inria, CNRS, I3S, France; **Philipp Cimiano**, Bielefeld University, Germany; and **Pascal Hitzler**, Kansas State University, Manhattan, Kansas, USA), with novel Semantic Web topics. Two industry sessions from Ultipa, USA and Building Digital Twin Association, Belgium were also organized.

The General and Program Committee chairs would like to thank the many people involved in making KGSWC 2024 a success. First, our thanks go to the four co-chairs of the main event and the more than 60 reviewers for ensuring a rigorous double-blind review process that led to an excellent scientific program, with an average of three reviews per article.

Further, we thank the kind support of all people from Université Paris Cité, Paris, France. We are thankful for the kind support of all the staff of Springer. We finally thank our sponsors and our community for their vital support of this edition of KGSWC.

The editors would like to close the preface with warm thanks to our supporting keynotes, the program committee for rigorous commitment in carrying out reviews, and last but not least, our enthusiastic authors who made this event truly International.

December 2024

Sanju Tiwari
Boris Villazón-Terrazas
Fernando Ortiz-Rodríguez
Soror Sahri

Organization

Chairs

Sanju Tiwari	Sharda University, India & TIB Hannover, Germany
Boris Villazón-Terrazas	Universidad Autónoma de Tamaulipas, Mexico
Fernando Ortiz-Rodríguez	Universidad Autónoma de Tamaulipas, Mexico
Soror Sahri	Université Paris Cité, France

Local Chairs

Sonia Guehis	Paris Nanterre University, France
Ioana Ileana	Université Paris Cité, France
Mourad Ouziri	Université Paris Cité, France

Workshop

Shishir Shandilya	VIT Bhopal University, India
Shikha Mehta	JIIT Noida, India

Sponsor

Fatiha Saïs	Université Paris-Saclay, France

Publicity

Riad Mokadem	IRIT, Paul Sabatier University, France
Ronak Panchal	Cognizant, India

Winter School

Sanju Tiwari   Sharda University, India & TIB Hannover, Germany
Fernando Ortiz-Rodríguez   Universidad Autónoma de Tamaulipas, Mexico

Program Committee Chairs

Sanju Tiwari   Sharda University, India TIB Hannover, Germany
Fernando Ortiz-Rodríguez   Universidad Autónoma de Tamaulipas, Mexico
Boris Villazón-Terrazas   Universidad Autónoma de Tamaulipas, Mexico

Program Committee

Adila Krisnadhi   Universitas Indonesia, Indonesia
Alejandro Rodríguez   Universidad Politécnica de Madrid, Spain
Andrea A. Alvarez   Universidad Politécnica de Madrid, Spain
Anisa Rula   University of Brescia, Italy
Antonella Carbonaro   University of Bologna, Italy
Adolfo Anton-Bravo   MPVD.es, Spain
Amed Abel Leiva Mederos   Universidad Central de las Villas, Cuba
Boris Villazón-Terrazas   Universidad Autónoma de Tamaulipas, Mexico
Carlos Bobed   everis/NTT Data - University of Zaragoza, Spain
Carlos F. Enguix   Universidad Autónoma de Tamaulipas, Mexico
Cogan Shimizu   Kansas State University, USA
Daniil Dobriy   Vienna University of Economics and Business, Austria
Dimitris Kontokostas   Medidata, Greece
Disha Purohit   TIB Hannover, Germany
Edelweis Rohrer   Universidad de la República, Uruguay
Edgard Marx   Leipzig University of Applied Sciences (HTWK), Germany
Erick Antezana   Norwegian University of Science and Tech, Norway
Eric Pardede   La Trobe University, Australia
Fatiha Saïs   Université Paris-Saclay, France
Fatima N. Al-Aswadi   Universiti Sains Malaysia, Malaysia; Hodeidah University, Yeman
Fatima Zahra Amara   University of Khenchela, Algeria
Federica Rollo   University of Modena and Reggio Emilia. Italy

Fernando Bobillo	University of Zaragoza, Spain
Fernando Ortiz-Rodríguez	Universidad Autónoma de Tamaulipas, Mexico
Ghislain Atemezing	European Union Agency for Railways (ERA), France
Gerardo Haces-Atondo	Universidad Autónoma de Tamaulipas, Mexico
Hassan Hussein	TIB, Hannover Germany
Hong Yung Yip	University of South Carolina, USA
Hussam Ghanem	Université de Bourgogne, France
Ioana IIeana	Université Paris Cité, France
Janneth Alexandra Chicaiza Espinosa	UTP de Loja, Ecuador
Jennifer D'Souza	TIB Hannover, Germany
Jose L. Martínez-Rodríguez	Universidad Autónoma de Tamaulipas, Mexico
Jose Melchor Medina-Quintero	Universidad Autónoma de Tamaulipas, Mexico
Jose Emilio Labra Gayo	Universidad de Oviedo, Spain
Marlene Goncalves	Universidad Simón Bolívar
Md Kamruzzaman Sarker	Bowie State University, USA
Meriem Djezzar	Khenchela University, Algeria
Miguel-Angel Sicilia	University of Alcalá, Spain
Mounir Hemam	Khenchela University, Algeria
Mourad Ouziri	Université Paris Cité, France
Nandana Mihindukulasooriya	IBM Research, USA
Pallavi Karanth karanth	TIB Hannover, Germany
Panos Alexopoulos	Textkernel B.V., The Netherlands
Pascal Hitzler	Kansas State University, USA
Patience U. Usip	University of Uyo, Nigeria
Ronak Panchal	Cognizant, India
Ronald O. Ojino	Co-operative University of Kenya, Kenya
Rosa M. Rodriguez	Universidad de Jaén, Spain
Sanju Tiwari	Sharda University, India & TIB Hannover, Germany
Shishir Shandilya	VIT Bhopal University, India
Shikha Mehta	JIIT Noida, India
Sonia Guehis	University Paris Nanterre, Paris, France
Soror Sahri	Université Paris Cité, Paris, France
Sven Groppe	University of Lübeck, Germany
Takanori Ugai	Fujitsu, Japan
Tek Raj Chhetri	MIT, USA
Tommaso Soru	Serendipity AI Ltd., UK
Valentina Janev	Institute Mihajlo Pupin, Serbia
Yusniel Hidalgo Delgado	Universidad de las Ciencias Informáticas, Cuba

Contents

ConfermentSampo – A Knowledge Graph, Data Service, and Semantic
Portal for Intangible Academic Cultural Heritage 1643–2023 in Finland 1
  *Eero Hyvönen, Patrik Boman, Heikki Rantala, Annastiina Ahola,
  and Petri Leskinen*

Towards Complex Ontology Alignment Using Large Language Models 17
  Reihaneh Amini, Sanaz Saki Norouzi, Pascal Hitzler, and Reza Amini

Enhancing Knowledge Graph Construction: Evaluating with Emphasis
on Hallucination, Omission, and Graph Similarity Metrics 32
  Hussam Ghanem and Christophe Cruz

Views, Semantic Data Catalog and Role-Based Access Control
for Ontologies and Knowledge Graphs 47
  Henrik Dibowski

Accelerating Medical Knowledge Discovery Through Automated
Knowledge Graph Generation and Enrichment 62
  *Mutahira Khalid, Raihana Rahman, Asim Abbas, Sushama Kumari,
  Iram Wajahat, and Syed Ahmad Chan Bukhari*

OntoSeer - A Recommendation System to Improve the Quality
of Ontologies 78
  Pramit Bhattacharyya, Samarth Chauhan, and Raghava Mutharaju

Adaptive Planning on the Web: Using LLMs and Affordances for Web
Agents 93
  Sebastian Schmid, Michael Freund, and Andreas Harth

Enriching RDF Data with LLM Based Named Entity Recognition
and Linking on Embedded Natural Language Annotations 109
  *Michael Freund, Rene Dorsch, Sebastian Schmid, Thomas Wehr,
  and Andreas Harth*

TEDME-KG Metrics Framework: A Metrics Framework for TEmporal
Data Modelling Evaluation in Knowledge Graphs 123
  *Sepideh Hooshafza, Beyza Yaman, Alex Randles, Mark Little,
  and Gaye Stephens*

SFARDE: A Knowledge-Centric Semantic Strategic Framework
for Heritage Artifact Recommendation Integrating Generative AI
and Differential Enrichment of Ontologies 139
   Archit Chadalawada and Gerard Deepak

Leveraging Graph Models for Comprehensive Visual Analytics of Equine
Heritage 153
   *Abdelkader Ouared, Noureddine Belarbi, Abdelhafid Chadli,
and Kebbal Seddik*

OWL2Vec4OA: Tailoring Knowledge Graph Embeddings for Ontology
Alignment 168
   *Sevinj Teymurova, Ernesto Jiménez-Ruiz, Tillman Weyde,
and Jiaoyan Chen*

Design of an Ontology-Driven Constraint Tester (ODCT) and Application
to SAREF and Smart Energy Appliances 183
   *Tareq Md Rabiul Hossain Chy, Henon Lamboro, Olivier Genest,
Antonio Kung, Cécile Rabrait, Dune Sebilleau, and Amélie Gyrard*

YOKO ONtO: You only KNIT One Ontology 199
   *Jorge Rodríguez-Revello, Cristóbal Barba-González, Maciej Rybinski,
and Ismael Navas-Delgado*

CoKGLM: Detecting Hallucinations Generated by Large Language
Models via Knowledge Graph Verification 212
   Rie Hasegawa and Ryutaro Ichise

AutOnto: Towards A Semi-Automated Ontology Engineering Methodology ... 225
   Kiara Marnitt Ascencion Arevalo, Shruti Ambre, and Rene Dorsch

Enhancing Question Answering Systems with Generative AI: A Study
of LLM Performance and Error Analysis 242
   Faiza Nuzhat, Kanchan Shivashankar, and Nadine Steinmetz

Disjointness Violations in Wikidata 259
   Ege Atacan Doğan and Peter F. Patel-Schneider

Enhancing WebProtégé with Version Control Systems 275
   Erhun Giray Tuncay, Nenad Krdzavac, and Felix Caspar Engel

Visual Presentation and Summarization of Linked Data Schemas 290
   *Lelde Lāce, Aiga Romāne-Ritmane, Mikus Grasmanis, Artūrs Sproģis,
Jūlija Ovčiņņikova, Uldis Bojārs, and Kārlis Čerāns*

A Proposed Ontology Evaluation Tool to Assist Ontology Engineers
in Selecting Ontologies During the Reuse Phase 306
    Lina Nachabe and Nushrat Jahan

Manufacturing Commonsense Knowledge 320
    *Muhammad Raza Naqvi, Arkopaul Sarkar, Farhad Ameri,
    Linda Elmhadhbi, and Mohamed Hedi Karray*

Construction and Canonicalization of Economic Knowledge Graphs
with LLMs 334
    *Hanieh Khorashadizadeh, Nandana Mihindukulasooriya,
    Nilufar Ranji, Morteza Ezzabady, Frédéric Ieng, Jinghua Groppe,
    Farah Benamara, and Sven Groppe*

Author Index 345

ConfermentSampo – A Knowledge Graph, Data Service, and Semantic Portal for Intangible Academic Cultural Heritage 1643–2023 in Finland

Eero Hyvönen[1,2](✉), Patrik Boman[1], Heikki Rantala[1], Annastiina Ahola[1], and Petri Leskinen[1]

[1] Semantic Computing Research Group (SeCo), Aalto University, Espoo, Finland
eero.hyvonen@aalto.fi
[2] Helsinki Centre for Digital Humanities (HELDIG), University of Helsinki, Helsinki, Finland

Abstract. This article presents a model for representing and studying academic intangible cultural heritage pertaining to conferment ceremonies organized by universities in Europe since the 1100's. A new Linked Open Data (LOD) service and semantic portal on top of it in-use called CONFERMENTSAMPO – *100 conferments of the Faculty of Philosophy at the University of Helsinki 1643–2023* is introduced. It allows data related to conferment celebrations, rituals, and academics involved in different roles to be published, stored, and researched using Semantic Web technologies. A goal of our work is to preserve and foster conferment traditions for the future generations of academics.

Keywords: Linked Data · Digital Humanities · Intangible Cultural Heritage

1 Publishing and Studying Intangible Cultural Heritage

The UNESCO Convention[1] identifies five types of intangible cultural heritage (ICH): 1) Oral traditions and expressions; 2) Performing arts; 3) Social practices, rituals, and festive events; 4) Knowledge and practices concerning nature and the universe; 5) Traditional craftsmanship. The focus of this paper is on the Social practices, rituals, and festive events pertaining to academic traditions.

University students get their master and doctoral degrees in conferment ceremonies of universities where new members are accepted into the academic community. Such ceremonies started at the first European University of Bologna in the 1100's and were soon commonly organized in Europe in the 1300's. The conferment tradition with its rituals have been preserved in exceptionally rich form in Finland, where both masters and doctors still participate in the proceedings as before. In particular, the Faculty of Philosophy at the University of Helsinki, originally the Royal Academy of Turku, has been the driving force here [7]. In

---

[1] UNESCO World Heritage Convention: https://whc.unesco.org/en/conventiontext/.

order to pass this tradition to new generations, conferment celebrations have been documented in student registers, anniversary books, photographs, on film, and related objects have been stored in collections of museums, libraries, and archives.

The Web offers new opportunities for fostering intangible cultural heritage: digital materials can be published and accessed conveniently regardless of time and place. With the help of the linked data of the Semantic Web, distributed cultural materials published by different actors can be aggregated into the same triplestore and under a single user interface, the data can be enriched with links both internally and externally, new data can be inferred with the help of artificial intelligence, and the data can be analyzed and visualized computationally [11,12] using methods of Digital Humanities [6].

This paper addresses the following research question: *How can ICH of ceremonies be published on the Web for human consumption through intelligent user interfaces and in a machine "understandable" way, so that the data can be used for Digital Humanities research analyses, too?* As a case study, we consider publishing and using data about the academic conferment ceremonies (1643–2023) of the Faculty of Philosophy at the University of Helsinki, Finland. As a solution approach, representing the data as a Knowledge Graph (KG) based on ontologies [27] and linked data [8] is proposed in order to enrich the data semantically from related linked data and other data sources. In order to test and evaluate this idea, a practical application called CONFERMENTSAMPO in use on the Web is presented. It includes a knowledge graph hosted at a Linked Open Data (LOD) service[2] and a semantic portal[3] based on a SPARQL endpoint.

In the following, related works are first discussed (Sect. 2), conferment ceremonies at the University of Helsinki are overviewed (Sect. 3), and a data model for representing them is presented (Sect. 4). After this, the methods for LOD publishing and implementing the CONFERMENTSAMPO web services are discussed in Sect. 5. Using the data service and semantic portal is then illustrated by examples (Sect. 6). In conclusion, contributions of the work are summarized and directions for further research suggested.

2 Representing Intangible Cultural Heritage Events

Linked Data for digital Intangible Cultural Heritage (ICH) presents new challenges and possibilities for memory organizations. Systems have been developed pertaining to the different categories of ICH using semantically informed conceptualizations and practices [31]. Linked data has been used for representing war history, biographical data of people [30], music [23], musical performances [1], etc., but to the best of our knowledge not for academic traditions. However, CONFERMENTSAMPO is a continuation of our earlier work at the beginning of the 2000's with the University Museum of the University of Helsinki, the current Science Museum Liekki[4], for the RDF(S)-based[5] Promoottori system [10,16],

---

[2] LOD service available at: https://ldf.fi/dataset/promootiosampo/.
[3] Portal available at: https://promootiosampo.ldf.fi.
[4] Liekki Museum: https://www.helsinki.fi/en/helsinki-university-museum-flame.
[5] RDF Schema recommendation of W3C: https://www.w3.org/TR/rdf-schema/.

whose goal was to publish and promote the conferment tradition of the University of Helsinki with the help of Semantic Web technologies. The innovation of this application was to create a formal data model, i.e., an ontology [27], describing conferment events, persons, places, and other concepts and entities. Based on the ontology, a "smart" semantic portal was created, which allowed the user to find photos, artefacts, and other tangible material related to the tradition and to browse related linked information. This application was used for several years in the museum's premises on a client terminal, but due to the copyright of the materials, it could not be published online. The system did not contain data about the different conferment events, the focus of this paper, but contained only a generic model of them and data about related tangible objects.

From a data modeling point of view, various event-based models [2], most notably the CIDOC CRM model[6] [5], have been developed for representing intangible phenomena. These include not only historical events and performance processes but also, e.g., acts of conserving and exhibiting artifacts in memory organizations.

3 Conferment Ceremonies at the University of Helsinki

In May 2023, the Faculty of Philosophy of the University of Helsinki organized its 100th anniversary conferment[7]; the first one was organized 380 years earlier in 1643. The ceremonies of the Faculty of Philosophy have become an exceptionally eventful tradition, even on a global international scale, and has been a model for the new universities founded in Finland in the 20th century.

The ceremonies last for four days. In addition, ceremonial old group dances are being practiced in advance to be presented at the conferment by participating masters and doctors separately.

1. **Floora's Day.** On May 13, the celebrations start and consist of four major rituals, such as *Inviting of the official wreath binder* and *Procession to the field of Kumtähti*, where the national anthem of Finland was first sang for the first time in 1848.
2. **Conferment Preparations.** On Thursday, last week of May, five major ritual are performed pertaining to the preparations for the actual conferment day (cf. below). For example, the spouses of the promoted masters bind during a special dinner laurel wreaths, symbols for the academic degree, to be used at the act of the conferment, and there is another event for sharpening the doctors' swords. At another dinner, the doctors are given as an insignia the permission to wear swords that symbolize the spirit for defending what is true, right, and good.
3. **Conferment Day.** Friday, last week of May, is the day for the act of conferment at the university Great Hall. During the act academic insignia are delivered, including laurel wreaths for the masters and special top hats and

---

[6] CIDOC Conceptual Reference Model (CRM): https://cidoc-crm.org.
[7] 110th jubilee conferment: https://www.helsinki.fi/fi/projektit/promootion-riemuvuosi/promootion-historiaa.

swords for the doctors. In total, the day includes more than a dozen additional smaller acts and rituals, such as various official speeches, and an academic procession to the church and a service there.

4. **Conferment Celebrations.** Saturday, last week of May, some 16 additional rituals and smaller acts are performed, such as the Conferment sailing trip and picnic to the archipelago, the great dancing ball, and finally the nocturnal procession with speeches to various statues of national heroes. The procession goes from the dancing ball to the university main building where the final act is a speech to the raising sun given typically by a new doctor in astronomy.

People in over 20 different official roles participate in these over 40 official rituals and acts of the conferment ceremonies. CONFERMENTSAMPO publishes information about the one hundred conferment ceremonies of the Faculty of Philosophy at the University of Helsinki from 1643 to 2023, with a focus on the academics who participated in them in different roles. CONFERMENTSAMPO is a new member in the series of over twenty Sampo systems in twenty years (2004–2024), which have had up to millions of users online[8] [13].

4 Creating the Knowledge Graph

This section explains the data models used in our case study and how the knowledge graph based on them was created and enriched from external data sources.

4.1 Data Models for Conferments and People

The core of CONFERMENTSAMPO data is a table of all 100 conferments of the Faculty of Philosophy of the University of Helsinki from 1643 to 2023. The University of Helsinki was originally the Royal Academy of Turku, later also the

Table 1. Conferment class: general properties. *seprs* refers to the namespace of the data.

General properties of the conferment class		
Property	Description	Label (in Finnish)
scprs:university	Name of the university	Yliopisto
scprs:year	Conferment year	Promootion vuosi
scprs:date	Date of the conferment	Tapahtumapäivä
scprs:conferment-Description	Short description of the events	Promootion kuvaus
scprs:conferment-DescriptionSource	Reference to data sources	Lähde promootion kuvaukselle
scprs:promovend	Number of students conferred	Promovendien lukumäärä
scprs:confermentPoem	Name of the conferement poem	Promootioruno
scprs:confermentCantata	Cantata presented at the conferment	Promootiokantaatti
scprs:externalLinks	Related externals links	Ulkoiset linkit
scprs:Image	URI of related person image	Kuva
scprs:ImageAttribution	License for the image	Valokuvan tekijän-oikeus-viittaukset
scprs:ImageDescription	Description of the image	Kuvaan liittyvä kuvaus
scprs:ImageSource	URI for the image source	Valokuvan lähdesivu

---

[8] Sampo systems: https://seco.cs.aalto.fi/applications/sampo/

Table 2. Role properties of the data model for the conferment class. The range of the properties is the Person class (cf. Table 3 below.)

Roles of people in conferments		
Property	Description	Label (in Finnish)
scprs:conferrer	Official conferrer	Promoottori
scprs:backupConferrer	Back-up conferrer	Varapromoottori
scprs:wreathWeaver	General wreath weaver	Seppeleensitoja
scprs:cantataPoet	Poet of the cantata	Kantaattirunoilija
scprs:cantataComposer	Composer of the cantata	Kantaattisäveltäjä
scprs:celebrationPreacher	Preacher of the celebration	Juhlasaarnaaja
scprs:confermentPoet	Poet of the conferment	Promootiorunoilija
scprs:danceMaster	Dance master	Tanssimestari
scprs:doctorPrimus	Primus doctor	Primustohtori
scprs:doctorUltimus	Ultimus doctor	Ultimustohtori
scprs:graphicDesigner	Grahic designer	Graafikko
scprs:goldsmith	Gold smith	Koruseppä
scprs:gratisti	Chairman of ceremonies	Gratisti (leader)
scprs:headMarshal	Head of ceremonies	Yliairut
scprs:honoraryDoctor	Honorary doctor	Kunniatohtori
scprs:jubileeDoctor	Jubilee doctor	Riemutohtori
scprs:jubileeGratisti	Jubilee gratisti	Riemugratisti
scprs:jubileeMagister	Jubilee magister	Riemumaisteri
scprs:jubileeWreathWeaver	Jubilee wreath weaver	Riemuseppeleensito-ja-tar
scprs:magisterPrimus	Primus magister	Primusmaisteri
scprs:magister-Ultimus	Ultimus magister	Ultimusmaisteri
scprs:doctorQuestioneer	Questioneer doctor	Tohtorikysymksen esittäjä
scprs:magister-Questioneer	Questioneer master	Maisterikysymyksen esittäjä
scprs:master-Of-Ceremonies	Master of the ceremonies	Juhlamenojen ohjaaja

Imperial Academy of Turku and the Imperial Alexander University of Finland during the time of the Grand Duchy of Finland (1809–1917).

A natural option for modelling data concerning ceremonial academic events would be to use an event-based ontology like CIDOC CRM. Its extensions have been used in various other Sampo systems, such as BiographySampo [15,28] and AcademySampo [20,21] whose data are re-used in CONFERMENTSAMPO. Here the lives of academics are represented as spatio-temporal sequences of biographical events participated by the people and other actors in different roles. However, the data related to conferment events is tabular in nature and it was decided that a simpler model would be enough in this case. Also presenting the data and maintaining it in the future would be easier in a tabular form than in CIDOC CRM.

The ontological model developed for the concept (class) of conferment ceremonies is therefore Dublin Core-like[9]. It consists of 38 metadata elements (properties). Firstly, each conferment is described in terms of 14 general properties,

---
[9] Dublin Core Metadata Initiative: https://dublincore.org.

such as the number and year of the conferment, a short description of the celebration events, a related image etc. (cf. Table 1). Secondly, each conferment class instance is linked to instances of the class people that acted in the ceremony in the 24 different roles listed in Table 2.

Table 3. Properties of the person class. *seprs* refers to the namespace of the data.

Person properties		
Property	Description	Label (in Finnish)
scprs:Image	Person image URI	Kuva
scprs:ImageAttribution	Copyright of the image	Valokuvan tekijän-oikeusviittaukset
scprs:ImageDescription	Image description	
scprs:ImageSource	URI of the image source	Valokuvan lähdesivu
scprs:matrikkeliLink	URL to student registry	Ylioppilasmatrikkeli linkki
scprs:Nation	Nation of the person	Osakunta
scprs:ParticipatedIn	Conferment participated	
scprs:personID	Person identifier	Henkilö ID
scprs:prefLabel	Person name	Henkilön nimi
scprs:role	Roles of the person	Rooli
scprs:title	Profession or rank	Ammatti tai arvo
scprs:wikiLink	Link to Wikipedia	Wikipedia linkki

Another key data table of the system consists of 1179 people who are known to have participated in the conferments in various official roles. This number excludes the thousands of students conferred during the events in the role of ordinary promovendei. The concept of a person is defined by the person's biographical properties (cf. Table 3). Resources belonging to the Finnish ontology infrastructure [14] are used as property values. With the help of the common infrastructure concepts, the data can be enriched and linked to other linked data publications. In the knowledge graph, instances of people are linked to instances of conferments by the explicit role properties of Table 2.

For the years 1643–1899 detailed information was also available on all conferred students (promovendei), thanks to the AcademySampo KG and LOD service. Their data and analyses are available by following links to the AcademySampo portal. Student-wise data of the 20th and 21st centuries is unfortunately not available, due to, e.g., personal data protection issues. Due to this shortcoming, the portal's application views and visualizations have been divided separately for the years 1643–1899 (conferments 1–70) and 1900–2023 (conferments 71–100). A separate but semantically similar conferment class is used for them to make the distinction.

4.2 Data Transformation, Data Enrichment, and Linking

The data was first harvested from various web pages into two CSV tables representing the 100 ceremonies and people involved.

In addition, these data was enriched manually and by links and related data of several external data sources:

1. AcademySampo.fi [20] based on the university's 28 000 student matriculation records.
2. Biografiasampo.fi [15,28], based on the 13 600 biographies of the Finnish Literature Society.
3. Wikidata/Wikipedia (e.g., paintings depicting the academics).
4. The national Finna.fi service[10] hosted by the National Library that aggregates and publishes collections of Finnish memory organizations.
5. Fennica – Finland's national bibliography[11] containing, e.g., data about the dissertations of the academics.

Furthermore, the national FIN-CLARIAH infrastructure[12] as part of the Research Council of Finland's research infrastructure roadmap[13] enabled the reuse of tools implemented in previous Sampo systems, e.g., visualizations with maps and timelines. At the same time, the data has been enriched also by reasoning as customary on the Semantic Web.

5 Methods

In our research, design science [9,22,24] methodology was used. The purpose of design science is to devise artifacts that are assessed against criteria of value or utility. The artifacts are created in an iterative fashion by building and evaluating them.

The data publication model of CONFERMENTSAMPO is based on the Sampo model [13] and the portal has been implemented with the Sampo-UI framework[14] [19,25] like other Sampos completed after NameSampo[15] published in 2018. The idea of the model is to gather mutually enriching data from different databases and sources, to harmonize the data using a shared infrastructure that includes shred data models (such as Dublin Core and CIDOC CRM) and ontological vocabularies (such as the ontological vocabularies of the National Library's Finto.fi service). The "standardized" principles and components of the Sampo-UI-based user interface included in the model makes the portal easy to use for end users on the one hand, and easy to implement for the application developers on the other hand.

The data aggregated and harmonized in a Sampo system is published as an open data service of linked data online in accordance with the standards and best practices of the W3C that coordinates the global Web infrastructure

---

[10] Finna portal: https://finna.fi.
[11] Fennica portal: https://kansalliskirjasto.finna.fi/Content/fennica.
[12] FIN-CLARIAH initiative, LOD part: https://seco.cs.aalto.fi/projects/fin-clariah/.
[13] Finnish roadmap of research infrastructures: https://www.aka.fi/tutkimusrahoitus/ohjelmat-ja-muut-rahoitusmuodot/tutkimusinfrastruktuurit/.
[14] Sampo-UI is openly available in Github: https://github.com/SemanticComputing/sampo-ui.
[15] NameSampo project page: https://seco.cs.aalto.fi/projects/nimisampo/.

development. For this purpose, a special Linked Data Finland platform LDF.fi[16] has been developed in Finland [17,18].

Once the linked data service has been published online, one can cost-efficiently use the data by SPARQL querying[17]. In this way it is possible, for example, to implement portals using the Sampo-UI tool [19,25] or use the data service directly in digital humanities studies. For example, in MMM-Sampo[18] (Mapping Manuscript Migrations) the triplestore related to medieval and Renaissance manuscripts of the University of Oxford, the Schönberg Institute of the University of Pennsylvania, and the French IRHT research center were combined into an open international data service and semantic portal utilizing international ontologies.

For CONFERMENTSAMPO, there was no ready-made database of conferments available, but the data had to be mined, e.g., from the website of the University of Helsinki[19] and online services related to the topic, and to supplement the data manually based on literature. An important literary source has been Tero Halonen's new book on the Finnish conferment traditions [7]. As a result of the data aggregation, tabular data was created and harmonized and transformed into a linked open data service in a similar way as in other database-based Sampos. In the future, the collected data can be supplemented and maintained, for example, in connection with future conferments by updating the tables. Alternatively, native linked data can be maintained using an editor. It is also possible to extend the system to conferment ceremonies in other faculties or universities.

The CONFERMENTSAMPO KG[20] is available on the Linked Data Finland[21] platform with open CC BY 4.0 license. The Apache Jena Fuseki[22] triplestore with Lucene[23] text indexing is used with a SPARQL server that is accessible from an open SPARQL endpoint[24]. The Varnish Cache web application accelerator[25] is used for routing URIs, content negotiation, and caching. Deployment of applications with a data service is based on a microservice architecture with Docker containers[26] where each individual component (the application, Varnish, and Fuseki) is run in its own dedicated container, making the deployment of the services easy due to installation of software dependencies in isolated environments.

---

[16] Linked Data Finland platform for LOD publishing: https://ldf.fi.
[17] SPARQL query language: https://www.w3.org/TR/sparql11-query/.
[18] MMM project homepage: https://seco.cs.aalto.fi/projects/mmm/.
[19] Conferments at the University of Helsinki: https://www.helsinki.fi/fi/tutustu-meihin/tama-helsingin-yliopisto/juhlat-ja-perinteet/promootio.
[20] ConfermentSampo data service: https://www.ldf.fi/dataset/promootiosampo.
[21] https://www.ldf.fi/.
[22] https://jena.apache.org/documentation/fuseki2/.
[23] https://lucene.apache.org/.
[24] https://ldf.fi/promootiosampo/sparql.
[25] https://varnish-cache.org.
[26] https://www.docker.com.

6 Using the Data Service and Portal

Based on the Sampo model, the data is published in an open SPARQL endpoint that can be used directly for data analyses using tools, such as the Yasgui editor [26] or Jupyter notebooks[27], or by applications created on top of the data service, such as the semantic Sampo portals[28] using the Sampo-UI framework [19,25].

Sampo-UI aims at "standardizing" UI development by proving a framework where faceted search and browsing is seamlessly integrated with data visualization and analysis tools. This makes using the UI easy for end users and application developer to implement. Developing a new Sampo portal can be started from an existing portal project whose configurations are modified in a declarative fashion to meet desiderata and the data models of the new application.

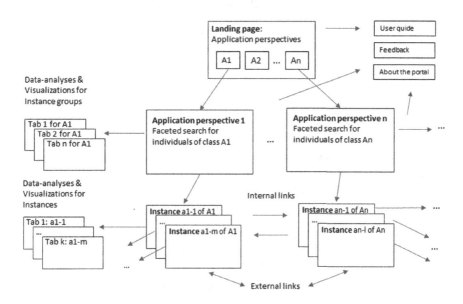

Fig. 1. The default navigational page structure of a portal based on Sampo-UI.

Figure 1 illustrates the navigational structure of using a Sampo-UI-based portal. The user first lands on the *landing page* with several *application perspectives* to the data. The perspectives are based on classes of the underlying KG. The usage cycle of each perspective can be divided into two steps: 1) filter and 2) analyze. The user first filters the data by using the faceted semantic search [29] tools provided by the portal. The results as well as the facet options are updated after each selection of a facet, making it possible for the user to precisely filter the end-result entities by different properties. After filtering the data to the wanted subset, the user can analyze the results set, i.e., a set of instances of the class

---
[27] Jypyter notebooks: https://jupyter.org/.
[28] Sampo portal series online: https://seco.cs.aalto.fi/applications/sampo/.

corresponding to the application perspective, with integrated data-analytic tools available as tabs on the application perspective page.

Fig. 2. CONFERMENTSAMPO's landing page with four application perspectives

The CONFERMENTSAMPO portal is based on Sampo-UI offering its user a landing page (cf. Fig. 2) from which different application perspectives can be selected for searching, browsing, and analyzing underlying linked data. In this case there are four perspectives for searching people, conferments 1643–1899, all conferments 1643–2023, and the last perspective offers a link to a page which explains the meanings of the official roles of the people participating in the promotion, as well as the course of ceremonial events, as explained in Sect. 3.

The perspectives are based on the key data types (classes) of the data model used, in this case the classes for conferments and people, and their properties described in Sect. 4. The top bar of the front page also contains links to additional information about the portal, a user manual, and a feedback channel for the end users to report possible errors and suggestion for further developments.

In the perspectives, one can search for individuals of the class related to the perspective, i.e., conferments and people, by using faceted search based on restrictions on the values of the class properties. In this way it is possible to limit the researched conferments only to, e.g., promotions held at the Royal Academy of Turku, or filter out people based on their roles. For example, one can find only promoters or cantata poets in the people application perspective.

In Fig. 3, the user has selected the perspective for conferments 1643–2023 and in the facet University the category Imperial Alexander University, in which case the search result will be 22 conferments. Facets are shown on the left and search results on the right. For each conferment, an image related to it in one way or another has been retrieved from different data sources, for example, a portrait of the promoter or an object related to the promotion, such as a sword.

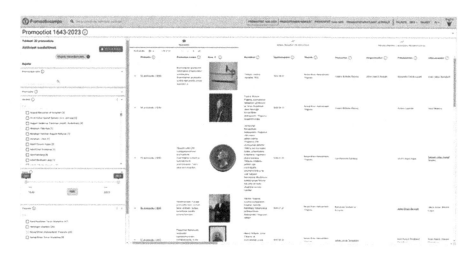

Fig. 3. Application perspective to search, browse, and analyze the conferment ceremonies 1643–2023

By clicking on the conferment link in the result set, one can access the homepage of the event, where various information and related links for further information have been collected. By clicking on the image, one gets additional information related to it.

By default, the search results are displayed as a table, but by changing the tab in the top bar, other data analytical visualizations are available for the search result. For example, in Fig. 3 there is a tab available for showing a histogram related to the number of people who participated in the promotions of the search result in different roles.

In faceted search, the information regarding the number of search hits for all facets is updated automatically after each selection. For example, the number of hits in the people facet tells how many times each person has been involved in an official role in different conferments. For instance, the total count for Professor Jakob Johan Wilhelm Lagus (1821–1909) is five; he was a central figure in developing the original student matriculation registry data behind AcademySampo and CONFERMENTSAMPO.

Figure 4 shows the perspective for people with search facets Name, Person, Roles, Department, Profession or rank, and Conferment on the left. The Roles facet has been opened, which shows the number of roles of the people in the search result. Here, for example, it is possible to see that there have been a total of 320 jubilee doctors and 194 jubilee masters[29] in different conferments.

In a similar way to the conferment perspective, the search result for people is shown as a table by default, but the result can also be visualized prosophographically on different tabs. For example, in this case histograms of roles can be

---

[29] The jubilee degree is given to people who were graduated in a conferment 50 years earlier.

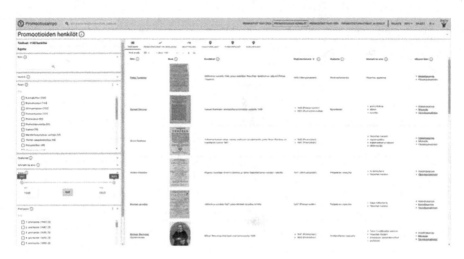

Fig. 4. The people application perspective of more than a thousand people in the official roles in 1963–2023

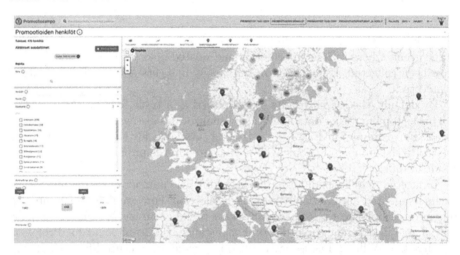

Fig. 5. Known events related to the persons of conferments 1643–1899, visualized on a map (© Mapbox, © OpenStreetMap). By clicking on the marker, you get information about the people associated with the place.

shown or maps used. Figure 5, for example, visualizes the career events related to the people in the result set on a map[30] during the period 1643–1899. This data originates from the underlying AcademySampo system based on CIDOC CRM events.

---

[30] The map is generated with Mapbox service with data from Mapbox and OpenStreetMap and their data sources. To learn more, visit https://www.mapbox.com/about/maps/ and http://www.openstreetmap.org/copyright.

Fig. 6. Homepage of Professor Michael Wexionius (1609–1670)

Each person (class instance) has an automatically generated homepage that can be accessed by clicking on the person's link in the search result. For example, Fig. 6 shows the homepage of the promoter of the 1647 and 1653 promotions, Michael Wexionius (1609–1670), ennobled later as Michael Gyldenstolpe. His data has been compiled and linked to diverse information about his life. For example, if a person has published a dissertation available in the National Library's *Fennica – Finland's national bibliography*, a link to it is provided for a reference.

A key source of data re-used in CONFERMENTSAMPO for academics 1640–1899 has been AcademySampo [20] which contains detailed biographical information on all approximately 28000 persons who received an academic education in Finland between 1640–1899. Most of them participated in one or more conferments in different roles. Based on CONFERMENTSAMPO, the most active conferment person has been Zacharias Topelius (1818–1898), who has been involved in official roles in eight different promotions. Three people have been participant with six official roles: Adolf Moberg (1829–1895), professor of physics and chemistry and rector of the Imperial Alexander University of Finland; the poet and Finnish writer Otto Manninen (1872–1950), who has participated as both a cantata poet and as an honorary doctor; the dance master Mrs. Sirpa Koivisto during the conferments 1986–2003.

Based on language agnostic semantic web technologies, the user interface of the portal is available in Finnish and English, the data only in Finnish.

7 Conclusions

This paper presented a new in-use application of linked open data, based on data harvested from web pages, enriched with related data sources, and published as an open knowledge graph in a SPARQL endpoint for 1) end users to use and 2) application developers to create applications for digital humanities.

The Sampo-UI framework was successfully tested and demonstrated for these two tasks. The portal UI has not been formally evaluated. However, previous implementations of several popular portals utilizing the Sampo-UI model [13] suggest empirically good usability and scalability of the model and tool [4]. From a software developer point of view, the Sampo-UI framework was deemed very useful by the portal main developer who was not involved in developing the tool and had never used it before [3].

Acknowledgement. Thanks to the Conferment Committee 2023 of the University of Helsinki for discussions. CSC – IT Center for Science provided computational resources for our projects.

References

1. Ahola, A., Hyvönen, E., Rantala, H., Kauppala, A.: Historical opera and music theatre performances on the semantic web: Operasampo 1830-1960. In: Knowledge Graphs in the Age of Language Models and Neuro-Symbolic AI: Proceedings of the 20th International Conference on Semantic Systems, Amsterdam, The Netherlands. IOS Press (2024)
2. Ali, A., Noah, S.A.M., Zakaria, L.Q.: Representation of event-based ontology models: a comparative study. IJCSNS Int. J. Comput. Sci. Netw. Secur. **22**(7), 147 (2022)
3. Boman, P.: Promootiosampo: Järjestelmä suomalaisen promootioperinteen kuvaamiseen, julkaisemiseen ja tutkimiseen semanttisessa webissä. Master's thesis, Aalto University, Department of Computer Science (2024). https://seco.cs.aalto.fi/publications/2024/boman-msc-2024.pdf
4. Burrows, T., Pinto, N.B., Cazals, M., Gaudin, A., Wijsman, H.: Evaluating a semantic portal for the "Mapping Manuscript Migrations" project. DigItalia **15**(2), 178–185 (2020)
5. Doerr, M.: The CIDOC CRM–an ontological approach to semantic interoperability of metadata. AI Mag. **24**(3), 75–92 (2003)
6. Gardiner, E., Musto, R.G.: The Digital Humanities: A Primer for Students and Scholars. Cambridge University Press, New York (2015)
7. Halonen, T.: Promootio: elävän yliopistoperinteen historiaa. Otava (2023)
8. Heath, T., Bizer, C.: Linked Data: Evolving the Web into a Global Data Space. Morgan & Claypool, Palo Alto (2011). http://linkeddatabook.com/editions/1.0/
9. Hevner, A.R., March, S.T., Park, J., Ram, S.: Design science in information systems research. MIS Q.: Manag. Inf. Syst. **28**(1), 75–105 (2004)
10. Hyvönen, E., Saarela, S., Viljanen, K.: Application of ontology techniques to view-based semantic search and browsing. In: Bussler, C.J., Davies, J., Fensel, D., Studer, R. (eds.) ESWS 2004. LNCS, vol. 3053, pp. 92–106. Springer, Heidelberg (2004). https://doi.org/10.1007/978-3-540-25956-5_7

11. Hyvönen, E.: Publishing and Using Cultural Heritage Linked Data on the Semantic Web. Morgan & Claypool, Palo Alto (2012)
12. Hyvönen, E.: Using the semantic web in digital humanities: shift from data publishing to data-analysis and serendipitous knowledge discovery. Semantic Web **11**(1), 187–193 (2020)
13. Hyvönen, E.: Digital humanities on the semantic web: Sampo model and portal series. Semantic Web **14**(4), 729–744 (2022)
14. Hyvönen, E.: How to create a national cross-domain ontology and linked data infrastructure and use it on the semantic web. Semantic Web (2024). https://doi.org/10.3233/SW-243468
15. Hyvönen, E., et al.: BiographySampo – publishing and enriching biographies on the semantic web for digital humanities research. In: Hitzler, P., et al. (eds.) ESWC 2019. LNCS, vol. 11503, pp. 574–589. Springer, Cham (2019). https://doi.org/10.1007/978-3-030-21348-0_37
16. Hyvönen, E., Styrman, A., Saarela, S.: Ontology-based image retrieval. In: Towards the Semantic Web and Web Services, Proceedings of XML Finland 2002 Conference, pp. 15–27, 21–22 October 2002. http://www.seco.hut.fi/publications/2002/hyvonen-styrman-saarela-ontology-based-image-retrieval-2002.pdf
17. Hyvönen, E., Tuominen, J.: 8-star linked open data model: extending the 5-star model for better reuse, quality, and trust of data. In: Posters, Demos, Workshops, and Tutorials of the 20th International Conference on Semantic Systems (SEMANTiCS 2024), vol. 3759. CEUR Workshop Proceedings (2024). https://ceur-ws.org/Vol-3759/paper4.pdf
18. Hyvönen, E., Tuominen, J., Alonen, M., Mäkelä, E.: Linked data Finland: a 7-star model and platform for publishing and re-using linked datasets. In: Presutti, V., Blomqvist, E., Troncy, R., Sack, H., Papadakis, I., Tordai, A. (eds.) ESWC 2014. LNCS, vol. 8798, pp. 226–230. Springer, Cham (2014). https://doi.org/10.1007/978-3-319-11955-7_24
19. Ikkala, E., Hyvönen, E., Rantala, H., Koho, M.: Sampo-UI: a full stack JavaScript framework for developing semantic portal user interfaces. Semantic Web **13**(1), 69–84 (2022)
20. Leskinen, P., Hyvönen, E.: Biographical and prosopographical analyses of Finnish academic people 1640–1899 based on linked open data. In: Proceedings of the Biographical Data in a Digital World 2022 (BD 2022), Tokyo. ZRC SAZU, Ljubljana (2024). https://doi.org/10.3986/9789610508120_7
21. Leskinen, P., Hyvönen, E.: Biographical and prosopographical analyses of Finnish academic people 1640–1899 based on linked open data. In: Proceedings of the Biographical Data in a Digital World 2022 (BD 2022), Tokyo. Institute of Cultural History, ZRC SAZU, Ljubljana, Slovenia, January 2024. https://doi.org/10.3986/9789610508120_7
22. March, S.T., Smith, G.F.: Design and natural science research on information technology. Decis. Support Syst. **15**(4), 251–266 (1995)
23. Pattuelli, M.C., Hwang, K., Miller, M.: Accidental discovery, intentional inquiry: leveraging linked data to uncover the women of jazz. Digit. Scholarsh. Humanit. **32**(4), 918–924 (2016). https://doi.org/10.1093/llc/fqw047
24. Peffers, K., Tuunanen, T., Rothenberger, M.A., Chatterjee, S.: A design science research methodology for information systems research. J. Manag. Inf. Syst. **24**(3), 45–77 (2007)
25. Rantala, H., Ahola, A., Ikkala, E., Hyvönen, E.: How to create easily a data analytic semantic portal on top of a SPARQL endpoint: introducing the configurable

Sampo-UI framework. In: VOILA! 2023 Visualization and Interaction for Ontologies, Linked Data and Knowledge Graphs 2023. CEUR Workshop Proceedings, vol. 3508 (2023). https://ceur-ws.org/Vol-3508/paper3.pdf
26. Rietveld, L., Hoekstra, R.: The YASGUI family of SPARQL clients. Semantic Web **8**(3), 373–383 (2017)
27. Staab, S., Studer, R. (eds.): Handbook on Ontologies, 2nd edn. Springer, Heidelberg (2009)
28. Tamper, M., Leskinen, P., Hyvönen, E., Valjus, R., Keravuori, K.: Analyzing biography collection historiographically as linked data: case national biography of Finland. Semantic Web **14**(2), 385–419 (2023). https://doi.org/10.3233/SW-222887
29. Tunkelang, D.: Faceted Search. Synthesis Lectures on Information Concepts, Retrieval, and Services, vol. 1, no. 1, pp. 1–80 (2009)
30. Warren, C.N.: Historiography's two voices: data infrastructure and history at scale in the Oxford Dictionary of National Biography (ODNB). J. Cult. Anal. **1**(2), 1–31 (2018)
31. Ziku, M.: Digital cultural heritage and linked data: semantically-informed conceptualisations and practices with a focus on intangible cultural heritage. LIBER Q. **30**, 1 (2020)

Towards Complex Ontology Alignment Using Large Language Models

Reihaneh Amini[1]($\boxtimes$), Sanaz Saki Norouzi[1], Pascal Hitzler[1], and Reza Amini[2]

[1] Kansas State University, Manhattan, KS, USA
{reihanea,sanazsn,hitzler}@ksu.edu
[2] Wright State University, Dayton, OH, USA
amini.4@wright.edu

Abstract. Ontology alignment, a critical process in the Semantic Web for detecting relationships between different ontologies, has traditionally focused on identifying so-called "simple" 1-to-1 relationships through class labels and properties comparison. The more practically useful exploration of more complex alignments remains a hard problem to automate, and as such is largely underexplored, i.e. in application practice it is usually done manually by ontology and domain experts. Recently, the surge in Natural Language Processing (NLP) capabilities, driven by advancements in Large Language Models (LLMs), presents new opportunities for enhancing ontology engineering practices, including ontology alignment tasks. This paper investigates the application of LLM technologies to tackle the complex ontology alignment challenge. Leveraging a prompt-based approach and integrating rich ontology content – so-called *modules* – our work constitutes a significant advance towards automating the complex alignment task.

Keywords: Complex Ontology Alignment · Ontology · Large Language Model · Knowledge Graph · Modular Ontology Modeling

1 Introduction

Ontology alignment (sometimes called ontology matching) [8] is the task of establishing mappings between different ontologies, and as a research field it is concerned with ways to automate or at least semi-automate this task. For those not familiar with the field: Ontologies, which are usually knowledge bases expressed using Description Logics [13] (including the W3C Web Ontology Language – OWL – standard [23]) in this case act as a type of data schema for data expressed as knowledge graphs [12], i.e. the establishing of mappings between ontologies is central for schema-based data integration purposes.

Ontology alignment [25] has been studied for over two decades, resulting in the development of many alignment approaches and systems. The majority of these systems are designed to detect only so-called "simple" 1-to-1 mappings between ontologies, primarily by establishing equivalence relationships between classes (unary predicates), or between properties (binary relationships); for example, one ontology may have a class called "Person" while another may have a class called "Human", and an ontology alignment mapping may state that these two classes are in fact equivalent. It has long been recognized in the Semantic Web and Ontologies community that such simple mappings are helpful but ultimately insufficient for data integration tasks for which mappings would need to be in the form of complex mapping rules[1] that can be expressed, e.g., as Datalog rules. However, detecting complex alignments between ontologies remains a very challenging and thus largely unexplored area with only few contributions that made progress in restricted settings. In current practice, establishing complex alignments between two or more ontologies requires domain experts to collaborate and manually generate the alignments, and this is usually a very work-intensive and thus expensive task. Any automation or semi-automation would have significant added value.

The Ontology Alignment Evaluation Initiative (OAEI)[2] is a long-standing coordinated international effort aimed at improving and evaluating ontology alignment and coreference resolution technologies.[3] It organizes annual evaluation campaigns [21] that provide a controlled environment where participants can test their ontology alignment systems using various benchmark tests. The benchmarks cover a range of complexity levels and real-world scenarios, aiming to simulate different aspects of the ontology alignment process. A complex alignment track was started in 2018, including evaluation of the GeoLink benchmark that we will make use of in this paper. It should be noted, however, that some datasets have ceased to be evaluated in recent years [2,21,29] as each track depends on volunteers to run it.

With significant advancements in the natural language processing (NLP) and natural language understanding (NLU) fields, spurred by Large Language Models (LLMs), it has become possible to extract meanings from text. OpenAI[4] has been at the forefront of this research, developing the Generative Pre-trained Transformer (GPT) series of models, which have attracted considerable attention from researchers, developers, and users. One of the most notable models, introduced in March 2022, is GPT-4 [1]. This transformer-based model is designed to predict the next token in generating text and has shown improvements in producing results that more closely align with user intent compared to its predecessor, GPT-3.5, on 70.2% of the prompts.[5]

Recent advancements in applying Large Language Models (LLMs) to Semantic Web and ontology engineering tasks have shown promising results, due to the importance of NLP for such tasks. A notable study [11] demonstrated the effectiveness of zero-shot and few-shot prompting with LLMs on various tasks within

---

[1] See Sect. 3 for an example.
[2] http://oaei.ontologymatching.org/.
[3] Co-reference resolution is about establishing equivalence and non-equivalence mappings between individuals, or constants, a related but different task.
[4] https://openai.com.
[5] See https://openai.com/blog/chatgpt and https://openai.com/blog/openai-api.

the Ontology Alignment Evaluation Initiative (OAEI), highlighting their potential in this area. The study was restricted to simple alignments. Likewise, [9] showcases how LLMs can effectively align diverse ontologies in knowledge engineering tasks, outperforming traditional Ontology Matching (OM) systems in simple alignment tasks.[6] As we will see later, a straightforward tasking of LLMs with the production of complex alignments does not quite work.

One of the difficulties with ontology alignment is that ontologies often tend to be underspecified and with little internal structure that may add some self-explainability. This can be seen for example by the considerable disagreement between humans as to "correct" alignments, even for the simple alignment task [6]. It has been posited that additional internal structure, e.g. in the form of conceptual "ontology modules" should aid with ontology engineering tasks that are hard to automate [24]. Following this hitherto merely conceptual argument, we made use of ontology modules in our approach to generate complex alignments, and as we will report below, our prompt-based approach for discovering complex alignments between ontologies yields significantly better results when richer content in the form of ontology modules is available.

Since ontologies are knowledge bases expressed using formal logic, and mapping rules are also expressed using formal logic and processed as such, ontology alignment is a key symbolic task that we are here addressing using by "neural" means (i.e. LLMs as artificial neural networks), as such contributing to the body of approaches and methods for neurosymbolic artificial intelligence [14].

The structure of this paper is as follows: Sect. 2 briefly describes our approach. Experiment details and quantitative evaluation can be found in Sect. 3, while Sect. 4 discusses the findings. Section 5 concludes.

An extended version of this paper is available on arXiv [4].

2 Complex Alignment by Large Language Model

Ontology Modules. In the Semantic Web community, the term **"module"** can mean many things. In our case, we understand the term in the way in which it is used in the Modular Ontology Modeling (MOMo) methodology [24]. It refers to a specific part of an ontology that encapsulates a principal concept and its main features, such as an Event module capturing details about location, date, organizer, etc. Modules serve dual roles: they are technical constructs that, on the one hand, demarcate parts of ontologies that group related classes and their interactions and, on the other hand, they do this in a way that aligns with domain experts' understanding. Despite potential overlaps and hierarchical structures within them, modules organize the ontology into a network of interrelated pieces, each mirroring the domain's conceptual framework as understood by experts.

Modules enable a strategic approach to ontology modeling by allowing the work to be broken down into manageable segments; initially focusing on individual modules before linking them together. This approach provides a clear way

---

[6] The unreviewed paper claims to contribute to complex alignment but lacks the data to support this.

to manage the complexities of large, cohesive ontologies by breaking down the process into understanding individual modules and their interconnections. This modular approach also aligns with the way domain experts conceptualize their fields, making both the ontology and its documentation more accessible and comprehensible. Since each module can be easily swapped out for another-perhaps one that offers a different level of detail-modifications are localized, making the entire system more adaptable [16].

In the research we present herein, we explore the impact of integrating module information on the effectiveness of identifying complex alignments. Specifically, we include descriptions, Core Axioms (where applicable), and alignment information as outlined in [15] that constitute the GeoLink Modular Ontology (GMO), that had been developed as an integrated schema for combining several large-scale ocean science data repositories [7]; GMO also underlies the OAEI GeoLink complex alignment benchmark. As we will see further below, the utilization of additional module information is of core importance in order to solve the complex alignment problem, in our setting.

Regarding the GeoLink complex alignment benchmark it contains two ontologies, the GeoLink Base Ontology (GBO) and the GMO already mentioned above. These are paired with a reference alignment created with help from domain specialists [5,29]. During these years, only two or three systems registered for evaluation in this track. Evaluation results show [3,19,20] that the systems that make use of instance data that is shared between the two ontologies perform relatively well, with high precision but low recall, i.e. many matches are missed. Furthermore, the availability of shared instance data, as for this benchmarking competition, while helpful for advancing the state of the art, is unrealistic for most practical application scenarios for ontology alignment: Usually, such shared data would not be available due to concerns over data privacy, the complexity of data collection, frequent changes in the data, and the demands of storage and processing. The new results that we present in this paper have been obtained without taking shared instance data into consideration, which, in our opinion, constitutes a major advance over the state of the art.

Design of the Prompting Process. There is a range of task-agnostic prompting techniques available for use with large language models.[7] We will in particular employ chain-of-thought prompting, which involves supplying the model with a series of thought process examples, helping it to navigate through reasoning steps to arrive at a conclusion [26].

Figure 1 illustrates our workflow, which begins by uploading the entire GMO file as an initial prompt as depicted in Fig. 2. Subsequently, we include specific entities from the second ontology, the GBO, to inquire about their alignment with the GMO. If the initial prompt successfully provides the relevant segments, further prompts

Fig. 1. Prompting workflow

---

[7] See https://www.promptingguide.ai/techniques and [17,27].

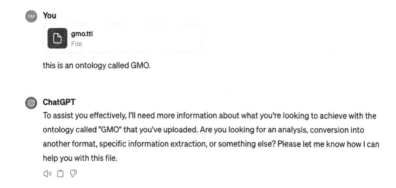

Fig. 2. Uploading GMO.ttl file as prompt.

in the chain-of-thought process are unnecessary. Otherwise, we can provide a list of all GMO module names and ask GPT to identify the most related modules for the GBO entities. Then, in a subsequent prompt, we supplement the inquiry with module information and request the related segments again.

3 Experiment and Evaluation

For the evaluation of our methodology, we use the GeoLink Complex Alignment dataset [28] from the OAEI. It comprises 109 complex alignment rules between GMO and GBO. List of all rules from GBO to GMO is available online.[8] Detailed findings from our study, including both the prompts used and the responses obtained, are available online.[9]

Prompting and Detailed Example. OpenAI's functionality includes the ability to load a prompt from a file, triggering backend processing to parse the uploaded data. We uploaded the entire GMO ontology RDF file in turtle (.ttl) format. OpenAI GPT-4 parsed the data, albeit with some latency, prompting a request for more specific tasks to be directed to it.

To give an example, consider the complex alignment rule

$$\text{Award}(x) \wedge \text{hasCoPrincipalInvestigator}(x, z) \leftrightarrow$$
$$\text{FundingAward}(x) \wedge \text{providesAgentRole}(x, y)$$
$$\wedge \text{CoPrincipalInvestigatorRole}(y) \wedge \text{performedBy}(y, z) \qquad (1)$$

where GBO entities hasCoPrincipalInvestigator (a relation or binary predicate or so-called object property) and Award (a class or unary predicate) can be found on the left-hand side, and the GMO entities on the right-hand side. This

---
[8] https://daselab.cs.ksu.edu/publications/alignment-rules-gbo-gmo.
[9] https://github.com/reddraa/complex_ontology_alignment.

rule is expressed in first-order predicate logic and our objective is to assess the effectiveness of detecting such rules by prompting GPT-4.

After uploading the GMO, we extract the GBO entities of the complex alignment rule (i.e., **"Award"** and **"hasCoPrincipalInvestigator"**) from the GBO RDF file in its original format. For our example, this looks as follows.

```
### http://gbo#Award
main:Award rdf:type owl:Class ;
        rdfs:comment "Funding provided by an Organization
        enabling Participation.";
        rdfs:label "Award" .
### http://gbo#hasCoPrincipalInvestigator
main:hasCoPrincipalInvestigator rdf:type owl:ObjectProperty ;
            owl:inverseOf main:isCoPrincipalInvestigatorOf ;
            rdfs:domain [ rdf:type owl:Class ;
                          owl:unionOf ( main:Award
                                        main:Program
                                      )
                        ] ;
            rdfs:range main:Person ;
            rdfs:label "hasCoPrincipalInvestigator" .
```

Next, our prompt instructs GPT-4 to examine the components in GMO that are associated with these elements in GBO, as depicted in Fig. 3.

What we observed from GPT-4's responses in this step is often not a comprehensive answer. Many times, it explicitly states that *It appears there were no results found in the ontology file "gmo.ttl" directly related to the terms you mentioned from the "GBO" ontology, such as "Award", "hasCo-PrincipalInvestigator", "Person", "Program", or "isCo-PrincipalInvestigatorOf"*. GPT-4 typically continues its response with further inquiries such as *To proceed, I can manually examine the "gmo.ttl"*

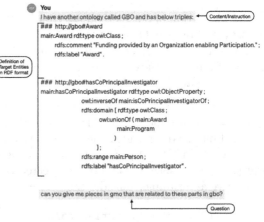

Fig. 3. GBO-related instruction in prompt

file to identify any related concepts or properties that might align with those in the "GBO" ontology. This process involves looking for classes or properties that serve similar purposes or are conceptually related, even if they are not directly named the same. Would you like me to perform this manual examination, or is there another way I can assist you with these ontologies?

Our consistent response to this question has always been "Yes", and the typical output we received from GPT is illustrated in Fig. 4. As depicted, the

> **ChatGPT**
>
> Upon examining the "GMO" ontology, I found several components that are related to the terms you mentioned from the "GBO" ontology. Here are the related parts:
>
> **Related to "Award":**
>
> - `http://gmo#AwardAmount`: Possibly refers to the financial aspect of an award.
> - `http://gmo#FundingAward`: This could be similar to the "Award" concept in GBO, referring to funding provided.
> - `http://gmo#hasAwardAmount`: A property that might relate to the amount associated with an award.
>
> **Related to "Person":**
>
> - `http://gmo#Person`: Directly relates to the concept of a person.
> - `http://gmo#PersonName`, `http://gmo#hasPersonName`: Properties related to a person's name.
> - `http://gmo#PersonalInfoItem`, `http://gmo#hasPersonalInfoItem`: May refer to various pieces of personal information about a person.
>
> **Related to "Program":**
>
> - `http://gmo#Program`: Directly corresponds to the concept of a program.
> - `http://gmo#associatedWithProgram`: A property that could indicate a relation or participation with a program.
> - `http://gmo#ProgramManagerRole`, `http://gmo#AgencyProgramManagerRole`: Roles that might be associated with managing or being involved in a program.
>
> There were no direct matches found for "hasCoPrincipalInvestigator" or "isCoPrincipalInvestigatorOf" in the "GMO" ontology, but the concepts related to "Award," "Person," and "Program" suggest that the "GMO" ontology includes entities and properties that could be conceptually aligned or used in conjunction with those in the "GBO" ontology, particularly around awards, people, and programs.
>
> If you need further analysis or have other specific queries about these ontologies, please let me know! ひ-ʻ

Fig. 4. GPT-4 response to our initial prompt

LLM successfully identifies a similar entity to **gbo#Award**, which in GMO is **gmo#AwardAmount** (highlighted in green). However, it struggles to conduct further analysis to identify alignments in GMO related to the relationship between "Award" and "hasCoPrincipalInvestigator" in GBO. Rather, it attempts to generate random information about recently detected entities such as "FundingAward", which hypothetically arises from its next-word-prediction logic (highlighted in yellow). However, the remaining generated information, such as "gmo#Program" (highlighted in red), is not relevant to this component in GBO. Ultimately, it states that *There were no direct matches found for hasCoPrincipalInvestigator or isCoPrincipalInvestigatorOf in the GMO ontology.*

We also attempted zero-shot prompting at this stage, combining both the GMO file and the second prompt into a single input to GPT-4. This approach resulted in increased latency and a confused response.

Our conclusion from this analysis is that GPT-4 failed to detect the complex alignment between the two ontologies. Instead, it provided partially related objects, including random information about different unrelated entity classes.

> This is a list of all the modules and patterns that are defined in GMO ontology:
>
> Agent, Agent Role, Event, Information Object, Identifier, Person, Personal Info Item pattern, Person Name, Organization, Funding Award, Program, Place, Cruise, Platform Pattern Stub, Vessel pattern, Physical Sample pattern, Property Value pattern.
>
> From this list, give me relevant module names that can help us find the most related GMO pieces.

Fig. 5. Prompt for module name suggestion

To proceed, let us revisit the GMO modules discussed in Sect. 2. The documentation [15] contains an informal description of the modules (in the documentation called *patterns*) in the GMO ontology, accompanied by visual depictions of these patterns. In the ongoing chain-of-thoughts prompting processes, we included the module names and asked for the most relevant module names related to the GBO pieces, as shown in Fig. 5. The result of this prompt is usually a single module or a list of suggested modules based on the information that the GPT has processed through the chain of prompts. In the next prompt, we included the descriptions of the suggested modules along with the question, as illustrated in the prompt below.

```
you couldn't give me all the pieces I need in GMO. here is more info that can
help: The Funding Award pattern describes the funding awards that fund all
kinds of ocean science research activities. We use the isFundedBy property
to connect anything to a funding award if the funding award funds it. Each
funding award has exactly one starting and ending date (aligned with
time:Instant). It provides at most one award amount, which is described via a
pair of decimal value and currency code. The currency code is not specified
here, but existing standards can be used, e.g., ISO 4217. There may be
people or organizations that have a role in a funding award. This is modeled
by re-using (and aligning with) the Agent Role pattern. In this version, we
include the following types of agent-roles, represented as classes:
SponsorRole, AgencyProgramManagerRole, PrincipalInvestigatorRole, and
CoPrincipalInvestigatorRole. Additional roles are possible in the future
versions. Each funding award is described by an InformationObject, which
when aligned to the Information Object pattern, allows one to represent
additional information such as identifier, description, etc.

ObjectProperty: isFundedBy
ObjectProperty: startsOnDate
ObjectProperty: endsOnDate
ObjectProperty: isDescribedBy
ObjectProperty: providesAgentRole
ObjectProperty: isPerformedBy
ObjectProperty: hasAwardAmount
ObjectProperty: hasCurrencyCode
DataProperty:   hasCurrencyValue
Class:  FundingAward
Class:  Agent
Class:  TimeInstant
Class:  AgentRole
Class:  SponsorRole
```

- **"FundingAward" (x)** ∧ **"providesAgentRole" (x, y)** ∧ **"CoPrincipalInvestigatorRole" (y)** ∧ **"isPerformedBy" (y, z)**

Here's the breakdown:

- **"FundingAward" (x)**: Directly equates to the "Award" concept in GBO, representing the funding award.
- **"providesAgentRole" (x, y)**: Connects a funding award to a role, which can include "CoPrincipalInvestigatorRole".
- **"CoPrincipalInvestigatorRole" (y)**: Specifies the role related to the funding award, equivalent to having a co-principal investigator in GBO.
- **"isPerformedBy" (y, z)**: Links the role (e.g., CoPrincipalInvestigatorRole) to the person or organization performing it, closing the equivalence to "hasCoPrincipalInvestigator" in GBO.

Fig. 6. GPT-4 response to our question with Module information included

```
Class:  PrincipalInvestigatorRole
Class:  CoPrincipalInvestigatorRole
Class:  AgencyProgramManagerRole
Class:  InformationObject
Class:  AwardAmount
Class:  CurrencyCode
```

By evaluating the outcome (see Fig. 6) of the prompt, it is evident that GPT-4 has a more informed and precise approach to conducting investigations and identifying semantically related components if it has the module information available. GPT-4 identifies all components related to the GMO as outlined in the alignment rule mentioned above. It further elaborates on each component and explains their interconnections. The full evaluation data of our study over all complex alignment rules is available online.[10]

Impact of Few-Shot vs Chain-of-Thought: Comparing few-shot and chain-of-thought approaches, we noted the differences between providing information in a single zero-shot prompt versus delivering it in a series of prompts. GPT tends to become confused about the question and the relevant information it needs to process in a zero-shot scenario. In contrast, introducing the prompt as part of a sequential chain of information clarifies the data pieces and their meanings for the model.

Impact of Adding Modules to the Prompt: We observed that while GPT-4 nearly grasps the query regarding the relevant components needed to fulfill the rule, it typically identifies only 10–20% of the required elements if no module information is given. However, if details about the module in the GMO are presented in a single prompt before posing the question again, GPT-4 is significantly more effective, clearly identifying the majority of the targeted components.

---

[10] https://github.com/reddraa/complex_ontology_alignment.

Table 1. Distribution of successful approaches for detecting complex alignment rules

detected pieces without module Information	5
detected pieces with module Information	104
total number of complex alignment rules	109

Quantitative Evaluation. To assess the effectiveness of our approach, we identified key entities within our complex alignment rules to serve as the basis for our metrics. As illustrated in Fig. 6, the entities detected as relevant by GPT-4 include FundingAwards(z), providesAgentRole(x,y), CoPrinciplaInvestigatorRole(y), isPerformedBy(y,z). It is important to note that response formats may vary, hence each response might be unique. We employed two key performance metrics: recall and precision. Recall, defined as the ratio of detected GMO-related instances to the total number of expected instances, evaluates our study's ability to identify all relevant GMO pieces comprehensively. Precision, on the other hand, measures the accuracy of our detection process by calculating the proportion of correctly identified GMO pieces out of all the instances flagged in our findings. Together, these metrics provide a holistic view of our study's identification capability, striking a balance between thoroughness and accuracy in detecting GMO-related instances of a complex alignment rule.

According to Table 1, in the study presented in the paper, an analysis of 109 complex alignment rules was conducted. The findings revealed that in only 4.5% of these alignments, GMO components were identified solely through the use of GBO entities (i.e., not using module information). Both identified complex alignment pieces, without the module, are actually simple 1 : 1 mappings, e.g., Program(x) ↔ Program(x), where 'Program' is an existing class in both the GMO and GBO ontology. However, in over 95% of the cases, the identification of GMO components was achieved through the application of information from the GMO module information, highlighting its significant role in the detection process.

In assessing GPT-4's performance on entity detection in the complex alignment, we analyzed the example alignment mentioned earlier in this paper. We compared the alignment pieces returned by GPT-4, shown in Fig. 6 for our running example based on rule (1), with four corresponding entities defined in the GMO ontology's alignment rule, i.e., FundingAward(x), providesAgentRole(x,y), CoPrincipalInvestigatorRole(y), performedBy(y,z). Note that we did not assess the actual return of alignment rules, but rather whether the relevant pieces (predicates) were detected. While the detection (without actual composition of the pieces into a rule) constitutes a simpler task than producing the rules, it nevertheless captures the core difficulty for complex ontology alignment. In fact, if the pieces are correct, the actual rule can easily be assembled by a human (or, in many cases, by a symbolic algorithm based on the ontology and example data).

In evaluating recall (coverage) in this setting, we found that all four expected GMO entities were accurately identified by GPT-4 for our running example, yielding a recall of 1.0 in this case. Additionally, for precision, we examined the entities returned by GPT-4 and found that aside from the four correct GMO entities, no irrelevant entities were detected, also resulting in a precision of 1.0. This indicates perfect alignment detection by GPT-4 in this instance. To further clarify the evaluation of recall, consider that if the prompt response from GPT-4 had included additional entities such as "Event" or "Place" which were not part of the expected entities, it would have negatively impacted the precision. These extraneous entities, not being included in the set of expected results, would reduce the precision as they represent incorrect identifications according to the specified alignment rule. In practical terms, the return of (in particular, many) superfluous entities by the system would make assembly of the actual rule by a human more difficult.

In Table 2 we see that for 73.3% of the complex alignment rules evaluated, the recall value exceeded 0.5. This indicates that more than half of the GMO entities involved in a complex alignment rule were successfully detected for this ratio in our population. Furthermore, the recall value surpasses 0.75 for approximately 62.3% of these rules, signifying a higher accuracy in detection. It's also noteworthy that *for 45% of the alignment rules recall is a perfect 1.0, and for 45.8% precision is also 1.0*, because of the integration of module information in the analysis process.

The precision metric indicates the accuracy of the responses in directing us toward expected entities within the GMO ontology. Our analysis found that responses achieved a precision higher than 0.75 for 59.6% of the evaluated records. Additionally, when the precision threshold is lowered to 0.5-implying that half of the entities suggested by the language model are the expected ones, while the other half may include relevant or irrelevant entities-the coverage of alignment rules increases to 69.7%.

Note that detection (recall) of half or more of the correct body predicates, paired with high precision, is already very helpful for human assembly of a rule.

The recall data exhibits a mean of 0.67 and a median of 0.75, indicating that half of the recall values exceed 0.75, while the other half fall below this threshold. This distribution highlights that the majority of our data points demonstrate recall values above 0.75, offering insight into the central tendency of our dataset. Similarly, for precision, both the median and mean values suggest promising results, with at least 50% of the records achieving a precision greater than 0.8. This underscores a generally reliable level of accuracy in the data.

4 Discussion and Future Work

The quantitative results we have just presented indicate that our setting—i.e., under inclusion of module information—produces high precision and recall values in many cases. Our results demonstrate a significant improvement compared to this baseline. While, absolutely speaking, the quantitative results are still

Table 2. Recall and Precision for detected GMO entities using module information

	Recall			Precision		
	≥0.5	≥0.75	= 1	≥0.5	≥0.75	= 1
with Module Information	73.3%	62.3%	45.0%	69.7%	59.6%	45.8%
mean	0.67			0.67		
median	0.75			0.80		
standard deviation	0.37			0.37		

moderate, we are in fact presenting the *very first* reasonably working approach for generating complex alignments that does not require shared individuals; as noted, shared individuals are far from a realistic setting in practice. In the latest evaluation of the participating systems in the OAEI complex alignment track, which was in 2021, most systems failed to detect any m:n complex alignments.[11] Only two systems proudced complex alignments,[12] but they required instance data. As such our contribution shows a path forward towards complex alignment in realistic settings, a challenge that had so far eluded researchers.

For our approach to work, we provided the LLM with module information, following the previously presented arguments that (1) ontologies without additional internal structure or meaningful additional information are often too ambiguous for automated complex alignment tasks, (2) module identification during the ontology design process can easily be provided by the ontology modelers at least as part of the documentation (while doing so post-hoc, by others, requires major efforts), and (3) providing such module information as part of ontologies (and/or their documentations) would likely significantly decrease the effort and cost of many ontology engineering tasks, including complex alignment [24]. As such, this is also a (repeated) call for improving ontology modeling methods to additionally provide module structure.

While our results, as presented, are very encouraging, substantial future investigations will be required to cast them into an ontology alignment system that can work autonomously at high precision. Intermediate steps could constitute human-in-the-loop approaches where a human ontology engineer receives suggestions from an LLM, e.g., as to the relevant modules for a question, postprocesses the LLM responses by manually checking the small number of suggestions, and feeding the correct suggestions back to the LLM for obtaining more complete responses. This is in line with the idea of an assisting system that limits the number of checks for the human, as opposed to the vast number of potential checks that would have to be done manually without such a limiting system.

In the future, we also intend to extend our approach to additional datasets featuring complex alignments for both evaluation and analytical objectives, e.g. [30]. Furthermore, we aim to explore alternative representations of modules to

---

[11] https://oaei.ontologymatching.org/2021/results/complex/geolink/index.html.
[12] https://oaei.ontologymatching.org/2021/results/complex/popgeolink/index.html.

LLMs and evaluate the model's performance with these variations. Fine-tuning existing LLMs, but also the integration of additional symbolic data or algorithms, e.g. pertaining to logical axioms that come with well-designed ontologies, and also the integration of traditional simple alignment algorithms to further assist with the complex alignment task are all on our path forward.

As for related work on LLMs for ontology alignment, for lack of space (see the extended arXiv version [4] for more details) we only briefly mention [10,11,18, 22], however, all these results do not go beyond simple alignment. In particular, no good results for complex alignment benchmarks have been reported yet.

5 Conclusion

We have presented the very first approach that is able to achieve good accuracy for complex ontology alignment without relying on shared individuals. The system is neurosymbolic in its nature as it addresses a symbolic task (complex ontology alignment), the output of which are alignment rules expressed in some logic, it furthermore makes decisive use of additional symbolic input in the form of ontology modules, and it uses an LLM as core processing engine. Our results suggest that further work bear the promise to result in strong complex ontology alignment systems, for ontologies that carry sufficient internal structure.

Acknowledgement. We acknowledge partial support by NSF award no. 2333532 "Proto-OKN Theme 3: An Education Gateway for the Proto-OKN" and by the Kansas State University GRIP program.

References

1. Achiam, J., et al.: GPT-4 technical report. arXiv preprint arXiv:2303.08774 (2023)
2. Algergawy, A., et al.: Results of the ontology alignment evaluation initiative 2018. In: 13th International Workshop on Ontology Matching Co-located with the 17th ISWC (OM 2018), vol. 2288, pp. 76–116 (2018)
3. Algergawy, A., et al.: Results of the ontology alignment evaluation initiative 2019. In: Shvaiko, P., Euzenat, J., Jiménez-Ruiz, E., Hassanzadeh, O., Trojahn, C. (eds.) Proceedings of the 14th International Workshop on Ontology Matching Co-located with the 18th International Semantic Web Conference (ISWC 2019), Auckland, New Zealand, 26 October 2019. CEUR Workshop Proceedings, vol. 2536, pp. 46–85. CEUR-WS.org (2019)
4. Amini, R., Norouzi, S.S., Hitzler, P., Amini, R.: Towards complex ontology alignment using large language models. arXiv preprint arXiv:2404.10329 (2024)
5. Amini, R., Zhou, L., Hitzler, P.: GeoLink cruises: a non-synthetic benchmark for co-reference resolution on knowledge graphs. In: Proceedings of the 29th ACM International Conference on Information & Knowledge Management, pp. 2959–2966 (2020)
6. Cheatham, M., Hitzler, P.: Conference v2.0: an uncertain version of the OAEI conference benchmark. In: Mika, P., et al. (eds.) ISWC 2014. LNCS, vol. 8797, pp. 33–48. Springer, Cham (2014). https://doi.org/10.1007/978-3-319-11915-1_3

7. Cheatham, M., et al.: The GeoLink knowledge graph. Big Earth Data **2**(2), 131–143 (2018)
8. Euzenat, J., Shvaiko, P.: Ontology Matching, 2nd edn. Springer, Heidelberg (2013)
9. Giglou, H.B., D'Souza, J., Auer, S.: LLMs4OM: Matching Ontologies with Large Language Models (2024)
10. He, Y., Chen, J., Dong, H., Horrocks, I.: Exploring large language models for ontology alignment. In: Fundulaki, I., Kozaki, K., Garijo, D., Gómez-Pérez, J.M. (eds.) Proceedings of the ISWC 2023 Posters, Demos and Industry Tracks: From Novel Ideas to Industrial Practice Co-located with 22nd International Semantic Web Conference (ISWC 2023), Athens, Greece, 6–10 November 2023. CEUR Workshop Proceedings, vol. 3632. CEUR-WS.org (2023)
11. Hertling, S., Paulheim, H.: OLaLa: ontology matching with large language models. In: Proceedings of the 12th Knowledge Capture Conference 2023, pp. 131–139 (2023)
12. Hitzler, P.: A review of the semantic web field. Commun. ACM **64**(2), 76–83 (2021)
13. Hitzler, P., Krötzsch, M., Rudolph, S.: Foundations of Semantic Web Technologies. Chapman and Hall/CRC Press (2010)
14. Hitzler, P., Kamruzzaman Sarker, Md., Eberhart, A. (eds.): Compendium of Neurosymbolic Artificial Intelligence. Frontiers in Artificial Intelligence and Applications, vol. 369. IOS Press (2023)
15. Krisnadhi, A., et al.: GeoLink core ontology design patterns. Technical report. https://people.cs.ksu.edu/~hitzler/pub2/gmo-tr.pdf
16. Krisnadhi, A., et al.: The GeoLink modular oceanography ontology. In: Arenas, M., et al. (eds.) ISWC 2015. LNCS, vol. 9367, pp. 301–309. Springer, Cham (2015). https://doi.org/10.1007/978-3-319-25010-6_19
17. Liu, P., Yuan, W., Jinlan, F., Jiang, Z., Hayashi, H., Neubig, G.: Pre-train, prompt, and predict: a systematic survey of prompting methods in natural language processing. ACM Comput. Surv. **55**(9), 1–35 (2023)
18. Norouzi, S.S., Mahdavinejad, M.S., Hitzler, P.: Conversational ontology alignment with ChatGPT. In: Shvaiko, P., Euzenat, J., Jiménez-Ruiz, E., Hassanzadeh, O., Trojahn, C. (eds.) Proceedings of the 18th International Workshop on Ontology Matching Co-located with the 22nd International Semantic Web Conference (ISWC 2023), Athens, Greece, 7 November 2023. CEUR Workshop Proceedings, vol. 3591, pp. 61–66. CEUR-WS.org (2023)
19. Pour, M.A.N., et al.: Results of the ontology alignment evaluation initiative 2020. In: Shvaiko, P., Euzenat, J., Jiménez-Ruiz, E., Hassanzadeh, O., Trojahn, C. (eds.) Proceedings of the 15th International Workshop on Ontology Matching Co-located with the 19th International Semantic Web Conference (ISWC 2020), Virtual Conference (originally planned to be in Athens, Greece), 2 November 2020. CEUR Workshop Proceedings, vol. 2788, pp. 92–138. CEUR-WS.org (2020)
20. Pour, M.A.N., et al.: Results of the ontology alignment evaluation initiative 2021. In: Shvaiko, P., Euzenat, J., Jiménez-Ruiz, E., Hassanzadeh, O., Trojahn, C. (eds.) Proceedings of the 16th International Workshop on Ontology Matching Co-located with the 20th International Semantic Web Conference (ISWC 2021), Virtual Conference, 25 October 2021. CEUR Workshop Proceedings, vol. 3063, pp. 62–108. CEUR-WS.org (2021)
21. Pour, M.A.N., et al.: Results of the ontology alignment evaluation initiative 2023. In: Shvaiko, P., Euzenat, J., Jiménez-Ruiz, E., Hassanzadeh, O., Trojahn, C. (eds.) Proceedings of the 18th International Workshop on Ontology Matching Co-located with the 22nd International Semantic Web Conference (ISWC 2023), Athens,

Greece, November 7, 2023. CEUR Workshop Proceedings, vol. 3591, pp. 97–139. CEUR-WS.org (2023)
22. Qiang, Z., Wang, W., Taylor, K.: Agent-OM: leveraging large language models for ontology matching. CoRR, abs/2312.00326 (2023)
23. Rudolph, S., Krötzsch, M., Patel-Schneider, P., Hitzler, P., Parsia, B.: OWL 2 web ontology language primer, 2nd edn. W3C recommendation, W3C, December 2012. https://www.w3.org/TR/2012/REC-owl2-primer-20121211/
24. Shimizu, C., Hammar, K., Hitzler, P.: Modular ontology modeling. Semantic Web **14**(3), 459–489 (2023)
25. Shvaiko, P., Euzenat, J.: Ontology matching: state of the art and future challenges. IEEE Trans. Knowl. Data Eng. **25**(1), 158–176 (2011)
26. Wei, J., et al.: Chain-of-thought prompting elicits reasoning in large language models. In: Advances in Neural Information Processing Systems, vol. 35, pp. 24824–24837 (2022)
27. White, J., et al.: A prompt pattern catalog to enhance prompt engineering with ChatGPT. arXiv preprint arXiv:2302.11382 (2023)
28. Zhou, L., Cheatham, M., Krisnadhi, A., Hitzler, P.: A complex alignment benchmark: GeoLink dataset. In: Vrandečić, D., et al. (eds.) ISWC 2018. LNCS, vol. 11137, pp. 273–288. Springer, Cham (2018). https://doi.org/10.1007/978-3-030-00668-6_17
29. Zhou, L., Cheatham, M., Krisnadhi, A., Hitzler, P.: GeoLink data set: a complex alignment benchmark from real-world ontology. Data Intell. **2**(3), 353–378 (2020)
30. Zhou, L., et al.: The enslaved dataset: a real-world complex ontology alignment benchmark using wikibase. In: d'Aquin, M., Dietze, S., Hauff, C., Curry, E., Cudré-Mauroux, P. (eds.) CIKM 2020: The 29th ACM International Conference on Information and Knowledge Management, Virtual Event, Ireland, 19–23 October 2020, pp. 3197–3204. ACM (2020)

Enhancing Knowledge Graph Construction: Evaluating with Emphasis on Hallucination, Omission, and Graph Similarity Metrics

Hussam Ghanem[(✉)] and Christophe Cruz

ICB, UMR 6306, CNRS, Université de Bourgogne, 21000 Dijon, France
hussam.ghanem@u-bourgogne.fr
https://icb.u-bourgogne.fr/

Abstract. Recent advancements in large language models have demonstrated significant potential in the automated construction of knowledge graphs from unstructured text. This paper builds upon our previous work [16], which evaluated various models using metrics like precision, recall, F1 score, triple matching, and graph matching, and introduces a refined approach to address the critical issues of hallucination and omission. We propose an enhanced evaluation framework incorporating BERTScore for graph similarity, setting a practical threshold of 95% for graph matching. Our experiments focus on the Mistral model, comparing its original and fine-tuned versions in zero-shot and few-shot settings. We further extend our experiments using examples from the KELM-sub training dataset, illustrating that the fine-tuned model significantly improves knowledge graph construction accuracy while reducing the exact hallucination and omission. However, our findings also reveal that the fine-tuned models perform worse in generalization tasks on the KELM-sub dataset. This study underscores the importance of comprehensive evaluation metrics in advancing the state-of-the-art in knowledge graph construction from textual data.

Keywords: Text-to-Knowledge Graph · Large Language Models · Zero-Shot Prompting · Few-Shot Prompting · Fine-Tuning · Hallucination

1 Introduction

Knowledge Graphs (KGs) play a crucial role in organizing complex information across diverse domains, such as question answering, recommendations, semantic search, etc. However, the ongoing challenge persists in constructing them, particularly as the primary sources of knowledge are embedded in unstructured textual data such as press articles, emails, and scientific journals. This challenge can be addressed by adopting an information extraction approach, sometimes implemented as a pipeline. It involves taking textual inputs, processing them using

Natural Language Processing (NLP) techniques, and leveraging the acquired knowledge to construct or enhance the KG.

In-context learning, as discussed by [7], coupled with prompt design, involves telling a model to execute a new task by presenting it with only a few demonstrations of input-output pairs during inference. Instruction fine-tuning methods, exemplified by InstructGPT [8] and Reinforcement Learning from Human Feedback (RLHF) [9], markedly enhance the model's ability to comprehend and follow a diverse range of written instructions. Numerous large language models (LLMs) have been introduced in the last year, as highlighted by [3], particularly within the ChatGPT [10] like models, which includes GPT-3 [11], LLaMA [12], Mistral [15], and Starling [17]. These models can be readily repurposed for KG construction from text by employing a prompt design that incorporates instructions and contextual information.

The task of converting textual information into structured KGs has gained significant traction with the advent of LLMs. These models offer unprecedented capabilities in understanding and generating human-like text, making them invaluable for a variety of NLP applications. Our previous work [16] explored different approaches to the Text-to-Knowledge Graph (T2KG) construction task, including Zero-Shot Prompting (ZSP) [19], Few-Shot Prompting (FSP) [6], and Fine-Tuning (FT) [4] of LLMs, employing models such as Llama2 [12], Mistral [15], and Starling [17]. In this work, we will include a little state of the art on contributions that use these three approaches (Sect. 2).

While traditional metrics like precision, recall, F1 score, triple matching, and graph matching provide a baseline for evaluating these models, they often overlook critical qualitative aspects of the generated graphs, such as hallucinations (incorrect or spurious triples) and omissions (missing relevant triples). Addressing these gaps, our current study introduces a refined evaluation framework that incorporates refined hallucination and omission metrics, and also incorporates BERTScore to measure the similarity between generated and ground truth graphs, setting an 95% similarity threshold for graph matching. This nuanced approach aims to provide a more comprehensive assessment of the models' performance in generating accurate and complete knowledge graphs.

In this paper, we specifically focus on comparing the original Mistral model and our finetuned Mistral (from our previous work) under zero-shot and few-shot settings. Additionally, we extend our experiments to include the KELM-sub dataset, utilizing few-shot examples to demonstrate that fine-tuning on a specific domain (WebNLG) significantly enhances performance when applied to related but distinct datasets with just few examples.

The present study is organized as follows, Sect. 2 presents a comprehensive overview of the current state-of-the-art approaches for Text to KG (T2KG) Construction and its evaluation metrics. In the Sect. 3, we present the general architecture of our proposed implementation (method), with datasets, metrics, and experiments. Section 4 then encapsulates the findings and discussions, presenting the culmination of results. Finally, Sect. 5 critically examines the strengths and limitations of these techniques.

2 Background

The current state of research on knowledge graph construction using LLMs is discussed. Three main approaches are identified: Zero-Shot, Few-Shot, and Fine-Tuning. Each approach has its own challenges, such as maintaining accuracy without specific training data or ensuring the robustness of models in diverse real-world scenarios. Evaluation metrics used to assess the quality of constructed KGs are also discussed, including semantic consistency and linguistic coherence. This section highlight methods and metrics to construct KGs and evaluate the result.

2.1 Zero Shot

Zero Shot methods enable KG construction without task-specific training data, leveraging the inherent capabilities of LLMs. [19] introduce an innovative approach using LLMs for knowledge graph construction, employing iterative zero-shot prompting for scalable and flexible KG construction. [20] evaluate the performance of LLMs, specifically GPT-4 and ChatGPT, in KG construction and reasoning tasks, introducing the Virtual Knowledge Extraction task and the VINE dataset, but they do not take into account open sourced LLMs as LLaMA [12]. [24] address the limitations of existing generative knowledge graph construction methods by leveraging large generative language models trained on structured data. The most of these approaches having the same limitation, which is the use of closed and huge LLMs as ChatGPT or GPT4 for this task. Challenges in this area include maintaining accuracy without specific training data and addressing nuanced relationships between entities in untrained domains.

2.2 Few Shot

Few Shot methods focus on constructing KGs with limited training examples, aiming to achieve accurate knowledge representation with minimal data. [6] introduce PiVe, a framework enhancing the graph-based generative capabilities of LLMs, and the authors create a verifier which is responsable to verifie the results of LLMs with multi-iteration type. [29] investigate LLMs' application in relation labeling for e-commerce Knowledge Graphs (KGs). As ZSP approaches, FSP approaches use closed and huge LLMs as ChatGPT or GPT4 [10] for this task. Challenges in this area include achieving high accuracy with minimal training data and ensuring the robustness of models in diverse real-world scenarios.

2.3 Fine-Tuning

Fine-Tuning methods involve adapting pre-trained language models to specific knowledge domains, enhancing their capabilities for constructing KGs tailored to particular contexts. [4] present a case study automating KG construction for compliance using BERT-based models. This study emphasizes the importance of machine learning models in interpreting rules for compliance automation. [31] propose Knowledge Graph-Enhanced Large Language Models (KGLLMs), enhancing LLMs with KGs for improved factual reasoning capabilities. These approaches that applied FT, they do not use new generations of LLMs, specially, decoder only LLMs as Llama, and Mistral. Challenges in this domain include ensuring the scalability, interpretability, and robustness of fine-tuned models across diverse knowledge domains.

2.4 Evaluation Metrics

As we employ LLMs to construct KGs, and given that LLMs function as Natural Language Generation (NLG) models, it becomes imperative to discuss NLG criteria. In NLG, two criteria [32] are used to assess the quality of the produced answers (triples in our context).

The first criterion is semantic consistency or Semantic Fidelity, which includes:

- **Hallucination**: Presence of information (facts) in the generated text that is absent in the input data.
- **Omission**: Omission of information present in the input data from the generated text.
- **Redundancy**: Repetition of information in the generated text (not considered in our evaluation).
- **Accuracy**: Exact match between the input and generated text without modification.
- **Ordering**: Sequence of information in the generated text differing from the input data (not considered in our evaluation).

The second criterion is linguistic coherence or Output Fluency, which evaluates the fluidity and linguistic correctness of the generated text. This criterion is not considered in our evaluation.

In their experiments, [3] calculated three hallucination metrics - subject hallucination, relation hallucination, and object hallucination - using preprocessing steps like stemming. They used the ground truth ontology and test sentence to determine if an entity or relation is present, considering any disparity between them as hallucination.

The authors of [6] evaluated their experiments using several metrics, including Triple Match F1 (T-F1), Graph Match F1 (G-F1), G-BERTScore (G-BS) from [33], and Graph Edit Distance (GED) from [35]. The GED metric measures the distance between the predicted and ground-truth graphs by calculating the number of edit operations needed to transform one into the other. To adhere to the semantic consistency criterion, we use the terms "omission" and "hallucination" instead of "addition" and "deletion," respectively.

3 Propositions

This section outlines our approach to evaluate the quality of generated KGs using metrics like T-F1, G-F1, G-BS, and GED. We also discuss the use of Optimal Edit Paths (OEP) to determine the precise number of operations needed to transform the predicted graph into an identical representation of the ground-truth graph. This method helps in calculating omissions and hallucinations in the generated graphs. Unlike our previous work where we marked a single hallucination or omission per generated graph, we now calculate the exact number of hallucinations and omissions for each generated graph (Fig. 1. Previously, we used examples from the WebNLG+2020 dataset [38] for testing with FSP techniques and trained LLMs using the FT technique. In this work, we take the best fine-tuned model (Mistral) from our previous work and apply zero/few-shot learning, comparing it with the original Mistral. Examples for few-shot learning are taken from WebNLG+2020 and the KELM-sub training dataset, and inference is applied on both datasets. We then compare these results with our previous work where models were applied on WebNLG+2020 and KELM-sub using examples from the WebNLG+2020 training dataset.

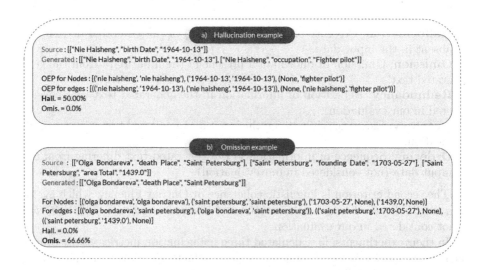

Fig. 1. Results examples

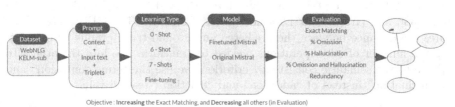

Fig. 2. Overall experimentation's process

3.1 Overall Experimentation's Process

In our previous work, we leveraged the WebNLG+2020 and KELM-sub datasets, specifically the version curated by [6]. Their preparation of graphs in lists of triples proves beneficial for evaluation purposes. We utilize these lists and employ NetworkX [39] to transform them back into graphs, facilitating evaluations on the resultant graphs. This step is instrumental in performing ZSP, FSP, and FT LLMs on these datasets. In this work, we will use examples from the training dataset of KELM-sub to do few-shot learning on the original and the finetuned (from our previous work) Mistral model.

Figure 2 illustrates the different stages of our experimentation process, including data preparation, model selection, training, validation, and evaluation. The process begins with data preparation, where the WEBNLG dataset is preprocessed and split into training, validation, and test sets. Next, the learning type is selected, and different models are trained using the training set. The trained models are then evaluated on the validation set to evaluate their performance. Finally, the best-performing model is selected and validated on the test set to estimate its generalization ability.

3.2 Prompting Learning

In this phase, we use ZSP and FSP techniques on LLMs to evaluate their proficiency in extracting triples for KG construction. We merge examples from the KELM-sub test dataset with our adapted prompt, strategically modified for contextual guidance without a support ontology description, as demonstrated by [3]. The prompts for ZSP and FSP are shown in Fig. 3(a) and Fig. 3(b).

For ZSP, we started with the method from [6], using the directive "Transform the text into a semantic graph" and enhanced it with additional sentences for our LLMs (Fig. 3(a)). For FSP, we used 6-shot learning, corresponding to the maximum KG size in KELM-sub, feeding the prompt with six examples of varying sizes (Fig. 3(b)).

3.3 Postprocessing

To evaluate the generated KGs against ground-truth KGs, we clean the LLM outputs by transforming generated graphs into organized lists of triples and transferring them to textual documents. This rule-based processing removes corrupted text outside the lists of triples, optimizing our evaluation process for metrics like G-F1, GED, and OEP (Sect. 3.4).

In our previous work, instructing LLMs to produce lists of triples sometimes resulted in unstructured text, which we addressed by substituting the generated text with an empty list of triples ('[["","",""]]'). This approach, however, underestimated hallucinations. In the current work, as illustrated in Fig. 1, we calculate the exact hallucination and omission for each generated graph through qualitative evaluation of two randomly generated graphs.

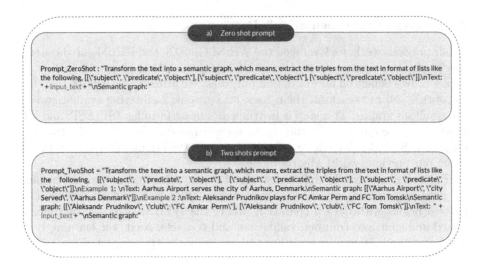

Fig. 3. Prompting examples

3.4 Experiment's Evaluation

To evaluate the generated graphs against ground-truth graphs, we use metrics such as T-F1, G-F1, G-BS [33], and GED [35] as in [6]. We also use Optimal Edit Paths (OEP) to calculate omissions and hallucinations in the generated graphs.

Our evaluation follows [6]'s methodology, especially in computing GED and G-F1, and involves constructing directed graphs from lists of triples using NetworkX [39]. Unlike [3], we do not use the ground truth test sentence of an ontology. Instead, we assess omissions and hallucinations using OEP, which provides the precise path of the edit, allowing exact quantification of these errors.

For example, Fig. 1 shows 2 omissions ('b)') and 1 hallucination 'a)' in using one of two paths "OEP for nodes" or "OEP for edges". Previously, we incremented the global hallucination metric for all graphs if ≥ 1 hallucinations or omissions were found. In the current work, we use OEP to detect the exact percentage of hallucination or omission in a generated graph, experimenting on 2 random examples from the WebNLG+2020 test dataset (Fig. 1).

Different from our previous work, our experiments are evaluated using examples from the KELM-sub test dataset (Table 2 and Table 1). Our primary goal is to improve G-F1, T-F1, G-BS and GM-GBS metrics, while reducing GED, hallucination, and omission.

3.5 Mathematical Representation of the Used Metrics

This study refines the metrics used for evaluating hallucinations and omissions in generated graphs and introduces a new metric, Graph Matching using Graph BERTScore (GM-GBS). In our previous work, we detailed the mathematical representation of all metrics used.

The G-BS metric evaluates graph matching by treating edges as sentences and using BERTScore to measure alignment between predicted and ground-truth edges. The F1 score for G-BS is calculated as follows:

$$R_{\text{BERT}} = \frac{1}{|x|} \sum_{x_i \in x} \max_{\hat{x}_j \in \hat{x}} x_i^T \hat{x}_j,$$

$$P_{\text{BERT}} = \frac{1}{|\hat{x}|} \sum_{\hat{x}_j \in \hat{x}} \max_{x_i \in x} x_i^T \hat{x}_j,$$

$$F1_{\text{BERT}} = \frac{2 \cdot P_{\text{BERT}} \cdot R_{\text{BERT}}}{P_{\text{BERT}} + R_{\text{BERT}}}.$$

where R_{BERT} is the recall, and P_{BERT} is the precision.

In this work, we use G-BS to compare generated graphs with ground-truth graphs, defining graph matching with a similarity threshold of 95% to introduce GM-GBS. This approach acknowledges that entities or relations in the generated graph may be synonymous with those in the ground truth graph. Results shown in Fig. 4 illustrate that even with 95% BERTScore similarity, the generated graph is nearly identical to the ground truth.

Source: [['Mitie', 'inception', '1987'], ['Mitie', 'instance of', 'Business'], ['Mitie', 'headquarters location', 'Bristol'], ['Mitie', 'stock exchange', 'London Stock Exchange']]
Generated : [['Mitie', 'instance of', 'Company'], ['Mitie', 'founded in', '1987'], ['Mitie', 'place of foundation', 'Bristol'], ['Mitie', 'country of foundation', 'United Kingdom']]
95.95%

Source: [['Ogert Muka', 'member of sports team', 'FK Dinamo Tirana']]
Generated: [['Ogert Muka', 'plays for', 'FK Dinamo Tirana']]
96.71%

Source: [['Seksyen 7 LRT station', 'country', 'Malaysia'], ['Seksyen 7 LRT station', 'owned by', 'Prasarana Malaysia']]
Generated: [['Seksyen 7 LRT station', 'owner', 'Prasarana Malaysia'], ['Seksyen 7 LRT station', 'country', 'Malaysia']]
99.18%

Fig. 4. Examples of the calculated GM-GBS

To calculate GM-GBS, we follow these steps: Given an array of F1 scores of G-BS $f_1, f_2, \ldots, f_n$ in `f1s_BS`, the fraction of F1 scores greater than 0.95 is calculated as follows:

1. Let $ToGrs$ be the total number of generated graphs.
2. Let f_m be the count of F1 scores that are greater than 0.95:

$$f_m = \sum_{i=1}^{N} \mathbf{1}(f_i > 0.95)$$

where $\mathbf{1}(\cdot)$ is the indicator function, which is 1 if the condition inside is true and 0 otherwise.

3. The fraction of F1 scores greater than 0.95 is given by: GM-GBS = $\frac{f_m}{N}$

For hallucinations and omissions, we use Optimal Edit Paths (OEP) to determine exact counts:

Hallucination: An edit operation is a hallucination if it adds an entity or relation not present in the gold graph. We previously used an overall hallucination metric **Hall.** = $\frac{hall}{ToGrs}$, where $hall$ is the number of graphs with hallucinations.

Omission: An edit operation is an omission if it deletes an entity or relation present in the gold graph. In the previous work the omission was computed by **Omis.** = $\frac{omiss}{ToGrs}$, where $omiss$ is the number of graphs with omissions.

In this work, we calculate exact percentages of hallucination and omission through qualitative evaluation.

Given a list of tuples lst = $[(g_1, p_1), (g_2, p_2), \ldots, (g_n, p_n)]$, where g_i represents a gold edge and p_i represents a predicted edge:

1. Let h be the number of hallucinations, where a hallucination is defined as g_i = None:

$$h = \sum_{i=1}^{n} 1(g_i = \text{None})$$

2. The exact hallucination rate is then calculated as: **Hall_Rate** = $\frac{h}{n}$

Where n is the total number of edges, and $1(\cdot)$ is the indicator function, which is 1 if the condition inside is true and 0 otherwise (Same for Omis_rate).

To calculate the exact omission rate:

1. Let o be the number of omissions, where an omission is defined as p_i = None:

$$o = \sum_{i=1}^{n} 1(p_i = \text{None})$$

2. The exact omission rate is then calculated as: **Omis_Rate** = $\frac{o}{n}$

4 Experiments

This section outlines the LLMs used in our experiments for ZSP and FSP and presents the experimental results.

We utilized the Mistral model from HuggingFace platform[1], specifically focusing on the finetuned Mistral model which showed the best results in our previous work. We also compared the finetuned model with the original Mistral model.

---
[1] Hugging Face: https://huggingface.co/.

- Original Mistral-7B-v0.1: A pretrained generative text model with 7 billion parameters introduced by [15], which outperforms Llama 2 13B in various benchmarks.
- Fine-tuned Mistral-7B-v0.1: Based on the original Mistral and fine-tuned on the WebNLG+2020 training dataset, this model outperformed other fine-tuned models like Llama2 (7b and 13b) and Starling in our previous work [16].

Our evaluation also considers hallucination and omission through a linguistic lens, unlike most studies which focus on precision, recall, F1 score, triple matching, or graph matching, except for [3] which includes hallucination evaluation.

Table 1 shows that the fine-tuned Mistral performs better in both ZSP and FSP compared to the original Mistral for the T2KG construction task. The performance improves with more examples (more shots), with both finetuned and original Mistral models. Seeing the fine-tuned Mistral, it has the best performance when given 7 shots, surpassing the original Mistral by a significant margin.

As mentioned in our previous work, to corroborate these findings, in this version of our study, we assess our fine-tuned models using KELM-sub dataset for few-shot. We see that even when we gave Mistral examples from KELM-sub, it works better than zero-shot for the test dataset of WebNLG.

As depicted in Fig. 2, Hall. represents Hallucinations, while Omis. denotes Omissions.

Table 1. Comparison of performance metrics and models on WebNLG test dataset. Lower values indicate better performance for GED, Hall., and Omis.

Model | Metric	G-F1	T-F1	G-BS	GED	Hall.	Omis.	GM-GBS
Mistral-0	2.30	3.27	77.87	15.84	20.35	31.31	33.27
Mistral-7	18.72	28.44	87.54	10.13	17.88	21.14	51.88
Mistral-FT-0	31.93	44.08	86.89	8.25	**13.55**	18.27	54.97
Mistral-FT-7	**34.68**	**49.11**	**91.99**	**6.69**	14.90	14.39	57.72
Mistral-6 (KELM-sub)	7.59	12.45	81.23	16.29	61.16	**7.64**	26.86
Mistral-FT-6 (KELM-sub)	31.37	47.49	91.27	7.51	27.37	8.26	**58.40**

The G-BS consistently remains high, indicating that LLMs frequently generate text with words (entities or relations) very similar to those in the ground truth graphs, which was one reason to use it for the GM-GBS metric. The finetuned Mistral with 7 shots achieves the highest G-F1, accurately generating approximately 35% of graphs identical to the ground truth. This model performs exceptionally well across various metrics, particularly in T-F1. Additionally, the finetuned Mistral with 6 examples from KELM-sub outperforms the finetuned Mistral with 7 examples from WebNLG+2020 using the GM-GBS metric.

In Table 2, we present the evaluation results of the original Mistral with 7-shot learning (using examples from WebNLG+2020) and the fine-tuned Mistral with zero-shot (Mistral-FT-0) and 7-shot (Mistral-FT-7) learning (also using examples from WebNLG+2020) on the KELM-sub test dataset, prepared by [6] and based on [40]. It is important to note that the experiments utilized the same prompts as previously described. The 7-shots experiments used examples from the WebNLG+2020 training dataset. These experiments aim to assess the generalization ability of the original LLMs with 7-shot learning and the fine-tuned LLMs with zero-shot and 7-shot learning across diverse domains in the T2KG construction task.

Another experiment was conducted using 6 random examples from the KELM-sub training dataset. We applied this prompt to both the original Mistral (Mistral-6) and our finetuned Mistral (Mistral-FT-6) models. As expected, Mistral-6 outperformed Mistral-7 because the examples were from the KELM-sub training dataset used in Mistral-6. However, it was interesting to observe that Mistral-FT-6 performed less effectively than Mistral-6 with the same examples. This suggests that finetuning on WebNLG domains reduces the generalizability of the LLMs.

The results in Table 2 indicate that the fine-tuned Mistral models perform less effectively than the original Mistral with 7 shots from WebNLG+2020 and with 6 shots from KELM-sub. Additionally, all fine-tuned versions of Mistral (Mistral-FT-7, Mistral-FT-0, and Mistral-FT-6) show inferior results on KELM-sub compared to WebNLG+2020. This disparity can be attributed to the presence of different relation types, with some types expressed differently in KELM-sub. To address this, we utilize G-BS to calculate the similarity between two graphs and consider them as synonyms if they are sufficiently similar (>95% of similarity). This metric, called GM-GBS (Graph Matching using Graph BERTScore), is the last metric presented in Table 2. GM-GBS indicates a higher value of graph matching. To assess the reliability of this metric, we conducted a qualitative evaluation as illustrated in Fig. 4.

Overall, unlike our previous work where we used examples from WebNLG with the original and fine-tuned models for few-shot learning, using examples from KELM-sub here shows that the results are relatively similar. This indicates that fine-tuning negatively affects the generalization capability of the models.

Table 2. Results on KELM-sub. Lower values indicate better performance for GED, Hall., and Omis.

Model	Metric	G-F1	T-F1	G-BS	GED	Hall.	Omis.	GM-GBS
Mistral-7	5.50	11.35	81.77	13.74	6.72	61.09	28.66
Mistral-FT-0	2.17	8.55	78.29	14.35	7.22	56.28	12.88
Mistral-FT-7	2.89	9.92	78.42	13.63	**6.22**	61.00	13.66
Mistral-6 (KELM-sub)	**12.00**	**31.08**	**85.49**	10.82	25.50	**32.44**	**38.88**
Mistral-FT-6 (KELM-sub)	4.00	17.66	84.30	12.50	11.06	48.17	36.22

Qualitative Results: As illustrated in Fig. 1, our metric precisely calculates the percentage of hallucinations and omissions in the generated graphs at the triple level. For example, if a generated graph contains 2 triples and 1 of them are not present in the ground truth graph, the hallucination rate is approximately 50%. Similarly, for omissions, if the generated graph is missing some triples present in the ground truth graph, the omission rate is calculated accordingly.

As previously mentioned, we use G-BS to calculate the similarity between generated and ground truth graphs. If the similarity value exceeds 95%, we consider it an exact match, based on the notion that entities or relations in the generated graph are very close to those in the ground truth graph, or what we refer to as synonyms. In Fig. 4, we present examples with varying levels of similarity, including one with approximately 95% similarity, to demonstrate that even with 95% similarity, the two graphs convey the same or very similar meanings.

5 Conclusion and Perspectives

In this study, we evaluated the performance of both the original and fine-tuned Mistral models for Text-to-Knowledge Graph (T2KG) construction tasks using Zero-Shot Prompting (ZSP) and Few-Shot Prompting (FSP). Our analysis incorporated a comprehensive set of metrics, including G-F1, T-F1, G-BS, GED, along with measures for hallucinations and omissions.

Our results demonstrate that the fine-tuned Mistral model generally outperforms the original Mistral, particularly in Few-Shot scenarios. The fine-tuned Mistral with seven shots achieved superior performance across most metrics, notably improving G-F1 and T-F1 scores, which indicates a higher fidelity in generating ground truth graphs, and reflects its improved ability to produce coherent and contextually relevant outputs.

Despite these improvements, we observed that fine-tuning on domain-specific data, such as WebNLG, can negatively impact the model's generalization capabilities. This was evident from the comparative performance of the fine-tuned models on the KELM-sub dataset, where the original Mistral model with 7 shots from WebNLG+2020 outperformed the fine-tuned variants. This finding highlights the importance of balancing domain-specific fine-tuning with maintaining broad generalization.

The inclusion of the GM-GBS metric provided valuable insights into the semantic similarity between generated and ground truth graphs. Our qualitative analysis of hallucinations and omissions further enhanced our understanding of model performance at the triple level.

Looking ahead, there are several promising avenues for further research. Refining evaluation metrics to account for synonyms of entities or relations in generated graphs could improve assessment accuracy. Additionally, leveraging LLMs for data augmentation in T2KG construction shows potential, as our experiments suggest that LLMs can maintain consistency in generating results and propose relevant triples.

Expanding evaluations to a broader range of domains and datasets can provide deeper insights into how various types of data influence model behavior and performance. Combining automated metrics with human evaluation could also offer a richer understanding of model quality, with domain experts providing valuable assessments of the relevance and accuracy of generated graphs. Exploring these directions will contribute to advancing the field of T2KG construction and enhancing the capabilities of language models in producing accurate and contextually appropriate knowledge graphs.

Acknowledgement. The authors thank the French company DAVI (Davi The Humanizers, Puteaux, France) for their support, and the French government for the plan France Relance funding.

References

1. Hogan, A., et al.: Knowledge graphs. ACM Comput. Surv. (CSUR) **54**, 1–37 (2021)
2. Noy, N., Gao, Y., Jain, A., Narayanan, A., Patterson, A., Taylor, J.: Industry-scale knowledge graphs: lessons and challenges: five diverse technology companies show how it's done. Queue **17**, 48–75 (2019)
3. Mihindukulasooriya, N., Tiwari, S., Enguix, C.F., Lata, K.: Text2kgbench: a benchmark for ontology-driven knowledge graph generation from text. In: International Semantic Web Conference, pp. 247–265. Springer, Heidelberg (2023). https://doi.org/10.1007/978-3-031-47243-5_14
4. Ershov, V.: A case study for compliance as code with graphs and language models: Public release of the regulatory knowledge graph. arXiv preprint arXiv:2302.01842 (2023)
5. Caufield, J.H., et al.: Structured prompt interrogation andrecursive extraction of semantics (spires): a method for populating knowledge bases using zero-shot learning. arXiv preprint arXiv:2304.02711 (2023)
6. Han, J., Collier, N., Buntine, W., Shareghi, E.: Pive: prompting with iterative verification improving graph-based generative capability of llms. arXiv preprint arXiv:2305.12392 (2023)
7. Min, S., et al.: Rethinking the role of demonstrations: what makes in-context learning work? arXiv preprint arXiv:2202.12837 (2022)
8. Ouyang, L., et al.: Training language models to follow instructions with human feedback. Adv. Neural. Inf. Process. Syst. **35**, 27730–27744 (2022)
9. Stiennon, N., et al.: Learning to summarize with human feedback. Adv. Neural. Inf. Process. Syst. **33**, 3008–3021 (2020)
10. R. OpenAI, Gpt-4 technical report, View in Article 2. arxiv:2303.08774 (2023)
11. Brown, T., et al.: Language models are few-shot learners. Adv. Neural. Inf. Process. Syst. **33**, 1877–1901 (2020)
12. Touvron, H., et al.: Llama: open and efficient foundation language models, arXiv preprint arXiv:2302.13971 (2023)
13. Workshop, B., et al.: Bloom: A 176b-parameter open-access multilingual language model. arXiv preprint arXiv:2211.05100 (2022)
14. Chowdhery, A., et al.: Palm: scaling language modeling with pathways. J. Mach. Learn. Res. **24**, 1–113 (2023)

15. Jiang, A.Q., et al.: Mistral 7b. arXiv preprint arXiv:2310.06825 (2023)
16. Ghanem, H., Cruz, C.: Fine-tuning vs. prompting: evaluating the knowledge graph construction with LLMs. In: International Workshop on Knowledge Graph Generation From Text (TEXT2KG), Co-located with the Extended Semantic Web Conference (ESWC) (2024)
17. Zhu, B., Frick, E., Wu, T., Zhu, H., Jiao, J.: Starling-7b: improving llm helpfulness & harmlessness with rlaif (2023)
18. Tunstall, L., et al.: Zephyr: direct distillation of lm alignment. arXiv preprint arXiv:2310.16944 (2023)
19. Carta, S., Giuliani, A., Piano, L., Podda, A.S., Pompianu, L., Tiddia, S.G.: Iterative zero-shot llm prompting for knowledge graph construction. arXiv preprint arXiv:2307.01128 (2023)
20. Zhu, Y., et al.: Llms for knowledge graph construction and reasoning: recent capabilities and future opportunities, arXiv preprint arXiv:2305.13168 (2023)
21. Li, B., et al.: Evaluating chatgpt's information extraction capabilities: An assessment of performance, explainability, calibration, and faithfulness.D arXiv preprint arXiv:2304.11633 (2023)
22. Wei, X., et al.: Zero-shot information extraction via chatting with chatgpt. arXiv preprint arXiv:2302.10205 (2023)
23. Jarnac, L., Couceiro, M., Monnin, P.: Relevant entity selection: knowledge graph bootstrapping via zero-shot analogical pruning. In: Proceedings of the 32nd ACM International Conference on Information and Knowledge Management, pp. 934–944 (2023)
24. Bi, Z., et al.: Codekgc: code language model for generative knowledge graph construction. ACM Trans. Asian Low-Res. Lang. Inf. Process. **23**, 1–16 (2024)
25. Yao, L., Peng, J., Mao, C., Luo, Y.: Exploring large language models for knowledge graph completion. arXiv preprint arXiv:2308.13916 (2023)
26. Khorashadizadeh, H., Mihindukulasooriya, N., Tiwari, S., Groppe, J., Groppe, S.: Exploring in-context learning capabilities of foundation models for generating knowledge graphs from text. arXiv preprint arXiv:2305.08804 (2023)
27. Deng, S., et al.: Construction and applications of billion-scale pre-trained multimodal business knowledge graph. In: 2023 IEEE 39th International Conference on Data Engineering (ICDE), pp. 2988–3002. IEEE (2023)
28. Trajanoska, M., Stojanov, R., Trajanov, D.: Enhancing knowledge graph construction using large language models. arXiv preprint arXiv:2305.04676 (2023)
29. Chen, J., et al.: Knowledge graph completion models are few-shot learners: an empirical study of relation labeling in e-commerce with llms. arXiv preprint arXiv:2305.09858 (2023)
30. Harnoune, A., Rhanoui, M., Mikram, M., Yousfi, S., Elkaimbillah, Z., El Asri, B.: Bert based clinical knowledge extraction for biomedical knowledge graph construction and analysis. Comput. Methods Prog. Biomed. Update **1**, 100042 (2021)
31. Yang, L., Chen, H., Li, Z., Ding, X., Wu, X.: Chatgpt is not enough: enhancing large language models with knowledge graphs for fact-aware language modeling. arXiv preprint arXiv:2306.11489 (2023)
32. Ferreira, T.C., van der Lee, C., Van Miltenburg, E., Krahmer, E.: Neural data-to-text generation: a comparison between pipeline and end-to-end architectures. arXiv preprint arXiv:1908.09022 (2019)
33. Saha, S., Yadav, P., Bauer, L., Bansal, M.: Explagraphs: an explanation graph generation task for structured commonsense reasoning. arXiv preprint arXiv:2104.07644 (2021)

34. Zhang, T., Kishore, V., Wu, F., Weinberger, K.Q., Artzi, Y.: Bertscore: evaluating text generation with bert. arXiv preprint arXiv:1904.09675 (2019)
35. Abu-Aisheh, Z., Raveaux, R., Ramel, J.-Y., Martineau, P.: An exact graph edit distance algorithm for solving pattern recognition problems. In: 4th International Conference on Pattern Recognition Applications and Methods 2015 (2015)
36. Papineni, K., Roukos, S., Ward, T., Zhu, W.-J.: Bleu: a method for automatic evaluation of machine translation. In: Proceedings of the 40th Annual Meeting of the Association for Computational Linguistics, pp. 311–318 (2002)
37. Lin, C.-Y.: Rouge: a package for automatic evaluation of summaries. In: Text Summarization Branches Out, pp. 74–81 (2004)
38. Gardent, C., Shimorina, A., Narayan, S., Perez-Beltrachini, L.: The webnlg challenge: generating text from rdf data. In: Proceedings of the 10th International Conference on Natural Language Generation, pp. 124–133 (2017)
39. Hagberg, A., Swart, P., Chult, D.S.: Exploring network structure, dynamics, and function using NetworkX, Technical Report, Los Alamos National Lab.(LANL), Los Alamos, NM (United States) (2008)
40. Agarwal, O., Ge, H., Shakeri, S., Al-Rfou, R.: Knowledge graph based synthetic corpus generation for knowledge-enhanced language model pre-training. arXiv preprint (2020)

Views, Semantic Data Catalog and Role-Based Access Control for Ontologies and Knowledge Graphs

Henrik Dibowski(✉)

Bosch Center for Artificial Intelligence, Robert Bosch GmbH, 71272 Renningen, Germany
henrik.dibowski@de.bosch.com

Abstract. Adequate means for viewing, browsing and searching knowledge graphs (KGs) and restricting the access of users are a crucial limiting factor for KGs. To address these issues, this paper presents a view concept and role-based access control (RBAC) for ontologies and KGs. The view concept allows for defining different lenses on a KG, which reduce the complexity to better readable subsets of a KG. Views can be defined with RDF and SHACL. Roles defined in a semantic data catalog can grant user groups access to views. SHACL-defined views combined with a RBAC approach are a novel concept. This combination can realize an expressive access control for KGs, which makes sure that user groups can only see and/or change the information that they are permitted to. It allows for a comprehensive and flexible access control, with different user groups having access to different subsets of a KG. Yet, it is simple to realize, apply and maintain.

The view concept and RBAC has been implemented at Bosch and is usable by more than 100,000 Bosch employees. It could successfully be demonstrated that views can significantly improve KG UIs by their different, simplified and specialized lenses that they provide.

Keywords: View Concept · Data Catalog · Role-based Access Control · Knowledge Graph · Ontology · SHACL

1 Introduction

Knowledge graphs (KGs) are on the rise and are spreading into more and more industrial use cases. Their strength of representing heterogeneous, highly connected information, of making it accessible via semantic search, and of inferring new facts via reasoning is superior to other data representations and data storage solutions.

A particular challenge of KGs consists in their much more distributed, heterogeneous structure, with many nodes and relations, which is generally more difficult to visualize than classical structured data formats, such as relational databases, tabular data or XML. For further increasing the acceptance and success of KGs, more mature software tools are needed for efficiently viewing, browsing and maintaining the contained information. Also, non-experts, with limited background in KGs and ontologies need to be able to use such tools intuitively and efficiently, which is rarely the case, as typical tools (e.g. Protégé

[1], WebProtégé [2], TopBraid Composer) require an in-depth knowledge of ontologies, their syntaxes, query languages etc. and are not meant for the broad audience.

Allowing more users to work with KGs furthermore requires new concepts for access control. The widespread simple access control concept that either grants full access or no access to a KG is insufficient, as in real-world scenarios it is rarely the case that every user is allowed to see everything. A rather dedicated access control is needed, which grants users of different groups access to different parts of the same KG. This is not only useful from the access control perspective, but also from the perspective of providing simplified views on a KG that make it easier for users to comprehend the complex graph structure.

This paper provides two main contributions that can improve the explained shortcomings. As first main contribution of this paper, a *view concept* for KGs is introduced. The view concept allows to define different lenses, the "views", on a KG. Each view can provide a reduced and specialized focus on a KG, and thus help to reduce the complexity when viewing KGs by only showing a subset of the defined classes and properties. This is related to views for relational databases, a concept still missing for KGs, since until today no standard way for defining views on KGs exists.

The view concept is combined with a *Role-Based Access Control* (RBAC), which is the second main contribution of the paper. The RBAC controls the access of users and applications to different KGs and their defined views. It applies an RDF-based semantic data catalog, which defines the assignment of views to roles via permissions. The combination of a view concept with a RBAC for KGs as proposed here is a novelty.

The paper is structured as follows: Sect. 2 describes the related work for views and access control for ontologies and KGs and their shortcomings. Section 3 introduces the view concept for KGs as first main contribution of the paper. Section 4 describes the realization of a role-based access control mechanism by a semantic data catalog as second main contribution. An evaluation is provided in Sect. 5, followed by a conclusion in Sect. 6.

2 Related Work

This section covers the related work for the main contributions of this paper, i.e. a view concept with a role-based access control mechanism for knowledge graphs. It summarizes existing work related to view concepts for ontologies and KGs in Sect. 2.1, and to access control mechanisms for KGs in Sect. 2.2.

2.1 Views for Ontologies and Knowledge Graphs

A yet unresolved topic for ontologies and KGs is the definition of views, i.e. different lenses that only show a relevant subset of the KG to certain groups of users. Views are an established technique for conventional relational databases but are missing for KGs. As KGs can reach a high complexity, comprising many different classes, properties and instances, displaying a KG in its entirety is not possible nor useful. Typically, different groups of persons require access to different sets of information. Furthermore, the maintenance of information shall be restricted, so that only dedicated users can edit dedicated subsets of a KG. Such a generic view concept providing different lenses on a KG and controlling the access of users is still missing. The closest solutions addressed in literature so far are summarized in the following.

Many different approaches can be found that handle the (semi-)automatic generation of UIs or Web pages from or with ontologies, such as [3–7] and [8]. They can generate static dedicated UIs from ontologies to render or even maintain the content. The visible parts are to be defined by a user a priory, or everything is included. A higher flexibility is offered by SPARQL Web Pages - an RDF-based framework to describe user interfaces for rendering Semantic Web data [9]. The defined user interfaces control what data of the KG is visible in what UI.

A few approaches for the specification of ontology views using a view definition language exist. [10] introduces RVL (RDF View Language), a view definition language based on the query language RQL (RDF Query Language). And [11] introduces CLOVE (Constraint Language for Ontology View Environments), a high-level constraint language that extends OWL constraints. However, they seem to be purely academic proprietary solutions.

Solutions for virtual KGs and ontology-based databases (OBDB) have proposed and implemented their own view concepts. The authors of [12] propose to extend the traditional OBDB model to support multiple descriptions of the ontological data under customized view ontologies as meaningful portions of the main ontology to allow different personalized views. The virtual KG system Ontop exposes the content of arbitrary relational databases as virtual KGs and supports the definition of lenses as relational views [13]. The KG platform Anzo follows yet a different approach and supports different types of lenses on the virtual KG, such as tables, lists, charts, web pages and others [14]. These view approaches however are proprietary solutions that are specific for the underlying tabular data and respective system.

Despite of several academic and commercial view approaches that appeared in the past, until today there is no standardized view approach for KGs and ontologies available.

2.2 Access Control Mechanisms for Knowledge Graphs

Access control is the selective restriction of access to a resource. This section summarizes the related work for access control for ontologies and KGs.

The most basic approach for access control is on a *per-user basis*. RDF triple stores usually support access control on a per-user basis, so that a particular user can either access the entire RDF triple store including all datasets, or specific datasets of the RDF triple store (dataset-level access). Additionally, the Oracle Database semantic data store supports triple-level security, which provides a thin layer of RDF-specific capabilities for security [15]. For documents, such as ontologies, Solid Pods constitute a way to control the access to them on a per-user basis [16]. On a per-user basis however means that access rights of each user need to be set, which can be challenging for large amounts of resources and/or users.

Role-Based Access Control (RBAC) is a more sophisticated and flexible solution for access control. Permissions are associated with roles, and users are made members of appropriate roles, thereby acquiring the roles' permissions [17]. RBAC support can be found in some RDF triple stores, for example Stardog [18], where re-usable roles can be defined and used for multiple users.

An alternative to RBAC in general is *Policy-Based Access Control* (PBAC). PBAC combines roles with attributes and determines access privileges via a set of rules or policies. There are first attempts that try to apply PBAC to KGs. A policy-based access control for an RDF store is proposed in [19]. It uses a collection of policy rules governing whether an action is permitted or prohibited. An ontology-based PBAC solution that can simplify and speed up the definition of rules in complex scenarios is introduced in [20].

An approach called *Ontology-Based Access Control* (OBAC) was proposed in [21]. It utilizes a domain ontology to describe raw data semantically and allow security policies to describe the information someone has access to. Other OBAC approaches are described in a comprehensive survey in [22]. The term *Knowledge-Based Access Control* (KBAC) is proposed in [23] as next level of authorization decision-making. It leverages the utility of KGs to determine permissions by examining how nodes are related in the graph.

Apart from RBAC, PBAC, OBAC and KBAC, other related work can be found. A fine-grained access control model based on logical descriptions of a KG and users is presented in [24, 25]. Recently, SHACL [26] has been proposed for implementing access control to RDF KGs. [27] presents a framework and architecture able to grant access to RDF sources for the execution of SPARQL queries using access control policies defined in SHACL.

2.3 Differentiation from State of the Art

Although handled separately so far, there is an overlap between views and access control. As views can constrain the visible (and hence accessible) information of a KG to a subset of it, they can control the access of users to a KG. This however requires the combination of a view approach with an access control mechanism, with the access control determining which users have access to what views. Such a combination of a view approach with an access control mechanism has been missing so far. Despite a variety of view approaches and access control mechanisms have been discussed in literature (see Sect. 2.1 and Sect. 2.2), none has covered the combination of generic KG views with a mature access control. Besides, until today there is no standardized view approach for KGs and ontologies, which is a large gap in the KG landscape of languages and standards of the W3C.

This gap is closed by the view and access control concept introduced in this paper as main contributions and will be explained in Sects. 3 and 4. This solution is especially elegant, as it makes use of RDF and SHACL, the standard languages for KGs, and as in combination of both complementary approaches it is expressive but still overall simple to realize, apply and maintain.

3 Knowledge Graph View Concept

KGs can reach a high complexity and comprise thousands of different classes and properties, which cover different aspects of a single or multiple domains. Various groups of persons or applications may need to use and access a KG, each of them with a different focus and interest and requiring to view or modify different parts of a KG.

For reducing the complexity and controlling the access to only the relevant subsets of a KG, a view concept has been developed and is introduced in this section as one of the main contributions of this paper. The view concept allows the definition of different views on a KG. Each view controls what parts of a KG are visible and modifiable in it. And to each view, different users can be granted access to.

The developed view concept can furthermore influence the appearance of the KG in UIs, such as by defining an ordering of the classes and properties, and by supporting multilinguality, so that the names and comments of classes and properties can appear in different languages, depending on the browser's language settings of the respective user. Design-related aspects of a UI, however, such as appearance, font size, positioning, color etc. are out of scope of the view concept.

3.1 Main Principle of the View Concept

This section introduces the view concept for KGs. A view is identified by its view name. A view can include one or more classes of a KG, and one or more properties of each included class, including both datatype and object properties. The classes and properties of a KG can be defined as viewable and modifiable in a specific view:

- *Viewable class*: the class and its instances are visible, i.e. they can be viewed and searched
- *Modifiable class*: the class and its instances are visible (see "viewable class") and modifiable, i.e. additionally, instances of the class can be created and deleted
- *Viewable property*: a property of a class and its values are visible
- *Modifiable property*: a property of a class and its values are visible (see "viewable property") and modifiable, i.e. additionally, the property values can be changed

In the following, the view concept is demonstrated by way of an KG example shown in Fig. 1. The upper part shows an extract of the ontology TBox with some of its classes. The lower part comprises the instance graph (ABox), consisting of instances of the classes and their object property relations and datatype property values (literals). This graph comprises the model of a system controller for heating systems, which is represented by the instance `d:ST350_CTR` of the class `s:SystemController`. Some datatype properties and its values are shown for the instance `d:ST350_CTR`, e.g. `rdfs:label (xsd:string)`, `s:isReleased (xsd:boolean)`, `s:typeNumber (xsd:int)` and `s:hasIPClass (xsd:int)`. Furthermore, two additional subgraphs expand from the instance via two object properties: 1) `s:hasSoftware` attaches a `s:SoftwareComponent` instance that models the software of the system controller; 2) `s:hasHardware` attaches a `s:HardwareComponent` instance that models its hardware.

For the graph given in Fig. 1, different aspects might be interesting for different personas or use cases. This can easily be accomplished by defining different views for the KG, for example a software-related view, and a hardware-related view. Currently, the software-related view is highlighted on the graph, comprising the colored classes and properties (only). Only these classes and properties are visible in the view. The grayed-out classes and properties, on the contrary, are not visible in the software-related view, but are only shown for the sake of completeness. They belong to the hardware-related view and are visible herein.

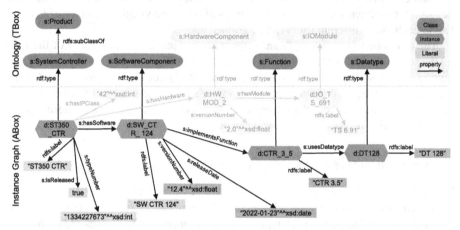

Fig. 1. Knowledge Graph example of a system controller model with a software-related view (colored classes and properties) and a hardware-related view (grayed-out part).

3.2 Definition of a Knowledge Graph View Layer

After the concept of views on KGs has been described in the previous section, the focus of this section is on how to define views. The visibility and modifiability of instances of certain classes and their properties is an additional aspect of a KG. In our opinion, it can be best defined as additional "view layer" on an existing KG, using SHACL and RDF.

The reasons for using SHACL are manifold: Many KGs nowadays use SHACL and not OWL for the definition of ontology axioms and constraints, and the dominance of SHACL over OWL is further growing. SHACL overcomes most of the complications and limitations of OWL and is becoming the industrial standard for KGs. It is well supported by the leading triple stores and ontology IDEs, allows for an extensive checking of completeness, is more expressive and supports far more complex constraints and rules than what can be formulated with OWL. Furthermore, SHACL purely relies on queries on a KG and does not require the computational overhead of a reasoning engine.

The view layer can be defined as extension of a KG itself. For that purpose, a new view ontology has been developed, which introduces a new "view" class and a few new properties. They can be seen in the UML class diagram in Fig. 2 using the prefix "ui". Figure 2 relates the view class ui:View via the annotation properties ui:isViewableIn and ui:isModifiableIn to the respective SHACL concepts, namely SHACL node shapes (sh:NodeShape) and SHACL property shapes (sh:PropertyShape).

A SHACL node shape, and hence the class it is defined for, can be defined as viewable or modifiable in a specific view by relating it to the respective view instance via one of the two properties. The class a SHACL node shape is defined for can either be expressed via the SHACL property sh:targetClass, or by declaring the class of interest to be both at the same time, a owl:Class and a SHACL node shape, which can be done by assigning both types via a respective rdf:type statement.

Also, a SHACL property shape, and hence the property it is defined for, can be defined as viewable or modifiable in a specific view by relating it to the respective view instance via the annotation properties ui:isViewableIn or ui:isModifiableIn.

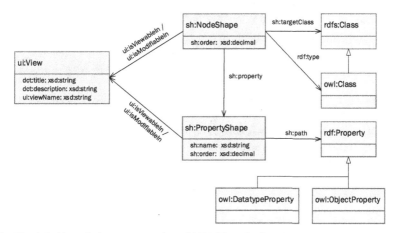

Fig. 2. The definition of views as extension of SHACL node shapes and SHACL property shapes.

Besides the visibility and modifiability declaration of classes and properties of a KG, a few more UI-related aspects can be defined additionally to the view. Via the SHACL property sh:name, display names can be defined for the properties, which are to be shown to the user in a UI, with support for multilinguality by using language tagged strings. And the SHACL property sh:order can be used for defining a sort sequence over the classes and properties of a KG by assigning ascending numbers to their SHACL node and property shapes.

To provide the largest benefit, the view concept can be combined with virtual properties, an approach proposed in [28, 29]. Virtual properties provide shortcuts on a KG that can enrich the scope of a class with other information beyond its direct neighborhood. With virtual properties, additional valuable information can be provided, which enriches the views. Virtual properties can be defined as visible in a view or not in the same way as standard properties (see above).

3.3 View Definition Example

This section provides an example of the view concept. Views are defined as extension of a KG using SHACL and RDF. This is shown in Fig. 3.

```
s:SystemController
  a owl:Class ;
  a sh:NodeShape ;
  ui:isViewableIn dcd:SoftwareView ;
  ui:isViewableIn dcd:HardwareView ;
  ui:isModifiableIn dcd:MasterDataView ;
  sh:order "15"^^xsd:decimal ;
  sh:property [
    a sh:PropertyShape ;
    ui:isViewableIn dcd:SoftwareView ;
    ui:isViewableIn dcd:HardwareView ;
    ui:isModifiableIn dcd:MasterDataView ;
    sh:order "2"^^xsd:decimal ;
    sh:path s:typeNumber ;
    sh:name "Type Number" ;
    sh:datatype xsd:float ;
  ] ;
  sh:property [
    a sh:PropertyShape ;
    ui:isViewableIn dcd:SoftwareView ;
    ui:isModifiableIn dcd:MasterDataView ;
    sh:order "4"^^xsd:decimal ;
    sh:path s:hasSoftware ;
    sh:name "Software" ;
    sh:class s:SoftwareComponent ;
  ] ;
  sh:property [
    a sh:PropertyShape ;
    ui:isViewableIn dcd:HardwareView ;
    ui:isModifiableIn dcd:MasterDataView ;
    sh:order "5"^^xsd:decimal ;
    sh:path s:hasHardware ;
    sh:name "Hardware" ;
    sh:class s:HardwareComponent ;
  ] ;
```

Fig. 3. Declaring classes and properties as visible and modifiable in different views by relating their SHACL node shapes and SHACL property shapes to the view instances (RDF turtle syntax).

Figure 3 shows the definition of the class s:SystemController, which was introduced in Fig. 1, and three of its properties, namely the float datatype property s:typeNumber and the two object properties s:hasSoftware and s:hasHardware. At the top of the RDF code, s:SystemController is defined therein as OWL class and SHACL node shape. The SHACL property sh:property is then used for attaching three SHACL property shape definitions to the SHACL node shape, one for each of the three properties.

A KG view is defined as instance of the class ui:View in the KG itself. In Fig. 3 overall three different view instances appear at multiple places: dcd:SoftwareView, dcd:HardwareView and dcd:MasterDataView. dcd:SoftwareView provides a software-related view on a system controller graph, whereas dcd:HardwareView provides a hardware-related view (see Sect. 3.1 and Fig. 1). dcd:MasterDataView defines a specific master data view dedicated for maintaining all data in the KG, comprising all classes and properties.

In its SHACL node shape definition, the s:SystemController class is declared to be visible in the views dcd:SoftwareView and dcd:HardwareView, and to be modifiable in the view dcd:MasterDataView. It furthermore has an order number of "15" defined, which determines its sort sequence among all classes defined in the KG.

By their SHACL property shapes, the three properties s:typeNumber, s:hasSoftware and s:hasHardware are given names (sh:name) and order numbers (sh:order). And most importantly, the properties are declared as being visible or modifiable in the three views. All properties are members of the view dcd:MasterDataView and are modifiable herein. Besides, s:typeNumber and s:hasSoftware are visible in the view dcd:SoftwareView, and s:typeNumber and s:hasHardware are visible in the view dcd:HardwareView.

4 Realizing Role-Based Access Control with a Semantic Data Catalog

The previous section introduced the view concept for KGs. This section explains how the view concept can be aligned with a previously introduced data catalog, in order to realize a role-based access control to the views of a KG. This is another main contribution of this paper.

In [30, 31] a *Data Catalog, Provenance, and Access Control* (DCPAC) ontology was introduced. The DCPAC ontology can be applied for defining a data catalog KG, which adds a semantic layer to a data lake and provides semantic description of the content, provenance, and access control permissions of the resources in a data lake. This ontology was created by combining several common, (predominantly) standardized ontology vocabularies and by aligning and extending them where necessary, most importantly the Data Catalog (DCAT) ontology (prefix "dcat"), the Provenance Ontology (PROV-O) (prefix "prov") and the Open Digital Rights Language (ODRL) ontology (prefix "odrl").

With the DCPAC ontology, a data catalog KG can be created that manages the datasets stored by a triple store. The semantic data catalog itself is stored in a dedicated dataset of the triple store, separated from the other KGs it describes and refers to. It is accessible by a dedicated backend only, which handles the authentication and role-based access control to the other KGs.

As another main contribution of this paper, the view ontology and the DCPAC ontology are aligned so that permissions to access certain views on certain KGs can be defined for users. This alignment between the DCPAC ontology and the view ontology is shown in Fig. 4: it consists of the new relationship `dcpac:permittedView` between the classes `odrl:Permission` and `ui:View`. Overall, the ontology allows for defining different roles (class `odrl:PartyCollection`), and the assignment of one or more roles to users and applications (class `prov:Agent`). A role is granted one or more permissions (class `odrl:Permission`), which hold for a certain KG (subclass `dcpac:GrapDatabaseDataset` of class `dcat:Dataset`). Via the permission, one or more views (object property `dcpac:permittedView`) can be associated with one or more KGs (object property `odrl:target`).

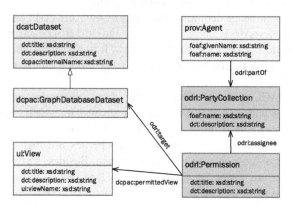

Fig. 4. Defining role-based access control to views of knowledge graphs with a data catalog ontology.

The membership of users or applications in a role can hence grant access to specific views on specific KGs. And via the view it is defined, which classes and properties of the KG are viewable and modifiable (see Sect. 3). In this way, a sophisticated RBAC for users and applications to resources in a KG via views can be realized, expressed as a (semantic data catalog) KG itself.

The proposed combination of a view concept with a RBAC utilizing a semantic data catalog is novel and elegant, as it makes use of RDF and SHACL, the standard languages for KGs. Despite both approaches are rather simple on their own, in combination they allow for an expressive access control for KGs, which is still overall simple to realize, apply and maintain. Users can get roles assigned, each of which can grant access to several KGs and several views on them. Each view controls the accessible classes and properties of the KG, and whether their instances can be viewed only, or also edited. Together, this allows for a comprehensive and flexible access control, where different user groups (aka roles) can have access to different subsets of KGs, with different access levels.

5 Evaluation

The main use cases of the view concept and RBAC are to control the access to the information in a KG, and to allow users to more easily view, browse and search a KG. The view concept therefore allows for defining and providing different lenses on a KG, each of which containing only a subset of classes and properties of the KG. Via roles defined in the semantic data catalog, permissions for different user groups are definable, so that each user group is allowed to access a certain view on the KG. The utilization of both concepts, e.g. in a UI, require a certain software architecture and tooling to be implemented. This is described in the following subsection.

The view approach and the RBAC via a semantic data catalog have been implemented at Bosch and are in use as functionality of the *Knowledge Graph Explorer*. The Knowledge Graph Explorer is a user interface for KGs and it was first introduced in [30] for a semantic data lake. It allows users to conveniently view, browse, search and even edit the data in a KG. This includes the roles and permissions in the semantic data catalog by the administrators as well. The Knowledge Graph Explorer, and with it the view approach and role-based access control, has been rolled out and is usable by more than 100,000 Bosch employees in a productive deployment, which proves the maturity and relevance of the approaches for Bosch.

5.1 Software Architecture

In this section, the solution and its software architecture are explained that realize the view concept and RBAC. The overall architecture and tooling is shown in Fig. 5. Herein, the *Knowledge Graph Explorer* is the user interface to the KG. Only the classes and properties of the view, which the given user is allowed to access, appear in it. The Knowledge Graph Explorer is realized as self-adaptive UI, which uses the view definition to dynamically render the permitted contents of the KG.

In between the triple store and the Knowledge Graph Explorer is a microservices backend that offers dedicated REST APIs and acts as interface. All data access, searches on the KGs and changes of KGs are controlled and handled via the *Knowledge Graph Engine* microservice of the backend. It has inbuilt support for the view concept in all its REST APIs, meaning that each request going to the backend needs to provide a view name. Here, the *Role-Based Access Control* comes into play, which is implemented as microservice on its own: For the given roles of a user or client application, it is checked in the data catalog whether a permission exists to access the view on the dataset. Furthermore, it is checked if the classes and properties to be accessed via a request are visible in the view, i.e. are included in it. Two REST APIs are utilized for the visibility checking: "Visible Classes API" returns all visible classes of a given view of a KG, and "Visible Properties API" returns all visible properties of a given class of a given view. Other REST APIs, which can fetch property values of instances ("Instance API"), or which can realize a KG search ("Search API"), apply the view concept as well and only allow to access, see and search the visible parts of a KG.

5.2 Quantitative Analysis

The view concept and RBAC have proven to be efficient and scalable, as the following quantitative analysis shows. The implementation shown in Fig. 5 is benchmarked with two different KG configurations, which are shown in Table 1: a smaller KG of Bosch Home Comfort for residential heating systems and heat pumps containing 39,569 instances and 511,000 triples in total, deployed on Apache Jena Fuseki; and a large KG

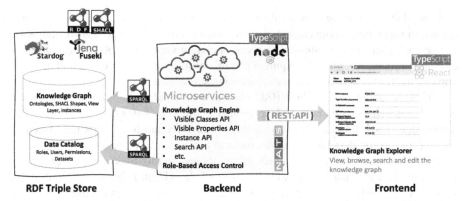

Fig. 5. Software architecture for the implementation of the view concept, semantic data catalog and role-based access control at Bosch.

containing Bosch master data with over 3.4 million instances and 69 million triples, deployed on Stardog.

As Table 2 shows, the SPARQL query for fetching all visible classes of a given view ("Visible Classes API") succeeds in a bit more than hundred milliseconds on Jena Fuseki, and in about 1.7 s on Stardog. A similar performance can be measured for a SPARQL query that fetches all visible properties of a given class ("Visible Properties API"), resulting in 114 ms and 1.817 s, respectively, which demonstrates a good scalability of the view concept.

Table 1. Knowledge graph examples and configurations from two different Bosch domains.

KG	Triple Store	# Classes	# Properties	# Instances	# Triples
Home Comfort	Jena Fuseki	209	315	39,569	511K
Master Data	Stardog	298	512	3.4M	69M

Table 2. SPARQL query execution times for fetching all visible classes and all visible properties of a class on the two different knowledge graphs.

KG	Triple Store	Visible classes query	Visible properties query
Home Comfort	Jena Fuseki	0.136s	0.114s
Master Data	Stardog	1.715s	1.817s

Overall, the backend and the Knowledge Graph Explorer have proven to be fast and responsive, even for large numbers of classes and for large KGs. Altogether, the responsiveness and scalability has ensured a good user experience and the viewing, browsing and searching of a KG has proven to be fast and responsive.

5.3 User Experience Study

From the productive usage and a conducted user experience study we could gain many insights and positive feedback about views and the RBAC and their benefit in the daily work of our employees. In the following, these points will be summarized.

The overall finding and feedback are that views can make a KG much better viewable, readable and searchable. The view approach proved to be the key enabler for realizing truly generic UIs that can self-adapt to the underlying information and lenses defined on it. Views can much reduce the complexity when viewing KGs by only showing a subset of the defined classes and properties. Via different views, different lenses can be defined on the same KG, allowing to have a reduced and specialized focus on a KG in each view. This helps users to find the required classes, properties and information in the UI more quickly and more easily, to more proficiently define searches on the KG, and it altogether improves the overall readability by compacting the content to the relevant only. Additionally, the definition of roles and view permissions via RDF in a data catalog have proven to be a powerful and flexible way for realizing a role-based access control on KGs. Roles and permissions can easily be defined and modified by administrators via the same generic UI, i.e. the Knowledge Graph Explorer, as they are a KG itself. It turned out to be very simple to create and maintain views and roles, to assign or remove roles to and from users, and to enlarge or limit the scope of what parts of a KG users can see and modify.

The expressiveness and flexibility of the approach has proven to be sufficient to meet the demands for an enterprise-level access control for KGs and views for a population of more than 100,000 employees.

6 Conclusion and Outlook

This paper presents a view concept and RBAC for ontologies and KGs. The view concept allows the definition of different views, i.e. lenses, on a KG. Views can reduce the complexity to relevant subsets of a KG and control the access to it. The views determine what classes and properties of a KG are visible and modifiable herein. Via a semantic data catalog realized as KG, roles can be defined and assigned to users, which can grant access to certain views on certain KGs. It allows the implementation of a powerful RBAC.

An approach has been shown how SHACL node and property shapes can be enhanced for defining views. A software architecture for a generic, self-adaptive, frame-based UI has been outlined, which can automatically generate user interfaces and masks for different views of a KG, thus allowing to view and search a KG by just using the structural information defined in the TBox and SHACL shapes.

The view approach has been implemented at Bosch and is usable by more than 100,000 Bosch employees in a productive deployment, which proves the maturity and relevance of the approach for Bosch. A quantitative analysis and user experience study have proven the scalability of the approach to large KGs also and the feasibility, scalability and significant value that views can provide.

The view concept is so far limited to defining which classes and properties are visible in a view, whether they are viewable or modifiable, what labels shall appear in what language, and how to order the entities in a UI. Out of scope are design-related aspects, such as font size, style and/or color to be used for rendering KG entities in the UI. This has been left for future research.

References

1. Musen, M.A., et al.: The Protégé project: a look back and a look forward. AI Matters **1**(4), 4–12 (2015)
2. Horridge, M., Tudorache, T., Nuylas, C., Vendetti, J., Noy, N.F., Musen, M.A.: WebProtégé: a collaborative Web-based platform for editing biomedical ontologies. Bioinformatics **30**(16), 2384–2385 (2014)
3. Hitz, M., Kessel, T.: Using application ontologies for the automatic generation of user interfaces for dialog-based applications. In: International Conference on Research and Practical Issues of Enterprise Information Systems (2016). https://doi.org/10.1007/978-3-319-499 44-4_2
4. Arndt, N., Zänker, S., Sejdiu, G., Tramp, S.: Jekyll RDF: template-based linked data publication with minimized effort and maximum scalability. In: International Conference on Web Engineering (ICWE 2019), pp. 331–346 (2019). https://doi.org/10.1007/978-3-030-19274-7_24
5. Khalili, A., Loizou, A., van Harmelen, F.: Adaptive linked data-driven web components: building flexible and reusable semantic web interfaces. In: International Conference on the Semantic Web (ESWC 2016), pp. 677–692 (2016). https://doi.org/10.1007/978-3-319-34129-3_41
6. Canadas, J., Palma, J., Tunez, S.: Model-driven rich user interface generation from ontologies for data-intensive web applications. In: 7th Workshop on Knowledge Engineering and Software Engineering, La Laguna, Spain (2011)
7. Gaulke, W., Ziegler, J.: Using profiled ontologies to leverage model driven user interface generation. In: Proceedings of 7th ACM SIGCHI Symposium on Engineering Interactive Computing Systems (EICS 2015), pp. 254–259 (2015)
8. Sahar, A., Armin, B., Shepherd, H., Lexing, L.: ActiveRaUL: automatically generated web interfaces for creating RDF data. In: Proceedings of the 12th International Semantic Web Conference (ISWC 2013), pp. 117–120 (2013)
9. Knublauch, H., Cyganiak, R.: SPARQL web pages – user guide (2019). https://uispin.org/ui.html
10. Magkanaraki, A., Tannen, V., Christophides, V., Plexousakis, D.: Viewing the semantic web through RVL lenses. Web Semant. **1**(4), 359–375 (2004)
11. Uceda-Sosa, R., Chen, C. X., Claypool, K. T.: CLOVE: a framework to design ontology views. In: Proceedings of the 13th International Conference on Conceptual Modeling (ER'04), vol. 3288 of Lecture Notes in Computer Science, Shanghai, China, pp. 844–849 (2004)
12. Bachtarzi, C., Fouzia, B.: View-OD: a view model for ontology-based databases. Int. J. Intell. Inf. Database Syst. **7**(4), 295–323 (2013)
13. Free University of Bozen-Bolzano, Ontop – Lenses. https://ontop-vkg.org/guide/advanced/lenses.html
14. Cambridge Semantics, Creating Lenses. https://docs.cambridgesemantics.com/anzo/v5.3/userdoc/create-lens.htm
15. Oracle: RDF Knowledge Graph Developer's Guide, 6. Fine-Grained Access Control for RDF Data. https://docs.oracle.com/en/database/oracle/oracle-database/19/rdfrm/fine-grained-access-control-rdf.html

16. Capadisli, S., Berners-lee, T., Verborgh, R., Kjernsmo, K.: Solid Protocol. W3C Solid Community Group Submission (2022). https://solidproject.org/TR/protocol
17. Sandhu, R.S.: Role-based access control. Adv. Comput. **46**, 237–286 (1998). https://doi.org/10.1016/S0065-2458(08)60206-5
18. Stardog Union, Stardog. https://www.stardog.com/
19. Reddivari, P., Finin, T., Joshi, A.: Policy-based access control for an RDF store. In: International Joint Conference on Artificial Intelligence (2005)
20. Imran-Daud, M., Sanchez, D., Viejo, A.: Ontology-based access control management: two use cases. In 8th International Conference on Agents and Artificial Intelligence (ICAART 2016), pp. 244–249, SCITEPRESS - Science and Technology Publications, Lda, Setubal, PRT (2016). https://doi.org/10.5220/0005777902440249
21. Brewster, C., Nouwt, B., Raaijmakers, S., Verhoosel, J.: Ontology-based access control for FAIR data. Data Intell. **2020**(2), 66–77 (2020). https://doi.org/10.1162/dint_a_00029
22. Gicquel, P.Y., Bouché-Pillon, J., Zaraté, P., Aussenac-Gilles, N., Chevalier, Y.: Ontologies and rules for access control: a feature oriented survey. In: 1st Workshop on Collaboration in Knowledge Discovery and Decision Making: Applications to Sustainable Agriculture, La Plata, Argentina, pp.1–12 (2022)
23. Tumy, B.: Implementing Knowledge-Based Access Control, Indykite (2022). https://www.indykite.com/blogs/implementing-knowledge-based-access-control
24. Maurino, A., Palmonari, M., Spahiu, B.: Towards an access control model for knowledge graphs. In: SEBD 2021 Italian Symposium on Advanced Database Systems - Proceedings of the 29th Italian Symposium on Advanced Database Systems, CEUR-WS (2021). https://ceur-ws.org/Vol-2994/paper37.pdf
25. Valzelli, M., Maurino, A., Palmonari, M.: A fine-grained access control model for knowledge graphs. In: 17th International Conference on Security and Cryptography (2020). https://www.scitepress.org/Papers/2020/98335/98335.pdf
26. Knublauch, H., Kontokostas, D.: Shapes constraint language (SHACL). W3C Recommendation (2017). https://www.w3.org/TR/shacl/
27. Rohde, P.D., Vidal, M.E., Iglesias, E.: SHACL-ACL: access control with SHACL. In: European Semantic Web Conference (ESWC 2023) (2023). https://2023.eswc-conferences.org/wp-content/uploads/2023/05/paper_Rohde_2023_SHACL-ACL.pdf
28. Dibowski, H.: Virtual properties for ontologies and knowledge graphs. In: International Workshop on Multilingual Semantic Web (IWMSW-2022), Co-located with the KGWSC-2022, pp. 11–25 (2022)
29. Dibowski, H.: Enhancing the viewing, browsing and searching of knowledge graphs with virtual properties. Int. J. Web Inf. Syst. (2024). https://doi.org/10.1108/IJWIS-02-2023-0027
30. Dibowski, H., Schmid, S.: Using knowledge graphs to manage a data lake. In: Reussner, R. H., Koziolek, A., Heinrich, R. (eds.) INFORMATIK 2020, Gesellschaft für Informatik, Bonn, pp. 41–50 (2020). https://doi.org/10.18420/inf2020_02
31. Dibowski, H., Schmid, S., Svetashova, Y., Henson, C., Tran, T.: using semantic technologies to manage a data lake: data catalog, provenance and access control. In: 13th International Workshop on Scalable Semantic Web Knowledge Base Systems (SSWS 2020) co-located with the 19th International Semantic Web Conference (ISWC 2020), Athens, Greece (2020)

Accelerating Medical Knowledge Discovery Through Automated Knowledge Graph Generation and Enrichment

Mutahira Khalid[1], Raihana Rahman[2], Asim Abbas[3], Sushama Kumari[4], Iram Wajahat[5], and Syed Ahmad Chan Bukhari[6](✉)

[1] TIB - Leibniz Information Centre for Science and Technology, 30165 Hannover, Lower Saxony, Germany
`Mutahira.Khalid@tib.eu`
[2] Department of Statistics, Harvard University, Cambridge, MA, USA
`rrahman@college.Harvard.edu`
[3] School of Computer Science, University of Birmingham, Edgbaston, Birmingham, UK
`axa2233@student.bham.ac.uk`
[4] Routing and Planning, Amazon, Bellevue, WA, USA
[5] Institute for Biotechnology, St. John's University, New York, USA
`wajahati@stjohns.edu`
[6] Division of Computer Science, Mathematics and Science, St. John's University, New York, USA
`bukharis@stjohns.edu`

Abstract. Knowledge graphs (KGs) serve as powerful tools for organizing and representing structured knowledge. While their utility is widely recognized, challenges persist in their automation and completeness. Despite efforts in automation and the utilization of expert-created ontologies, gaps in connectivity remain prevalent within KGs. In response to these challenges, we propose an innovative approach termed "Medical Knowledge Graph Automation (M-KGA)". M-KGA leverages user-provided medical concepts and enriches them semantically using BioPortal ontologies, thereby enhancing the completeness of knowledge graphs through the integration of pre-trained embeddings. Our approach introduces two distinct methodologies for uncovering hidden connections within the knowledge graph: a cluster-based approach and a node-based approach. Through rigorous testing involving 300 frequently occurring medical concepts in Electronic Health Records (EHRs), our M-KGA framework demonstrates promising results obtained 50% accuracy, 57% F1-score, 50% recall and 65% precision on a cluster base approach. Similarly, we achieve 85% accuracy, 87% F1-score, 89% recall and 88% precision on a node base approach indicating its potential to address the limitations of existing knowledge graph automation techniques.

Keywords: Knowledge Graphs · Ontologies · Healthcare · Decision Support systems

1 Introduction

While once considered a relic of early Artificial Intelligence (AI) research [1], knowledge graphs (KGs) have experienced a remarkable resurgence in recent years. Knowledge graphs, which serve as the foundation of symbolic AI, consist of interconnected knowledge pertaining to many domains like medical, finance, commerce, and education [2]. Particularly in medicine, KGs have emerged as indispensable tool.

KGs offer numerous advantages over traditional relational databases, primarily stemming from their diverse array of nodes and the ability to establish connections between them. This versatility lends itself to applications ranging from search engine optimization to recommendation systems, knowledge discovery, and research facilitation. However, the process of constructing KGs is inherently labor-intensive, especially in the intricate domain of medicine, despite its profound significance. At its core, a KG comprises facts, often represented as triplets, each consisting of a relationship and two nodes. With KGs consisting of millions to billions of these triplets, their aggregation holds immense potential for information discovery, data integration, and effective management. Yet, crafting KGs, particularly within the medical domain, presents formidable challenges due to the complexity of medical concepts and relationships between them. Compounding these challenges is the prevalence of unstructured medical data, further complicating the KG creation process.

Various methodologies for graph creation have emerged in recent years, ranging from automatic to semi-automated and manual approaches [3]. While these methods addresses some challenges. However they often suffer from significant deficiencies such as they lack standardized platforms or code for graph creation despite offering graphical methodologies. Furthermore, some approaches utilize hospital notes to generate nodes and relations, they overlook the potential benefits of data augmentation, resulting in incomplete graphs. Additionally, there is currently no promising technology capable of generating graphs in real-time, further impeding the process.

In response to these challenges, our proposed approach, Medical Knowledge Graph Automation (M-KGA), effectively addresses these obstacles by seamlessly processing both structured and unstructured data in real-time. The preprocessing steps involve Named Entity Recognition (NER)-based keyword extraction from unstructured data using the SciSpacy library [4], tailored for scientific and biomedical content. Subsequently, a knowledge filtration phase eliminates duplicates and extraneous terms before rapidly generating the knowledge graph using Neo4j's query language, 'Cypher'. Furthermore, we leverage Bioportal [5] for data augmentation, enriching medical terms semantically via incorporating metadata such as definitions, synonyms, and hierarchies. Following data augmentation, a semantic information filtration phase removes duplicates and non-English terms, enhancing the quality of the knowledge graph. To uncover concealed linkages and associations between medical terms, we utilize the pre-trained contextual word embedding model Clinical BERT [6], trained on the MIMIC-III dataset. This facilitates the discovery of valuable insights within the data and contributes to the creation of a comprehensive knowledge graph. We proposed cluster-based and node-based comparison methods to unveil hidden relationships by exploiting Clinical BERT within the knowledge graph.

Furthermore, our proposed approach enables users to effortlessly navigate complex features and generate autonomous knowledge graphs. Consequently, users can efficiently generate KGs based on input data, eliminating the need for prolonged waiting periods. Additionally, users have access to the generated files for further study and analysis. Ultimately, the discovery of hidden connections through our approach aids clinicians in gaining a deeper understanding of patient symptoms. Also it is benefiting insurance companies in identifying fraudulent claims and examining inaccurate forecasts of medical codes.

In summary, our research makes the following contributions:

i. Proposed a significant approach for automating the construction of a Medical Knowledge Graph, known as Medical Knowledge Graph Automation (M-KGA).
ii. Utilization of node-based and cluster-based comparisons for KG completion.
iii. Conducting rigorous evaluation to demonstrate the efficiency of our technique and the resulting knowledge graph.

Further, the paper is arranged as: Related work on KG automation in the medical and other fields is discussed in Sect. 2. The proposed methodology to create KG is presented in Sect. 3. The result and assessment is describe in Sect. 4, conducted on several medical use cases. Finally, we have comprehensively presented the limitation, future direction and conclusion in Sect. 5.

2 Related Work

Over the past few years, numerous knowledge graph (KG) automation techniques have emerged across sectors like business, healthcare, finance, and education, each designed for specific use cases. KG creation in healthcare, however, faces unique challenges due to the data's diversity, volume, and complexity. While KG completion methods are valuable for uncovering connections using machine learning, they are rarely applied. Additionally, there is no platform that provides a fully automated KG for any specific area within the medical field.

Medical KG are built using a variety of approaches, including human, semi-automatic, and automated methods, as well as modern and conventional procedures. An important example of a semi-automated graph generation procedure that made use of scientific literature and pre-existing datasets is the COVID-19 graph [7]. Evidence mining, hypothesis ranking, and relation extraction were carried out using hierarchical spherical embeddings, ontology-enriched text embeddings, and cross-media semantic-structure representation. A mechanism for producing reports and responding to inquiries was also devised. 7,230 diseases, 9,123 chemicals, and 50,864 genes are included in the final KG. There are 1,725,518 chemical-gene relationships, 5,556,670 chemical-disease ties, and 77,844,574 gene-disease links.

Another method for creating healthcare knowledge graphs was available; it consisted of eight modules and produced a KG semi-automatically utilizing 16,217,270 de-identified clinical visit data from 3,767,198 patients [8]. Entity recognition, entity normalization, relation extraction, property calculation, graph cleaning, related-entity ranking, and graph embedding are among the processes in the process. To store extra context in healthcare graphs, the quadruplet form is used instead of the traditional triplet. A medical KG of 22,508 entities and 579,094 quadruplets with nine distinct entity kinds was the end result.

Knowledge graphs are widely used in fields such as medicine, diagnosis, and fraud detection. A Fraud, Waste, and Abuse (FWA) detection system, utilizing a Chinese medical knowledge graph, was developed to identify fraudulent claims in insurance [9]. Entities were extracted using deep learning techniques, with human validation ensuring accuracy. The graph, built from medical texts and drug labels from the Chinese Food and Drug Administration, comprised 1,616,549 nodes and 5,963,444 relations, achieving a 70% accuracy in detecting fraud claims.

Knowledge Graphs are often constructed for text data, which makes up 80% of all available data. The Semi-automated KG Construction and Application (SAKA) framework [10] is an interesting use case that utilizes both auditory and structured data for KG generation. The audio-based KG Information Extraction (AGIE) technique employs Voice Activity Detection (VAD), Speaker Diarization (SD), and the Medical Information Extractor (MIE) model to extract entities. Additionally, a system was developed to handle user inquiries and ensure the data remains relevant and updated. Testing on datasets like LibriSpeech, VoxCeleb, and doctor-patient dialogues showed promising results. An automated medical knowledge graph for "Subarachnoid hemorrhage" used over a thousand case records, enhanced by Bioportal ontologies across several layers, including semantic, statistical, and predictive knowledge layers [11]. Although the source code is publicly available, adapting it for other medical fields is challenging, as switching ontologies alone isn't enough, and word embeddings need retraining. The authors also overlooked knowledge graph (KG) completion. Another automation method for evidence-based medicine focused on cerebral aneurysm and COVID-19, using peer-reviewed ontologies and clustering models with deep learning techniques like RNN and BioBERT [12]. Accuracy reached 93% for COVID and 82% for aneurysm datasets.

The majority of approaches depict the KG creation process in a manual or semi-automated manner, as the literature demonstrates. The majority of them are tailored to specific use cases and the medical field, and they rarely incorporate the KG completion technique, which can harness the potential of BIG data to uncover facts. Although the generated KGs are indeed helpful in many ways, they cannot be applied generally. Furthermore, there isn't a platform that can handle user requests to validate specific use cases and create the appropriate KG in a matter of minutes or seconds. Our method, which offers automation, is distinct and creative in that it uses expert-created ontologies to produce a full and comprehensive KG while satisfying the user's request for a specific KG generation. People can benefit from the approach in a variety of fields, and the KGs produced by our method can be used to enhance the research of others.

3 Methods

Our proposed approach is designed in mutiple steps. Figure 1 illustrates the entire workflow of the Medical-Knowledge Graph Automation (M-KGA). This approach acquire data in two formats: structured and unstructured. It then applies various natural language processing (NLP) techniques to process the data. Initially, Bioportal is utilized to identify and enhance medical concepts with semantic information. The fetched data is filtered and used to create nodes in a knowledge graph (KG) along with their relationships. A pre-trained contextual word embedding model Clinical BERT is leveraged to discover hidden connections for KG completion. Finally, a Cypher query file is generated to facilitate the creation of the semantically enriched KG in Neo4j. The details of each individual stage are outlined below.

3.1 User Input

The M-KGA technique allows users to input medical data in two distinct formats: structured and unstructured. When we say "structured," we mean that the user defines the medical terms with precision. The data does not contain any interconnected notions. Here is an example of a text that is organized in a structured manner:

> Structured Input Example:
> *['fever', 'diarrhea', 'insomnia', 'severe acute respiratory syndrome', 'diabetes']*

Unstructured text, on the other hand, is free natural language text that is understandable to people but not to computers. It is the text written by a medical professional for instance diagnosis of a patient. An illustration is:

> Unstructured Input Example:
> *["If you have a condition called polyuria, it's because your body makes more pee than normal. Adults usually make about 3 liters of urine per day. But with polyuria, you could make up to 15 liters per day. It's a classic sign of diabetes."]*

The developed code can take data in both formats; if structured text is needed, it will ask for the data numerous times. If unstructured formatted text is needed, it will accept it all at once and find out the concepts on its own.

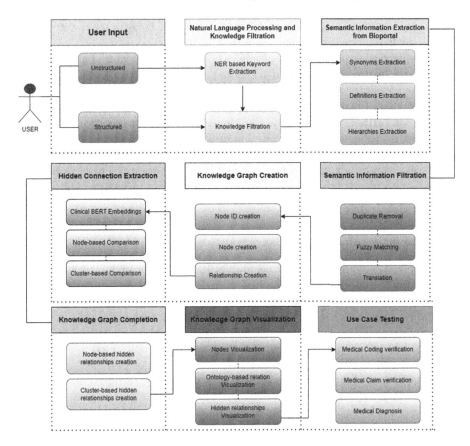

Fig. 1. Medical-Knowledge Graph Automation Approach (M-KGA).

3.2 NLP-Based Knowledge Filtration

We propose two NLP-based approaches for knowledge filtration: i) NER-based keyword extraction, and ii) knowledge filtration. The NER-based extraction identifies medical concepts from unstructured text and converts them into a structured format. Then, the knowledge filtration process selects the most relevant information from the extracted data. Each process is detailed below:

3.2.1 NER-Based Keyword Extraction and Knowledge Filtration

An unstructured document is input to extract clinical entities using an NER-based keyword function. This process populates nodes in the medical knowledge graph, integrating entities like diseases, treatments, and clinical concepts. We used the SciSpacy library [6], an extension of spaCy tailored for scientific and biomedical text, specifically leveraging the en_core_sci_sm model optimized for this domain. The NER-based function processes the unstructured text to extract clinical concepts, producing a list of entities that can be seamlessly integrated into the medical knowledge graph.

Subsequently, acquiring a list of clinical or medical concepts, knowledge filtration function would be applied to choose only relevant and prominent concepts. The concepts that are previously extracted may contain some redundancies. Knowledge Filtration looks for duplicates in the data by Fuzzy-matching and filters it further. Additionally, the knowledge filtration assists the medical practice in decision making towards building a comprehensive KG by including and excluding certain concepts. The concept that is extracted during the NER-based keyword extraction process may be not relevant to current problem, disease diagnosis or treatment. By mapping these words to Bioportal Ontologies, more filtering is applied.

3.3 Semantic Information Extraction

The organized, enhanced, and sifted material from the preceding stage is used in the semantic information extraction step. Mapping these terms to expert-created Bioportal ontologies [5] allows you to retrieve the semantically enriched information using the Bioportal REST API. Our approach's strength is rooted in the notion that M-KGA is not exclusive to any particular medical condition or illness. Instead, it attempts to enrich data without being limited to particular ontologies.

Different sorts of semantically enriched information are retrieved from ontologies Such as Synonyms words, phrases, or morphemes that share the same meaning as the term being mapped, Definition which is a group of terms or phrases that provide a longer description of the term that is mapped. In this process, the two types of semantic information extraction took place.

3.4 Semantic Knowledge Filtration

Since our method is not limited to a specific medical condition or issue, M-KGA anticipates a high degree of data diversity, variation, and redundancy. In this step, data that has been semantically enhanced is filtered using a variety of techniques. The method attempts to translate data from many languages to English, eliminate duplicates from the retrieved data, and then use fuzzy-matching to further filter the findings.

3.4.1 Translation, Duplicate Removal and Fuzzy Matching
The data used in the enrichment stage is multilingual and comes from many ontologies. To translate this material, we exploited the 'translate' library in python. We need to take this action since, in the absence of translation, we will lose some important information. Text that we identified as non-English was translated into English; if the language cannot be identified or there are any exceptions, the text is eliminated. The subsequent stage does not include this deleted portion. Since the non-English text cannot be used in any further steps,

it has been removed. Pre-trained model employed to find hidden connections are unable to comprehend the data, which will result in problems. Aside from that, this step will address the limitations imposed by Neo4j on the creation of a Cypher Node ID.

This duplicate removal step takes the enriched data from the previous stage and tries to remove the duplicates. As we are fetching data from 1000+ Bioportal ontologies, we will likely data massive duplications. In this step we used semantic information, changed all synonyms and definitions to lower case and then used simple set operation on python to remove redundant entries.

An expansion of the duplicate removal process is fuzzy matching. Certain enriched data may contain semantically comparable text that cannot be removed with set procedures. In an effort to maintain the content's uniqueness, we employed this strategy. This also fixes problems in the ID creation stage and greatly aids in the removal of semantically duplicate items.

3.5 Knowledge Graph Creation

Creating a KG is a challenging endeavor in and of itself because it requires extreme caution while creating nodes and interactions. The KG was developed in Neo4j's Cypher query language. The format has unique limitations. The node ID in Cypher ought to begin with a character rather than a number, special character, non-English phrase, etc. Taking these factors into account, we produced graphs. This will be further explained in the steps following.

3.5.1 Node ID Creation

Nodes ID was developed with the understanding that hundreds of connections—both hidden and provided by the ontology—must be made between nodes. We translated the Node content or enriched data into ID by adhering to the ID requirements for different KG formats in order to reduce the amount of computing resources required for ID retrieval for comparison and connection formation. As previously stated, the Node ID in Cpyher only accepts data in English format; special characters are not permitted, etc. Using this method, the enriched data on *polyuria* is transformed into IDs such as *excessivesecretionofurine* from the definition of 'excessive secretion of urine'. Therefore, we don't have to go look for the ID connected to that Node every time we need to establish a connection. All we had to do was apply our function and turn the content into ID.

3.5.2 Nodes and Relationships Creation

This step builds the nodes for the structured and unstructured (converted to structured) data, as well as the semantically enriched data, according to the ID creation technique previously outlined. Different kinds of nodes have been created. Synonyms, medical concepts, definitions and so on are among the categories. Depending on its kind, every node in the graph is represented by a distinct color. The node displays the content. All KG nodes are constructed in this step.

After node creation, relationships among different nodes based on the expert crafted ontology-provided connections are created. With each iteration, a semantically enriched node is created with ID, the connection creation step, uses the ID and connects the node with the main medical concept. Same is the case for all semantically enriched data. Here, relationships are also of different types such as synonyms, definitions, and so on. Relationships are labeled and directed.

3.6 Hidden Connection Extraction

The earlier processes collect user data, filter it, obtain enriched data from ontologies produced by experts, and produce a knowledge graph. In addition to the links supplied by experts, our method looks for hidden connections absent from ontologies. The ontologies offer richer medical terminology, but it might be challenging to determine whether or not these concepts are related to one another. Exist any connections that could be omitted to improve the analysis of the medical data? We made knowledge graphs, but how do we complete them?

To address these problems, we attempted using KG embeddings for our method, which can predict links given KG triplets. Sadly, these methods are ineffective for tiny graphs. Our method can create both large and tiny KG in response to user requests; nevertheless, KG embeddings are unable to function on small networks because these models need thousands of triplets. Therefore, in order to create connections, we took advantage of word embeddings to determine a word's meaning and relationships with other words. It should be highlighted that our method looks for connections with other medical concepts and their enriched content rather than trying to establish links with its semantic enriched nodes, which are all already connected.

3.6.1 Clinical BERT Embeddings
We took advantage of Clinical BERT embeddings to extract vector representation of medical concepts and their contextual meta-data. We utilized the Clinical BERT embeddings, which are trained on a sizable medical corpus, in place of creating our own model. Medical Information Mart for Intensive Care III (MIMIC-III) is used to train the model. We took use of their pretrained nature and open-source nature to comprehend medical concepts and their interrelationships. We calculated the degree of similarity between various terms and built relations that led to knowledge graph completion based on the distance and user-defined threshold.

3.6.2 Cluster-Based Comparison
We offered two methods for locating the links that are buried in knowledge graphs. We treated every medical concept and its semantically enriched data as a cluster in a cluster-based method. Using all available semantic information, we composed a paragraph and then used the Clinical BERT model to look for embeddings. Clinical BERT implementation is not scalable and introduces mistakes on big clusters. In order to address this method, we segmented the

paragraph into chunks, handled exceptions, fetched embeddings for each chunk and then divided by the total number of pieces. A cluster is mapped with other clusters according to a user-defined threshold. The threshold and the degree to which users require specific or general connections are key factors here. In actuality, the threshold is the separation between the clusters. To determine whether clusters have strong relationships or not, users can choose the lower threshold. Cluster-based comparisons or connections lead to KG completion quickly and at low computing cost. This step introduces further relationships named 'embedding_match_cluster' in the KG. Figure 2 is an example of Cluster-based comparison method.

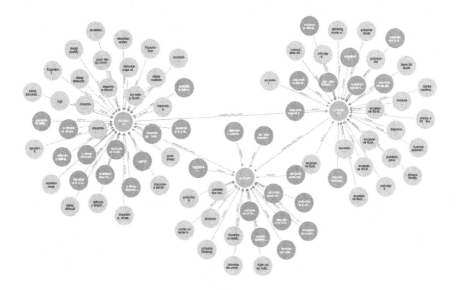

Fig. 2. Cluster-based comparison on Medical diagnosis use-case.

3.6.3 Node-Based Comparison

In contrast to the cluster-based approach, the node-based technique looks for connections with other nodes in the cluster. Using word embeddings, this compares a single node to every other cluster's node based on nodes. The lack of a large amount of text on the nodes means that scalability is not an issue. Furthermore, this method requires around n2 time and is computationally costly, as opposed to the cluster-based method. Because the node-based technique allows us to determine the exact match of the link, it is much easier to understand. Here, connections are also established according to user-specified thresholds. Depending on the size of the graph, connecting nodes takes minutes. Figure 3 shows the result depiction of this method. This adds relationship named 'embedding_match_node' in the KG.

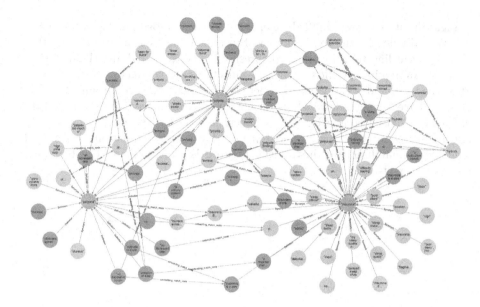

Fig. 3. Node-based comparison on Medical diagnosis use-case.

3.7 Use Case Testing

To determine the efficacy of our method, we ran three separate use cases through it. Three use cases were selected: medical claims, medical diagnostics, and medical coding. We used the dataset "CodiEsp" [13] to test each of these use cases. Experts annotate CodiEsp data with ICD-10 codes. To test our strategy, we used the discharge summaries along with their annotations.

3.7.1 Medical Coding, Claim Verification and Medical Diagnosis
We used several synopses, medical terms, and related medical codes We extracted the medical terms contained in the descriptions of the medical codes after converting them into descriptions. We then used our Medical-Knowledge Graph Automation (M-KGA) technique to see the outcomes after passing all the medical concepts-such as discharge summaries or descriptions of diagnostic codes—through it. True positive and true negative cases were used in our tests. We took the summaries, applied our method to their annotations, annotated the summaries once again with fictitious examples, and retested the method. The approach demonstrated its usefulness through visualization and proved satisfactory in all cases. We also experimented with various thresholds.

We also tested the method for medical diagnosis using CodiEsp data. Each medical summary's knowledge graph was made using the concepts that were extracted from the summaries using the NER-based keyword extraction stage. We applied both node-based and cluster-based comparisons, and we generated a complete KG. We presume that any medical ideas included in a summary must be related to one another; this relationship will confirm the usefulness of our method for completing graphs and offer a far more profound comprehension of the relationships taken from ontologies. Using our node-based and cluster-based comparisons technique, the strategy demonstrated significance in all experiments and most medical terms within the same summary generated links. To better understand the approach's operation, we also put it to the test with negative cases as well.

4 Results, Evaluation and Discussion

In this section, we present the results obtained from the implementation and testing of our proposed Medical Knowledge Graph Automation (M-KGA) approach. The evaluation aims to assess the effectiveness and efficiency of M-KGA in constructing comprehensive knowledge graphs from medical concepts provided by users. We conducted experiments using a diverse set of 300 medical concepts to evaluate the performance of our approach across various domains within healthcare.

Table 1. Inter Annotator Agreement (IAA) Among Domain Experts for Node-based Annotation

% of Agreement				Cohen's Kappa Value	
	Expert1	Expert2	Expert3	IAA between two Expert	
Expert1	1	94.25%	93.79%	Expert1, Expert2	0.86
Expert2	94.25%	1	94.55%	Expert1, Expert3	0.87
Expert3	93.79%	94.55%	1	Expert2, Expert3	0.86

Table 2. Inter Annotator Agreement (IAA) Among Domain Experts for Cluster-based Annotation

% of Agreement				Cohen's Kappa Value	
	Expert1	Expert2	Expert3	IAA between two Expert	
Expert1	1	92.56%	91.27%	Expert1, Expert2	0.85
Expert2	92.56%	1	91.85%	Expert1, Expert3	0.84
Expert3	91.27%	91.85%	1	Expert2, Expert3	0.84

In our evaluation, we partitioned the 300 medical concepts into two sets: 150 for assessing the cluster-based comparison method and another 150 for evaluating the node-based comparison approach. Each set underwent pairing, facilitated by the GPT-3.5 model, to create pairs of medical concepts. These pairs were

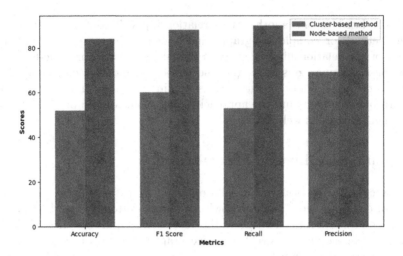

Fig. 4. Node-based verses Cluster-based comparison on Medical diagnosis use-case.

compiled into an Excel file for annotation by human medical experts. Owing to this, we calculated Inter-Annotator Agreement (IAA) assesses how consistently multiple annotators (domain experts) can make identical annotation decisions for a specific category, playing a vital role in validating and reproducing annotation results. Similarly, We utilized a well- known statistical measurements for the evaluation purposes Cohen's Kappa which is a statistical coefficient that quantifies the degree of agreement and reliability in statistical classification. It measures the concordance between two annotators (Domain Experts), each categorizing items into exclusive categories. Kappa value greater than 0.81% presented almost perfect agreements, where as the proposed framework achieve almost perfect annotation agreements of over 90% and Cohen's Kappa over 81% among the three domain experts both at Node-based and Cluster-based annotation as shown in Table 1 and Table 2. This indicates the robustness and reliability of the "M-KGA" framework in facilitating agreement among domain experts in knowledge graph recommendations.

The experts were tasked with annotating each pair based on measures of True Positive (TP), False Positive (FP), True Negative (TN), and False Negative (FN), providing valuable insights into the accuracy and performance of our approach. Following the annotation process, we applied both the cluster-based and node-based comparison methods to the pairs of medical concepts with threshold=4. Utilizing these methods, we constructed knowledge graphs for each pair and analyzed whether they successfully identified connections as annotated by the experts. Given that traditional ontologies often struggle to find connections among certain medical concepts, our objective was to determine if our approach could uncover hidden connections that might otherwise remain undiscovered. This analysis aimed to ascertain the efficacy of our proposed methods in augmenting existing knowledge and revealing previously unrecognized relationships within the medical domain.

In Fig. 4, the metrics constructed using True Positive (TP), False Positive (FP), True Negative (TN), and False Negative (FN) are presented. Our analysis focused on evaluating accuracy, F1 score, recall, and precision based on these metrics. The comparison depicted in the Fig. 4 highlights the performance disparity between the node-based and cluster-based methods. Notably, the node-based method emerges as the clear frontrunner, exhibiting significantly higher accuracy, F1 score, recall, and precision compared to the cluster-based approach. This observation underscores the effectiveness of the node-based method in accurately capturing connections within the knowledge graph, ultimately leading to superior performance across all evaluated metrics. The observed limitations in accuracy, F1 score, recall, and precision of the cluster-based method can be attributed to the utilization of Clinical BERT. While Clinical BERT is a powerful pre-trained model, its efficacy is constrained by practical considerations such as computational resources and sample size limitations. Due to the vast scale of clusters within the knowledge graph, it becomes necessary to divide them into smaller, manageable chunks for processing. However, this segmentation introduces a challenge: the loss of contextual coherence across multiple chunks. As a consequence, the embeddings derived from fragmented clusters may lack the holistic context necessary for accurate representation and inference, resulting in diminished performance metrics. This phenomenon underscores the importance of considering both the capabilities and limitations of pre-trained models when designing and implementing knowledge graph construction methodologies.

In addition to evaluation measures, transparency and time are critical factors in assessing the effectiveness of the M-KGA approach. As depicted in Figs. 2 and 3, transparency refers to the clarity and comprehensibility of the constructed knowledge graph. The node-based method excels in transparency by establishing direct connections between nodes, thereby presenting a clear and intuitive representation of relationships. In contrast, the cluster-based approach may exhibit less transparency, as it tends to add fewer relationships, resulting in a less explicit depiction of connections. Time, on the other hand, pertains to the efficiency of the knowledge graph construction process. The cluster-based approach demonstrates an advantage in terms of time efficiency, requiring less computational resources and processing time compared to the node-based method. However, this efficiency comes at a cost, as the cluster-based approach may sacrifice performance metrics such as accuracy, F1 score, recall, and precision, as previously discussed. While the cluster-based approach offers a quicker construction process, it may compromise transparency and performance. In contrast, the node-based method prioritizes transparency and performance, albeit at the expense of increased computational complexity and time consumption. Thus, the choice between these approaches should be carefully considered based on the specific requirements and priorities of the knowledge graph application.

5 Conclusion and Future Work

In conclusion, this study introduces the Medical Knowledge Graph Automation (M-KGA) approach, which aims to address the challenges associated with automating the construction of knowledge graphs (KGs) and enhancing their completeness. Leveraging user-provided medical concepts and BioPortal ontologies, M-KGA enriches the semantic content of KGs using pre-trained embeddings, thereby facilitating a more comprehensive representation of structured medical knowledge. Our approach incorporates two distinct methodologies, namely a cluster-based approach and a node-based approach, to uncover hidden connections within the knowledge graph. Through rigorous testing involving 300 medical concepts, our M-KGA framework demonstrates promising results, showcasing its potential to overcome the limitations of existing knowledge graph automation techniques. The performance metrics and graph visualizations presented in this study underscore the effectiveness of our approach in enhancing the transparency and accuracy of knowledge graphs, particularly in the medical domain. Looking ahead, future work will focus on addressing scalability issues associated with the cluster-based method, aiming to improve its performance. Additionally, we plan to explore retrieval augmented generation (RAG) with Large Language Models (LLMs), for knowledge graph development and performance comparison with our current approach. By continuing to innovate and refine our approach, we aim to further advance the field of knowledge graph automation and contribute to the development of more comprehensive and accurate representations of structured knowledge in healthcare domain.

Acknowledgement. This work is supported by the National Science Foundation under Grant No. 2431840.

References

1. Smolensky, P.: Connectionist ai, symbolic ai, and the brain. Artif. Intell. Rev. **1**(2), 95–109 (1987)
2. Zou, X.: A survey on application of knowledge graph. In: Journal of Physics: Conference Series, vol. 1487, p. 012016. IOP Publishing (2020)
3. Hao, X., Ji, Z., Li, X., Yin, L., Liu, L., Sun, M., Liu, Q., Yang, R.: Construction and application of a knowledge graph. Remote Sens. **13**(13), 2511 (2021)
4. Neumann, M., King, D., Beltagy, I., Ammar, W.: Scispacy: fast and robust models for biomedical natural language processing. arXiv preprint arXiv:1902.07669 (2019)
5. Noy, N.F., et al.: Bioportal: ontologies and integrated data resources at the click of a mouse. Nucleic Acids Res. **37**(suppl_2), 170–173 (2009)
6. Alsentzer, E., et al.: Publicly available clinical bert embeddings. arXiv preprint arXiv:1904.03323 (2019)
7. Wang, Q., et al.: Covid-19 literature knowledge graph construction and drug repurposing report generation. arXiv preprint arXiv:2007.00576 (2020)
8. Li, L., et al.: Real-world data medical knowledge graph: construction and applications. Artif. Intell. Med. **103**, 101817 (2020)

9. Sun, H., et al.: Medical knowledge graph to enhance fraud, waste, and abuse detection on claim data: model development and performance evaluation. JMIR Med. Inf. **8**(7), 17653 (2020)
10. Zhang, H., Wang, X., Pan, J., Wang, H.: Saka: an intelligent platform for semi-automated knowledge graph construction and application. In: Service Oriented Computing and Applications, pp. 1–12 (2023)
11. Malik, K.M., Krishnamurthy, M., Alobaidi, M., Hussain, M., Alam, F., Malik, G.: Automated domain-specific healthcare knowledge graph curation framework: subarachnoid hemorrhage as phenotype. Expert Syst. Appl. **145**, 113120 (2020)
12. Alam, F., Giglou, H.B., Malik, K.M.: Automated clinical knowledge graph generation framework for evidence based medicine. Expert Syst. Appl. **233**, 120964 (2023)
13. Miranda-Escalada, A., Gonzalez-Agirre, A., Armengol-Estapé, J., Krallinger, M.: Overview of automatic clinical coding: annotations, guidelines, and solutions for non-english clinical cases at codiesp track of clef ehealth 2020. In: CLEF (Working Notes), vol. 2020 (2020)

Ontoseer - A Recommendation System to Improve the Quality of Ontologies

Pramit Bhattacharyya[1](✉)[iD], Samarth Chauhan[2][iD], and Raghava Mutharaju[2][iD]

[1] Indian Institute of Technology, Kanpur, Kanpur, India
pramitb@cse.iitk.ac.in
[2] Indraprastha Institute of Information Technology, Delhi, New Delhi, India

Abstract. Building an ontology is time-consuming and confusing, especially for beginners and the inexperienced. Although ontology developers can take the help of domain experts in building an ontology, they are not readily available in several cases for various reasons. Ontology developers must grapple with several questions about the choice of classes, properties, and axioms that should be included. Apart from this, aspects such as modularity and reusability should be considered. From among the thousands of publicly available ontologies and vocabularies in repositories such as Linked Open Vocabularies (LOV) and BioPortal, it is hard to know the terms (classes and properties) that can be reused in the development of an ontology. A similar problem exists in implementing the right ontology design patterns (ODPs) from among the several available. Generally, ontology developers use their experience in handling these issues, and the inexperienced ones have a hard time. To bridge this gap, we developed a tool named OntoSeer that monitors the ontology development process and provides suggestions in real time to improve the quality of the ontology under development. It can provide suggestions on the naming conventions, vocabulary to reuse, ODPs to implement, and axioms to be added to the ontology. OntoSeer has been implemented as a Protégé plug-in. We conducted a user study of the tool to evaluate the quality of the recommendations. Almost all the users are satisfied with the recommendations provided by OntoSeer, and most agreed that OntoSeer reduces their modelling time. The source code and the plug-in instructions are available under Apache License 2.0 at The source code and the plug-in instructions are available under Apache License 2.0 at https://github.com/kracr/ontoseer. A short video demonstrating the use of OntoSeer is available at https://youtu.be/iNQOJGZkZKQ.

Keywords: Ontology engineering · Ontology quality · Ontology design patterns · Recommendation system · Protégé plug-in

1 Introduction

Ontology development is generally a group activity where domain experts, ontology developers, and other stakeholders meet to discuss and develop an ontology.

This makes ontology development time-consuming and expensive, especially if the scope of the ontology is broad. It is not always possible to take the help of domain experts due to various reasons such as their unavailability and cost. Experienced developers face issues while building ontologies, and this problem only magnifies in the case of inexperienced ontology developers. They will have to deal with several questions, such as the classes and properties to reuse, the appropriate class hierarchy to build, the axioms and the ontology design patterns (ODPs) [9] to use to make the ontology modular. Experienced ontology developers may be able to answer some of these questions and make better design decisions. But even for them, keeping track of all the new vocabularies, ontologies, and ODPs published in the repositories will be hard. Our tool, named *OntoSeer*, is a Protégé plug-in that works with the ontology developer during the ontology development process and provides suggestions in real-time that can lead to better quality ontologies. The contributions of this work are as follows.

- A tool that recommends classes, properties, axioms, and ODPs to (re)use based on the description, competency questions, and the ontology under development.
- A mechanism to provide suggestions of names to use during the ontology development. These suggestions follow the naming conventions.
- A process to validate the class hierarchy based on the inputs provided by the ontology developer.
- Integration with Protégé (available as a plug-in) to improve the ontology development experience.

2 Related Work

Apart from OntoSeer, there have been a few efforts in this direction. Tools that are similar to OntoSeer are OntoCheck [16], and OOPS! [14]. We briefly discuss the functionality of these tools and the way in which they differ from OntoSeer. OntoCheck [16] is a plugin for Protégé that indicates the class names that violate the naming convention. For example, if there are classes such as HumanBeing and nitrogenoxide, then OntoCheck reports that 50% of the classes do not follow naming conventions along with the class nitrogenoxide. In contrast, OntoSeer indicates the violation and recommends possible class names such as NitrogenOxide to the user. OOPS! stands for Ontology Pitfall Scanner. It is an online tool that detects pitfalls or common modelling errors that ontology developers make. The pitfalls are divided into structural, functional, and usability-profiling.

OntoClean [7] provides guidelines for properties and class hierarchies. It introduced terms such as essence, rigidity, identity, and unity. ODEClean plug-in of WebODE [5] devised a way to validate class hierarchy using OntoClean's guidelines. Unlike OntoSeer, ODEClean is a stand-alone web plugin. Also, ODEClean assumes that the user will know the characteristics of each property, which may not be true in the case of novice developers. OntoSeer, on the other hand, asks the user simple questions and validates the class hierarchy based on the answers. OnToology [1] is a web-based system that integrates existing tools for documentation, evaluation, and publishing ontologies. It encourages the automation of these support activities through a Git-based environment.

One of the main differences between OntoClean, OOPS!, OnToology and OntoSeer is that OntoSeer works with the ontology developer to create good quality ontologies. It does not evaluate ontologies post creation. Another significant difference is that OntoSeer recommends terms (classes, properties, instances) that can be reused. It does this by checking the similarity of the terms in the ontology being built with existing ontologies. OntoSeer, like OnToology, is also compatible with all the well-known ontology development methodologies like METHONTOLOGY [6], On-To-Knowledge [17], and the NeOn Methodology [18] since OntoSeer does not mandate the use of any specific ontology development methodology.

An essential functionality of OntoSeer is to encourage reusability by using existing ontologies and ODPs. An ODP helps domain experts in ontology engineering by packaging reusable best practices into small blocks of ontology functionality [8]. Since there are several hundred ODPs, it is challenging to select the relevant ones. OntoSeer helps users select ODPs by using the classes and properties of the ontology under construction and asking a few simple questions to the user. This feature is one of the major differences between the existing tools described earlier and OntoSeer.

3 Preliminaries

Ontology. Description Logics [10] is the formal underpinning for OWL 2 ontologies. So, there is a very close correspondence between Description Logics and OWL 2, the W3C standard to create ontologies. The signature Σ for Description Logics (and hence ontologies) is defined as $\Sigma = \langle N_C, N_R, N_I \rangle$, where N_C, N_R, and N_I are countably infinite, mutually disjoint sets of class names, property names, and instance names respectively. Ontologies consist of statements called axioms that capture the relationship between classes, properties and instances. Although not part of the signature, competency questions, denoted by CQ and ontology domain, denoted by D, are often associated with an ontology. The ontology under construction, i.e., the ontology for which the recommendations are sought, is denoted by O' and its signature $\Sigma' = \langle N'_C, N'_R, N'_I \rangle$, with CQ' and D', as the competency questions and domain of O'.

Inverted Index. An inverted index [12] is a data structure that efficiently retrieves documents containing a specific term or set of terms. In an inverted index, the index is organized by terms (words), and each term points to a list of documents containing that term. Search engines and database systems use inverted indices for efficient text search. OntoSeer creates an inverted index on the classes and properties of the ontologies. It queries this index to get a ranked list of the classes and properties to reuse, along with the ontology they are part of.

Jaro-Winkler Distance. Jaro-Winkler distance [3] is a string metric for measuring the edit distance between two sequences, referred to as the Jaro distance. It is the minimum number of single-character transpositions required to change one word into the other. The Jaro Distance between two sequences s1 and s2 is defined by

$$\text{Jaro Similarity} = \begin{cases} 0, & \text{if } m = 0. \\ 1/3 * ((m/|s1|) + \\ (m/|s2|) + \\ ((m-t)/m)), & \text{otherwise.} \end{cases} \quad (1)$$

where,

- m is the number of matching characters (characters that appear in s1 and s2)
- t is half the number of transpositions (compare the i-th character of s1 and the i-th character of s2 divided by 2)
- |s1| is the length of the first string
- |s2| is the length of the second string

The Jaro-Winkler distance uses a prefix scale, giving more favourable ratings to strings that match a set prefix length. As it gives more importance to the words with an identical prefix, the Jaro-Winkler distance is appropriate to our use case of syntactic matching. The Jaro-Winkler distance is calculated using

$$S_w = S_j + P * L * (1 - S_j) \quad (2)$$

where S_w is the Jaro- Winkler similarity, S_j is Jaro similarity obtained from Eq. 1, P is the scaling factor (0.1 is set as default), and L is the length of the matching prefix. The Jaro-Winkler similarity ranges between 0 and 1, where 1 denotes two strings being exactly similar and 0 indicates that the two strings are completely dissimilar.

4 Approach

In this section, we discuss the various features of OntoSeer and the implementation details for each. Since OntoSeer uses the existing ontologies and vocabularies in making the recommendations, we first describe the datasets used and then describe the method used to make the recommendations. OntoSeer makes use of four types of datasets. Competency questions provide hints about the potential classes and properties that make up the ontology. We make use of these hints in recommending ODPs to reuse. The other three types of datasets (ontologies, ODPs and vocabularies) are indexed and used to recommend ontologies, ODPs and axioms.

1. **Competency questions (CQs)**. 92 CQs and their corresponding ontologies are available at Software Ontology[1]. 52 CQs and ontologies are available

---

[1] https://softwareontology.wordpress.com/2011/04/01/user-sourced-competency-questions-for-software.

at ArCo[2]. Several CQs and their associated ontologies are available from CORAL [4] and [15].
2. **Ontologies.** We collected ontologies from several repositories such as NCBO BioPortal[3], Manchester OWL Corpus[4], Oxford OWL Repository[5], and Protégé Ontology Library[6].
3. **Ontology Design Patterns (ODPs).** ODPs are available at the ODP repository[7]. Since no bulk download option exists, we used a web scraper to collect all the ODPs across six categories. Each ODP has several fields associated with it, such as intent, domain, competency questions, elements and description. These fields are used in making the recommendations.
4. **Vocabularies.** Several vocabularies are indexed at the Linked Open Vocabularies (LOV)[8]. These vocabularies can be accessed using the SPARQL endpoint or the LOV API.

4.1 Class, Property and Vocabulary Recommendation

A recommended best practice in building an ontology is to reuse existing ontologies. However, there are several ontologies spread across many repositories. This makes it hard to find the right ones to reuse. OntoSeer creates an inverted index on N_C and N_R of each ontology, O, in the dataset. It queries this index using N'_C, N'_R of O' to get a ranked list of the classes and properties to reuse in O'. Figure 1 shows OntoSeer's recommendations for a class Book. The first recommendation is of FOAF along with its IRI. The second description represents the vocabulary *Comic Book Ontology* along with its IRI. In both these vocabularies, a class named Book is present and the user can choose to reuse this class from one of the two vocabularies instead of creating a new one.

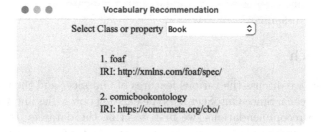

Fig. 1. Vocabulary recommendation for the Book class

---

[2] https://github.com/ICCD-MiBACT/ArCo.
[3] https://bioportal.bioontology.org/ontologies.
[4] http://mowlrepo.cs.manchester.ac.uk/datasets/mowlcorp/.
[5] https://www.cs.ox.ac.uk/isg/ontologies/.
[6] https://protegewiki.stanford.edu/wiki/Protege_Ontology_Library.
[7] http://ontologydesignpatterns.org/wiki/Main_Page.
[8] https://lov.linkeddata.es/dataset/lov/.

4.2 Class and Property Name Recommendation

Following the appropriate naming conventions for classes and properties helps improve the readability and maintenance of an ontology. However, it is a challenge to abide by the naming conventions, especially for a beginner, because they may not be familiar with the conventions. OntoSeer supports the class and property naming recommendations [13] using regular expressions. They can be summarized as follows.

- Use of numbers in the name is discouraged. If there is a class named Human1234_being in the ontology, a suggestion to use HumanBeing is given. However, names like RFID_000017 are used in the biomedical domain. The occurrence of digits in these terms is generally at the end of the name. To accommodate these types of names, we do not remove the digits if they occur at the end of a name. Thus, if there is a class named RFID_000017, Ontoseer does not suggest any changes to the name.
- Use of any special character other than an underscore is not recommended.
- Use of upper camelcase for naming classes and lower camelcase for properties and instances is encouraged. If a class is named Human_being, the recommendation would be to use HumanBeing.

4.3 Axiom Recommendation

Ontology developers often forget to include the axioms in the ontology. OntoSeer retrieves axioms from the ontology corpus that closely match N'_C, N'_R of O'. It will recommend axioms such as disjointedness among class siblings and property characteristics (inverse, symmetric, transitive, etc.). The axioms are extracted from the ontology corpus. The similarity between the given N'_C, N'_R and the axioms is computed using the Jaro-Winkler distance. The axioms are ranked based on this score, and the top ones are provided as recommendations. In Fig. 2, the top two axiom recommendations for the Person class are shown. The similarity threshold has been varied between 0 to 1 with an increment of 0.05 and has been fixed at 0.85 after observing the recommendations on 150 different instances.

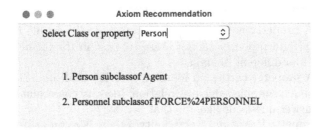

Fig. 2. Axiom recommendation for the Person class

4.4 ODP Recommendation

Ontology Design Patterns help make an ontology modular and reduce the effort in building an ontology since the design solutions approved by the community can be reused. However, finding the right ODP from among the hundreds of available ODPs is hard. OntoSeer helps ease this process by recommending appropriate ODPs by considering N'_C, N'_R, and CQ' of O'. It uses the Jaro-Winkler distance metric to compute the similarity between N'_C, N'_R, CQ' of O' and their corresponding N_C, N_R, CQ for each ODP, O, in the corpus. Due to its semantic matching capability, we tried Doc2Vec [11] to compute similarity. But deeplearning4j[9] is incompatible with Protégé. So, we used the Jaro-Winkler distance for computing similarity. However, the results obtained using this method are close to the ones obtained using the vector-based approach, as discussed in [2].

More weightage was given to the similarity scores obtained from class and property matching, followed by the scores for the description of the ontology, domain of the ontology, and competency questions. The total score, thus, can be represented as, $ODP_{score} = w_1 * s_1 + w_2 * s_2 + w_3 * s_3 + w_4 * s_4$, where s_1 is the similarity score obtained from term (class and property) matching, s_2 is the score obtained from matching the description of the ontology, s_3 is the score obtained from matching the domain of the ontology, and s_4 is the score obtained from matching the competency questions. The weightage for components ($w_1 = 5$, $w_2 = 3$, $w_3 = 2$, $w_4 = 3$) and a similarity threshold of 0.65 has been set after running the experiments for more than 150 iterations.

For an ontology with classes **Person**, **Professor**, **School** and the description as *College*, the recommended ODPs are *TimeIndexedPersonRole*, *AgentRole*, *BiologicalEntities* and *RelativeRelationship*. On analyzing the recommendations, we noticed that these ODPs and their descriptions have the same or similar terms (classes and properties) as the ones found in the ontology.

4.5 Class Hierarchy Validation

Determining the class hierarchy is confusing, especially for inexperienced ontology developers. This often leads to improper hierarchy. OntoSeer makes use of the rigidity (R), identity (I) and unity (U) characteristics [7] of the classes and takes the user input based on these characteristics to validate the class hierarchy. These three characteristics are defined as follows.

1. **Rigidity:** A property is rigid if it is essential to all possible instances. An instance of a rigid property cannot cease to exist in the future or fail to be its instance in a different domain.
2. **Identity:** A property carries an identity criterion if all its instances can be identified using a suitable *sameness* relation. Identity constraint is the criteria we use to answer questions like, "Is that my dog?".
3. **Unity:** A property P is said to carry unity if there is a common unifying relation R such that all the instances of P are wholes under R. Unity constraint checks whether properties have wholes as instances.

---

[9] https://deeplearning4j.org/.

OntoSeer - A Recommendation System to Improve the Quality of Ontologies

OntoSeer interacts with the ontology developer and determines the characteristics of classes. The questions asked by OntoSeer are as follows.

1. Do the properties of the class cease to exist in the future? For example, `Person` will always be a person, but `Student` can cease to exist to be a student after a certain time.
2. Are the properties of superclass and subclass identical? For example, two one-hour duration time intervals are identical, but an hour interval on Wednesday is not identical to an hour on Friday.
3. Is the property of the subclass part of the properties of the superclass? For example, a lump of clay is part of the "amount of matter", but the "amount of matter" is not part of a lump of clay.

Table 1 shows the superclass and its accepted subclass characteristics for validating the class hierarchy. Consider the first row of the table. If the superclass satisfies the Identity criteria, the Identity constraint for the hierarchy will be maintained only when the subclass also satisfies the Identity criteria. The rest of the rows of Table 1 can be interpreted similarly.

Table 1. The rigidity, identity and unity values for class hierarchy validation.

Rule	Superclass value	Subclass value
Identity	Positive	Positive
Identity	Negative	Negative
Rigidity	Positive	Negative, Positive
Rigidity	Negative	Negative
Unity	Positive	Positive
Unity	Negative	Negative

5 User Study

We evaluated OntoSeer through user study by considering the following questions.

1. Is OntoSeer useful for the inexperienced and the experienced ontology developers?
2. Does OntoSeer reduce the ontology development time?
3. Are ontology developers able to make richer ontologies due to the axiom recommendations of OntoSeer?
4. Are the ontology developers reusing existing ontologies more than they generally do?
5. How easy is it to install and use OntoSeer?

6. Are the ontology developers able to create modular ontologies by incorporating ODPs?
7. Are the ontology developers able to build ontologies that follow the naming conventions?

We advertised OntoSeer across several relevant mailing lists[10] and contacted some of the research groups that work on ontology design. We requested them to use OntoSeer and provide an anonymous response to a questionnaire. It had questions on the modelling experience of the users, their familiarity with ODPs, the ease of installing OntoSeer, the relevance of the recommendations provided by OntoSeer, the modelling experience with and without using OntoSeer and whether OntoSeer helps save modelling time.

There were twenty-one respondents. Of the twenty-one, twelve have identified themselves as having basic proficiency in ontology modelling, seven are at intermediate level, and two indicated that they are very good (advanced level) at ontology modelling. Nine out of twenty-one respondents indicated they are unaware of LOV and other ontology repositories. Ten out of twenty-one respondents said they do not use ODPs during ontology modelling. To evaluate OntoSeer, users can either build an ontology or import an existing one. We use the Likert scale in our user study. The ratings are on a scale of 1 to 5, with 1 being the lowest (not satisfied) and 5 being the highest (very satisfied).

Twelve respondents said that the plugin's installation was easy and the instructions were clear, whereas seven found it moderately difficult to install the plugin. Table 2 represents the cumulative user response for vocabulary, ODP and axiom recommendation with and without the CQs. As expected, providing CQs improves the recommendation. OntoSeer can use the additional information in the CQs to provide better recommendations. However, the performance of OntoSeer does not drop drastically in the absence of CQs. Table 2 show that the average satisfaction of the intermediate users decreases minimally when compared with the basic user response (advanced user response is not considered significant since there were only two users). With more modelling experience, users become aware of LOV and ODPs. But even for experienced modellers, OntoSeer can provide helpful recommendations.

Figures 3, 4, 5, 6 and 7 are the graphical representations of the user response to the different features of OntoSeer. The average user satisfaction index for name recommendation and class hierarchy validation is given in Table 3. Considering Table 2, it is clear that users are happy with Ontoseer's recommendations.

An essential aspect in evaluating OntoSeer is the quality of the developed ontology and the amount of effort in terms of the time taken to develop the ontology. On average, there is an 8.33% improvement in modelling experience for users with basic proficiency when using OntoSeer. This increases to 25.72% for users with intermediate proficiency. The two advanced proficiency level users chose to be neutral, i.e., they said that it neither improves nor deteriorates their modelling experience. Fourteen, that is, 66.67% of users believed that OntoSeer saves modelling time while the remaining seven chose to be neutral.

---

[10] Protégé users, ODP users, Semantic Web, OWL API.

Table 2. Average satisfactory index with and without CQs in each user category

Proficiency Type	Vocabulary Recommendation	ODP Recommendation	Axiom Recommendation
Basic	4.25	4	4.083
	4	3.67	4
Intermediate	3.86	3.86	4.143
	3.571	3.571	3.714
Advanced	3	3.5	3.5
	3	3.5	3

Table 3. Average user satisfactory index for naming recommendation and class hierarchy validation

Proficiency Type	Naming Recommendation	Class Hierarchy Validation
Basic	4.08	3.92
Intermediate	4.14	4
Advanced	4	3.5

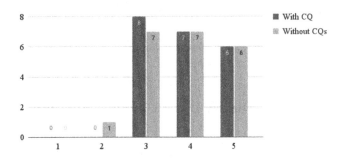

Fig. 3. User response on the vocabulary recommendation with and without CQs. The X-axis is the Likert scale, and the Y-axis is the number of respondents.

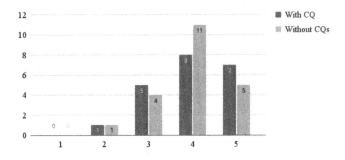

Fig. 4. User response on the axiom recommendation with and without CQs. The X-axis is the Likert scale, and the Y-axis is the number of respondents.

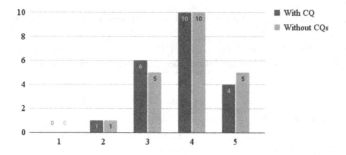

Fig. 5. User response on the ODP recommendation with and without CQs. The X-axis is the Likert scale, and the Y-axis is the number of respondents.

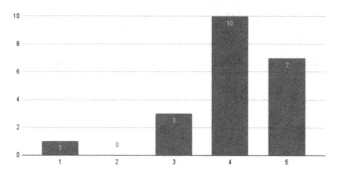

Fig. 6. User response on the name recommendation. The X-axis is the Likert scale, and the Y-axis is the number of respondents.

6 Evaluating the Quality of OntoSeer's Recommendations

Using the user study, we evaluated the utility and the effectiveness of OntoSeer's recommendations in real-time, i.e., when it was used as a plug-in while building an ontology in Protégé. There exists the possibility of a bias here because OntoSeer already provided the recommendations, and the users were asked to evaluate them. To avoid this, we evaluate OntoSeer against the users' recommendations, i.e., we first ask the users to make the recommendations, which are then compared with OntoSeer's recommendations. We provided the users with two simple use cases in the form of English language descriptions. Along with the descriptions, a few ontologies and ODPs are also provided. Among them, some are relevant to the use case, and some are not. Four users were asked to build an ontology (without OntoSeer) by reviewing each use case and reusing the classes, properties, and axioms from the provided ontologies and ODPs. Users worked individually and in a group of two while building ontologies for the two use cases. The ontologies built by the users, along with the reused classes, properties, and axioms from the provided ontologies and ODPs were collected. OntoSeer is then run on the same ontologies, and its recommendations are compared against the ones from the users. To mitigate the effects of building the ontologies individually

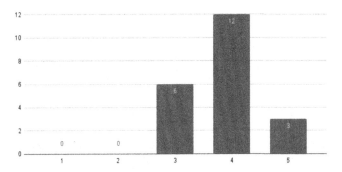

Fig. 7. User response on the class hierarchy validation. The X-axis is the Likert scale, and the Y-axis is the number of respondents.

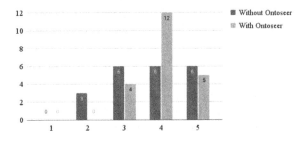

Fig. 8. Overall user satisfaction in building ontologies without OntoSeer and with OntoSeer. The X-axis is the Likert scale, and the Y-axis is the number of respondents.

and in a group (since the same set of users will be participating in the individual and group activities, it may not be a fresh start of the ontology design each time), we took the union of the recommendations for each use case and compared them with OntoSeer's recommendations. We made these comparisons in two settings. In the exact setting, OntoSeer used only the ontologies and the ODPs provided to the users to produce the recommendations. In the general setting, OntoSeer used the entire repository of ontologies and ODPs that were available to it to generate the recommendations. Table 4 shows the precision and recall of OntoSeer for use case 1 in the general and exact setting. Table 5 shows the precision and recall of OntoSeer for use case 2 in the general and exact setting. The precision@k and recall@k metrics indicate the number and the proportion of relevant recommendations in top k. From this evaluation, it is evident that OntoSeer's recommendations are relevant and of good quality.

7 Sustainability Plan

The index that OntoSeer uses may become stale over a period of time. Users can add more ontologies and regenerate the indexes. We will keep updating the index regularly. For the ODP recommendation, OntoSeer expects each ODP

Table 4. Precision and recall (PR) for use case 1 in the general and exact setting.

Features	General Setting			Exact Setting		
	PR@3	PR@5	PR@7	PR@3	PR@5	PR@7
ODP	(0.66, 0.4)	(0.6, 0.6)	(0.71, 1)	(0.66, 0.4)	(0.8, 0.8)	(0.71, 1)
Vocabularies	(0.66, 0.33)	(0.6, 0.5)	(0.71, 0.83)	(0.66, 0.33)	(0.66, 0.66)	(0.857, 1)
Axioms	(0.33, 0.2)	(0.4, 0.4)	(0.43, 0.6)	(0.66, 0.4)	(0.6, 0.6)	(0.571, 0.8))

Table 5. Precision and recall (PR) for use case 2 in the general and exact setting.

Features	General Setting			Exact Setting		
	PR@3	PR@5	PR@7	PR@3	PR@5	PR@7
ODP	(0.66, 0.4)	(0.8, 0.8)	(0.571, 0.8)	(0.33, 0.2)	(0.6, 0.6)	(0.71, 0.8)
Vocabularies	(0.66, 0.4)	(0.8, 0.8)	(0.71, 1)	(0.66, 0.4)	(0.4, 0.4)	(0.57, 0.66)
Axioms	(0.66, 0.33)	(0.6, 0.5)	(0.71, 0.83)	(0.33, 0.17)	(0.4, 0.33)	(0.29, 0.5)

(including its description, CQs, axioms, etc.) to be in a separate text file. So the new additions to the ODP repository need to be made available as text files. Naming recommendations and class hierarchy validation will not go stale. Based on the user response and the initial interest shown by the community on the code repository (11 users have *starred* the repository), we expect an increase in the uptake of this tool by the community, and this, in turn, will help with the sustainability. We plan to keep developing OntoSeer as a Protégé plugin and as an independent tool. The latter enables us to use libraries and techniques that are not compatible with Protégé and may improve the performance of OntoSeer (Fig. 8).

8 Conclusion and Future Work

Developing an ontology is often confusing and time-consuming, especially for the inexperienced. They have many choices to make, such as the classes and properties to create or reuse, ODPs to use, the different axioms to include in the ontology and the naming conventions to follow. We developed a Protégé plugin named OntoSeer to ease the process of building an ontology. Our tool recommends classes, properties, axioms, and ontology design patterns in real-time while a user is developing the ontology. We conducted a user study of the tool. There were 21 respondents. Almost all the users are satisfied with the recommendations provided by OntoSeer. Most users (14 out of 21) said that OntoSeer also helps them save modelling time. Our evaluation also shows that the recommendations generated by OntoSeer are relevant and of good quality. The source code and the instructions to install and use the plug-in is available at https://github.com/kracr/ontoseer.

We plan to have several extensions to OntoSeer. Along with providing the recommendations, it would be very helpful to incorporate the suggested changes in the ontology. Decoupling OntoSeer from Protégé and having a standalone version in the form of an API would make it possible for OntoSeer to be used outside of the Protégé environment. This would also enable using machine learning libraries incompatible with Protégé to improve the recommendations. We also plan to implement an interactive dialogue system to engage in real-time with the ontology developer to resolve confusing issues, such as a term being a class vs. a property vs. an instance. A dialogue with the user also helps check whether the ontology can answer the competency questions.

Acknowledgement. This work has partially been supported by the Infosys Centre for Artificial Intelligence (CAI), IIIT-Delhi, India.

References

1. Alobaid, A., Garijo, D., Poveda-Villalón, M., Santana-Perez, I., Fernández-Izquierdo, A., Corcho, O.: Automating ontology engineering support activities with OnToology. J. Web Semant. **57**, 100472 (2019). https://doi.org/10.1016/j.websem.2018.09.003
2. Cahyono, S.C.: Comparison of document similarity measurements in scientific writing using Jaro-Winkler Distance method and Paragraph Vector method. In: IOP Conference Series: Materials Science and Engineering, vol. 662, no. 5, p. 052016 (2019). https://doi.org/10.1088/1757-899x/662/5/052016
3. Dreßler, K., Ngonga Ngomo, A.C.: On the efficient execution of bounded Jaro-Winkler distances. Semant. Web **8**(2), 185–196 (2017). https://doi.org/10.3233/SW-150209
4. Fernández-Izquierdo, A., Poveda-Villalón, M., García-Castro, R.: CORAL: a corpus of ontological requirements annotated with lexico-syntactic patterns. In: Hitzler, P., et al. (eds.) ESWC 2019. LNCS, vol. 11503, pp. 443–458. Springer, Cham (2019). https://doi.org/10.1007/978-3-030-21348-0_29
5. Fernández-López, M., Gómez-Pérez, A.: The integration of OntoClean in WebODE. In: Angele, J., Sure, Y. (eds.) EON2002, Proceedings of the OntoWeb-SIG3 Workshop at the 13th International Conference on Knowledge Engineering and Knowledge Management EKAW 2002, Siguenza (Spain), 30th September 2002. CEUR Workshop Proceedings, vol. 62, pp. 38–52. CEUR-WS.org (2002)
6. Fernández-López, M., Gómez-Pérez, A., Juristo, N.: Methontology: from ontological art towards ontological engineering. In: Proceedings of the AAAI97 Spring Symposium, Stanford, USA, pp. 33–40 (1997)
7. Guarino, N., et al.: An overview of OntoClean. In: Staab, S., Studer, R. (eds.) Handbook on Ontologies. INFOSYS, pp. 201–220. Springer, Heidelberg (2009). https://doi.org/10.1007/978-3-540-24750-0_8
8. Hammar, K.: Ontology design patterns in use - lessons learnt from an ontology engineering case. In: Blomqvist, E., Gangemi, A., Hammar, K., Suárez-Figueroa, M.C. (eds.) Proceedings of the 3rd Workshop on Ontology Patterns, Boston, USA, 12 November 2012. CEUR-WS.org (2012)
9. Janowicz, K., Gangemi, A., Hitzler, P., Krisnadhi, A., Presutti, V.: Introduction: Ontology Design Patterns in a Nutshell. IOS Press (2016)
10. Krötzsch, M., Simančík, F., Horrocks, I.: Description logics. IEEE Intell. Syst. **29**(1), 12–19 (2014). https://doi.org/10.1109/MIS.2013.123

11. Le, Q.V., Mikolov, T.: Distributed representations of sentences and documents. In: Proceedings of the 31th International Conference on Machine Learning, ICML 2014, Beijing, China, 21–26 June 2014, pp. 1188–1196. JMLR.org (2014)
12. Manning, C.D., Raghavan, P., Schütze, H.: Introduction to Information Retrieval. Cambridge University Press, Cambridge (2008)
13. Noy, N., Mcguinness, D.: Ontology Development 101: A Guide to Creating Your First Ontology. Knowledge Systems Laboratory (2001)
14. Poveda-Villalón, M., et al.: OOPS! (OntOlogy Pitfall Scanner!): an on-line tool for ontology evaluation. Int. J. Semant. Web Inf. Syst. (IJSWIS) 7–34 (2014). https://doi.org/10.4018/ijswis.2014040102
15. Ren, Y., Parvizi, A., Mellish, C., Pan, J.Z., van Deemter, K., Stevens, R.: Towards competency question-driven ontology authoring. In: Presutti, V., d'Amato, C., Gandon, F., d'Aquin, M., Staab, S., Tordai, A. (eds.) ESWC 2014. LNCS, vol. 8465, pp. 752–767. Springer, Cham (2014). https://doi.org/10.1007/978-3-319-07443-6_50
16. Schober, D., et al.: Ontocheck: verifying ontology naming conventions and metadata completeness in Protégé 4. J. Biomed. Semant. (2012)
17. Staab, S., Studer, R., Schnurr, H.P., Sure, Y.: Knowledge processes and ontologies. IEEE Intell. Syst. **16**(1), 26–34 (2001). https://doi.org/10.1109/5254.912382
18. Suárez-Figueroa, M.C., Gómez-Pérez, A., Fernández-López, M.: The neon methodology framework: a scenario-based methodology for ontology development. Appl. Ontol. **10**, 107–145 (2015). https://doi.org/10.3233/AO-150145

Adaptive Planning on the Web: Using LLMs and Affordances for Web Agents

Sebastian Schmid[1](✉), Michael Freund[2], and Andreas Harth[1,2]

[1] Chair of Technical Information Systems, Friedrich-Alexander-Universität Erlangen-Nürnberg, Nuremberg, Germany
sebastian.schmid@fau.de

[2] Division Data Spaces and IoT Solutions, Fraunhofer IIS, Nuremberg, Germany
{michael.freund,andreas.harth}@iis.fraunhofer.de

Abstract. We investigate the adaption of agents using plans on the Web despite its large and dynamic nature, as well as agents' constrained perception. Based on Semantic Web technologies and affordances, we compare how agents choose appropriate actions to adapt to their environment by condition-action rules or suggested actions of large language models. We conduct experiments on execution cost and plan stability distance to see whether agents choose appropriate actions to adapt their plans. We find that cost and stability of rule-based and LLMs for adaptation with affordances are close together, while performance differs greatly.

Keywords: Web Agents · Adaption · Dynamic Environments

1 Introduction

The Web has been long discussed as an environment where software agents act for humans [16]. Agents look for a course of action to reach a goal, that is, by using a plan [36]. The Web, however, is inherently dynamic [33] and large scale so agents are constrained by local perception [3].

Limited by local perception, using plans is especially hard in the dynamic Web environment, as plans will easily get outdated and agents detect errors only during execution, e.g. the environment does not look like expected or the agent has to perform an unforeseen action to proceed. So, how can agents use plans on the Web and choose appropriate actions on deviations, when only local information is reliable?

For dynamic environments, solutions using agents [11,36] range from replanning and plan adaption [30] to plan repair [18], which are suitable for agents that have reliable information to adapt but not for agents on the Web [15]. On the other hand, crawling every resource quickly enough might lead to having all information, but as the Web is an almost infinitely big maze the agent runs easily astray. Current research focuses heavily on Web agents using large language

---

This work is partially funded by the German Federal Ministry of Education and Research via the MANDAT project (FKZ 16DTM107A).

models (LLM) [26,34], which have successfully interacted with complex simulated environments. Direct interactions with the Web for information gathering or interactions are possible, but difficulties like hallucinations, a lack of robustness against prompt deviations, and a slow inference speed make the application for planned actions on the Web difficult [34]. Additionally, models are often pre-trained and fine-tuned for limited, specific use cases [26]. A promising idea is to use the environment itself for support: affordances, that is interactions offered by the environment for state transitions [24], can help agents in their decision-making on a local level. Here, the Semantic Web offers a uniform interface for interaction between agents and the environment to share data with a common understanding by using established web technologies, as also the rising recent interest in the combination of autonomous agents and the Semantic Web shows [6]. While the prevalent rule-based web agents [15] need to rely on rigid assumptions for affordances, we now see a chance to unify the approaches and mitigate the disadvantages of LLMs.

We combine existing agent architectures with LLMs and affordances for accurate and context-aware decision-making. We use a hybrid memory approach where agents use long-term memory, including a plan for execution, and short-term memory, using local percepts, [26], and derive injectable knowledge for LLMs without additional training [22] to suggest an action for adaption.

Based on the Web's analogy as a maze, we address our question with an experiment in an extended testbed of *Mike's maze* [1], an environment of connected web resources to be escaped by following links, to include *unforeseen actions* defined by affordances. We assume that all data is available in RDF[1] and accessible via a RESTful Read-Write Linked Data interface.

We compare two different approaches for agents, based on a model-based agent [30], to choose appropriate actions after a plan to escape Mike's Maze fails a) use condition-action rules, or b) use an LLM with local perception and knowledge injection. We implement the approaches and measure their effectiveness with respect to the distance and cost of an ideal plan and discuss the application in three maze environment sizes: a small maze with 25 web resources, a medium maze with 250, and a big maze with over 28,500. We summarize our contributions:

- We present a formalized and implemented hybrid agent architecture that integrates agent planning with LLMs and affordances to enhance robust and flexible decision-making in dynamic Web environments
- We measure the cost, stability, and performance to adapt of our different agent approaches in dynamic scenarios built on Mike's maze and discuss their implications

2 Running Example: Mike's Maze

We follow the agent A1 through a small part of the maze M (Fig. 1). Mazes are represented by cell entities that have four direction links, north, south, east, and

[1] Resource Description Framework: https://www.w3.org/TR/rdf11-concepts/.

west, that lead to other cells or end in walls. Agents can use safe HTTP requests (GET) to access cells and need to perform actions with unsafe requests (POST) to unlock "locks" to eventually reach their goal, where each lock needs a key.

Example 1. A1 shall traverse the maze M and reach the exit. (1) Accessing M to find the entrance cell, A1 gets a link to cell 0 (marked as entrance) and the useful information of a map of M. Using the map, A1 may conclude that east of cell 0 is cell 5, east of cell 5 is cell 10, etc., and that cell 24 leads to the exit, resulting in a plan for how to traverse the maze. (2) A1 executes its plan, accesses cell 0, and leaves via the east connection. (3) At cell 5, A1 notes that **no** east connection to cell 10 exists, so A1's plan *fails*. (4) Looking in its environment for a way to proceed, A1 evaluates if cell 5 describes an affordance to perform an action. Indeed, cell 5 is *locked* and requests a *red key*. To proceed, A1 has to choose appropriate actions that is to *acquire* information about a red key and then *act* on the affordance of cell 5. (5) By choosing an appropriate action, e.g. with a given rules set, A1 finds the red key at cell 0. (6) A1 returns to cell 5 and posts the red key to the locked cell, revealing an east connection to cell 10 as planned before. A1 may proceed, follow its initial plan and arrive at the exit.

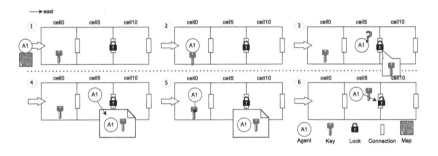

Fig. 1. Illustration of A1's interaction with M, cf. Example 1

We focus on the implications of steps (4)–(6): confronted with a failed plan, the agent has to recognize that an action is needed, what is needed to attempt the action, and finally how to perform the action. The way that these actions are suggested is with an affordance. Our maze present two kinds of obstacles to be overcome with affordances: locked cells that need keys, and dead-ends where special links show the correct path.

3 Theoretical and Technical Background

Agents in Uncertain Environments: Agents' behavior in uncertain environments is traditionally of high interest [30]. Current architectures like the Belief-Desire-Intention model, however, rely on stable environments [28]. In dynamic environments, plans can easily become incorrect during execution, so many solutions exist [32] that range from fixed actions for deterministic environments to time-intensive operations like alternative planning [35] or probabilistic modeling [27]. Kirrane [15] presents an exhaustive overview of Web agents and concludes that most approaches rely on simple event-condition-action rules.

Affordances: *Affordances* are the offered use or interaction with the environment via an operation [24]. In the sense of Hypermedia as the Engine of Application State (HATEOAS) [29], affordances build on the RESTful interfaces such that agents perform state transitions on resources using actions that are suggested by the environment itself [20]. For hypermedia affordances, such a transition can either be a *safe* action, e.g. by following a suggested link via GET, or an unsafe action by manipulating the environment, e.g. with a POSTing a form.

Agents and LLMs: LLMs, as a base for autonomous agents, gained a huge amount of interest, where the applications are plentiful [34], as the abilities for tool usage in LLM agents are generally deemed high [13,23]. Here, we want to focus on the interplay of LLMs, agents, and structured Web data: Yao et al. [37] train LLMs generate actions to order products according to a given description by consuming HTML and using observed human performances. In AgentBench, Liu et al. [21] provide eight testbeds to test different open and closed-source LLMs, including a task to produce answers based on a knowledge graph (KG). Liu et al. find that open-source LLMs have generally worse performance in web based scenarios. In HPTSA [8], Fang et al. use a multi-agent scenario where a hierarchical planner provides plans to a fleet of specialized agents with fixed tools to exploit vulnerabilities; the plan is formed on the consumption of plain HTML to guess interactions with which agents proceed in a trial and error fashion. Threats remain for LLMs i.e. hallucinations, a slow inference speed, and a limited input context length. Here, Pan et al. find in a recent survey that a synergy between LLM and KG leads to enhanced performance [26].

Agents and Artifacts: We give the definitions for RDF as data model, the maze environments, and our agents, derived from the definitions of [6,25,31].

Definition 1 (RDF Graph, triple, RDF dataset). *Let U, B, L be the sets of all URIs, blank Nodes, and literals and the set of all triples $T = (U \cup B) \times U \times (U \cup B \cup L)$, then a set $G \subseteq T$ is an RDF Graph. $\mathcal{G} = 2^T$ is the set of all RDF Graphs. We call the tuple $\langle s, p, o \rangle \in T$ a triple with the subject s, predicate p and object o of the triple. Let $U \times \mathcal{G}$ be the set of named RDF graphs. We call the pair $\langle u, g \rangle \in U \times \mathcal{G}$ a named RDF graph with its name u and the graph g. We call a set of named graphs an RDF dataset.*

Definition 2 (HTTP operations). *Let $OP = \{GET, POST\}$ be the set of HTTP methods, U the set of URIs and $\mathcal{G}$ the set of all RDF Graphs. Thus, the* set of HTTP operations is $O = OP \times U \times \mathcal{G}$. We call the tuple $\langle op, u, b \rangle \in O$ an HTTP operation with its operator $op \in OP$, target URI $u \in U$ and body $b \in \mathcal{G}$.

We call the set of operations $O_{GET} = \{o \in O | o = \langle GET, u, \emptyset \rangle, \forall u \in U\}$ the set of all GET requests that are used to retrieve the representation of the current state of u (read only). As operations $o \in O_{GET}$ do not change the resource state, we call these operations "safe". We call the set of operations $O_{POST} = \{o \in O | o = \langle POST, u, b \rangle, \forall u \in U, b \in \mathcal{G}\}$ the set of all POST requests that are used to append the data in b to an existing representation of a resource u. As operations $o \in O_{POST}$ change the resource state, we call these operations "unsafe".

Definition 3 (Environment, affordance). Let $U_M \subseteq U$ be the set of URIs of resources in the environment and $T_M \subseteq T$ the set of triples that describe resources in the environment. Let $K \subset T$ be the set of triples that describe affordances. Let the set $D \subset U_M \times (T_M \cup K)$ be the RDF dataset of the environment such that $D = \{\langle u_1, t_1 \rangle, \langle u_2, t_2 \rangle, \dots\}$. Let $\mathcal{D}$ be the set of all RDF datasets. Let σ be a function over $\mathcal{D}$ such that $\sigma(u_n, D)$ denotes $\{\langle u_n, t_n \rangle\}$, the set containing the unique element $\langle u, t \rangle \in D$ with $u = u_n$. Let $D' \subset D$ be the set of finite percepts. Thus, the environment is $M = \langle D, D_0, O, transfer, update, evolve \rangle$ with the initial state $D_0 \subset D$, and

$$transfer: O_{GET} \times \mathcal{D} \to \mathcal{D} \text{ with } transfer(\langle GET, u, \emptyset \rangle, D) = \sigma(u, D) = \{\langle u, t \rangle\}$$

$$update: O_{POST} \times \mathcal{D} \to \mathcal{D} \text{ to apply unsafe HTTP operations with}$$

$$update(\langle POST, u, b \rangle, D) = \begin{cases} (D \setminus \{\langle u, t \rangle\}) \cup \{\langle u, t \cup b \rangle\} & \text{if } \exists t \in T: \langle u, t \rangle \in D \\ D \cup \{\langle u, b \rangle\} & \text{otherwise} \end{cases}$$

$evolve: \mathcal{D} \to \mathcal{D}$, for the environment's dynamic state change

An operation $o \in O_{POST}$ with a body b as described by K (an affordance) lets the environment evolve D, e.g. by creating new triples.

Definition 4 (Model-based agents). Let A be a model-based agent that makes decisions on its own via a perception-thought-action cycle. Then the tuple $A = \langle G, perceive, apply, act \rangle$ where

$$G \subset \mathcal{G}, \text{ the agent's possible knowledge}$$
$$perceive: \mathcal{D} \times \mathcal{G} \to \mathcal{G} \text{ with } perceive(D', g) = g \cup D' \text{ as perception function}$$
$$apply: \mathcal{G} \to \mathcal{G}, \text{ function for inference rules,}$$
$$act: \mathcal{G} \to O, \text{ function for decision making for HTTP operations,}$$

where $D' \subset D$ is a finite percept (cf. Definition 3). The agent's current knowledge $g \in G$ contains sets of triples describing the currently perceived resource u possibly including triples with affordances, the triple Γ describing the goal and a plan π.

Definition 5 (Plan, planning problem). *Let $o \in O$ be an operation (Definition 2) on a URI u and let $t \in T$ be a triple. Then π is a plan of length n with $\pi = \langle \langle o_1, t_1 \rangle, \ldots, \langle o_n, t_n \rangle \rangle$ where each pair $1 \leq k < n$ contains a triple $t_k = \langle u_k, p_k, u_{k+1} \rangle$. In each step $\langle o, t \rangle$, agents perform an operation and need $t \in g$ to proceed in the plan, so t acts as a post-condition. We call the triple $t_n = \Gamma$, describing the goal. If $\Gamma \in g$, the goal is reached (Definition 4). A planning problem is $\mathcal{P} = \langle m, A, \varepsilon, \gamma, \pi \rangle$ with $m \in \mathcal{M}$, the set of all environments (Definition 3), and $\varepsilon, \gamma \in U$ as entry and exit URI.*

Definition 6 (Language model). *Let V be a vocabulary with tokens t_i and $\mathcal{T}$ the set of sequences of tokens $T_i = t_1, \ldots, t_n$. Given a corpus $\mathcal{C} \subseteq \mathcal{T}$, a language model $\mathcal{L}_\mathcal{C}$ is a probabilistic model that defines a distribution over sequence of tokens $\mathcal{L}_\mathcal{C}(T_i) = p(t_1, \ldots, t_n)$, which is an estimate of the probability of a sequence T_i given C. A prompt $P \subseteq \mathcal{T}$ is a sequence of tokens that may include tokens not in C (out-of-vocabulary tokens). Given a prompt P, a language model $\mathcal{L}_\mathcal{C}$ generates a sequence of tokens that maximizes the conditional probability under $\mathcal{L}_\mathcal{C}$ to generate the output token sequence $T_{out} = \arg\max_T \mathcal{L}_\mathcal{C}(T|P)$.*

Data Model and Maze. The maze is an environment after Definition 3 with $C \subset U$ as set of cell URIs, $W \subset U$ the set of wall URIs, $N \subset U$ the set of direction URIs, so $T_M = C \times N \times (C \cup W)$ describes all cells' directions to other cells and walls. We use the `maze` ontology[2] to describe `maze:Maze`, `maze:Cell`, `maze:Wall`, the directions `maze:north`, `maze:south`, `maze:east`, `maze:west`, and `maze:green`. `xhv`[3]`:start` and `maze:exit` mark the entry and exit cell (cf. ε and γ in Definition 5). We created the DYNMAZE vocabulary[4] to introduce `dyn:Lock` which has a `dyn:state` that is either `dyn:locked` or `dyn:unlocked`. If it is `dyn:locked`, the `dyn:Lock` hides all but one connection to a connected cell. As affordance, a `dyn:Lock` defines via `dyn:needsAction` a `dyn:Actions` with needed `dyn:Keys`.

Affordances K describes actions to "unlock" a cell with `dyn:Action` that uses the `http:`[5] vocabulary to define an HTTP method, request URI together with the properties `dyn:foundAt` and `dyn:needsProperty` that give hints what entities are sought (e.g. a `dyn:RedKey`) and what properties of these entities (e.g. `dyn:keyValue`) shall be submitted in the body. When the `dyn:keyValue` is POSTed to the respective `dyn:Lock`, the lock's `dyn:state` changes to `dyn:unlocked` and reveals the path (cf. Sect. 4.3).

Thus, the maze environment is $M = \langle D, D_0, O, transfer, update, evolve \rangle$ with the initial cells, directions, and affordances $D_0 \subset D$. Agents perceive and "traverse" the maze with operations from O_{GET} (e.g. moving to cell u), and manipulate the maze with operations in O_{POST} (e.g. put a key in a locked cell).

---

[2] `maze`: https://kaefer3000.github.io/2021-02-dagstuhl/vocab#.
[3] `xhv`: http://www.w3.org/1999/xhtml/vocab#.
[4] `dyn`: https://paul.ti.rw.fau.de/~am52etar/dynmaze/dynmaze.ttl#.
[5] `http`: https://www.w3.org/2011/http.

Example 2. The dyn:Cell with the URI /cells/5 is locked and specifies an affordance with information on a dyn:RedKey (cf. Example 1):

```
</cells/5>  a         maze:Cell, dyn:Lock ;
      maze:north   maze:Wall ;  maze:south   maze:Wall ;
      maze:west    </cells/0> ; #note the missing maze:east!
      dyn:state dyn:locked ;
      dyn:needsAction [
            dyn:hasStatus dyn:open ; http:requestURI </cells/5> ;
            http:mthd httpm:POST ;  dyn:foundAt dyn:RedKey ;
            dyn:needsProperty dyn:keyValue ] .
```

4 Agent Architecture and Approach

4.1 Planning and Plan Execution

Our agents use plans (Definition 5) to reach the maze's exit γ described by Γ. Agents may construct plans themselves or use provided plans, e.g. by other agents, a plan library, or the environment itself etc. [2,10]. Here, we opt for our agents to use a simple breadth first search algorithm [30, p. 374] from entry ε to exit γ based on the map provided by the environment.

Example 3. A1 GETs the URI of maze M and retrieves the representation describing M. A link to a resource of type dyn:Map hints at a useful description of the cells, so A1 GETs the map, too. Using the overview of maze:Cells, the directions maze:north, maze:south, maze:east, and maze:west, and the marked xhv:start and maze:exit relations, A1 constructs π, with the first steps:
$\pi = \langle \langle o_1, t_1 \rangle, \langle o_2, t_2 \rangle, \ldots \rangle = \langle \langle \langle GET, \text{/cells/0}, \emptyset \rangle, \langle \text{/cells/0}, \text{maze:east}, \text{/cells/5} \rangle \rangle, \langle \langle GET, \text{/cells/5}, \emptyset \rangle, \langle \text{/cells/5}, \text{maze:east}, \text{/cells/10} \rangle \rangle, \ldots \rangle$

Algorithm 1 shows the agent loop with relation to Definition 4. Using its plan π, the agent executes the next step $\langle o, t \rangle$ with operation o (cf. Definition 5), tries to match the defined condition triple t in $g \in G$ according to the plan: if the triple exists in g, the plan is still valid and the agent continues with the next step. If the agent notices a deviation, that is $t \notin g$, the plan fails.

For perception, the agent evaluates the current step in π to GET a URI u, e.g. a cell u, and transfers the perceived environment d' into the current knowledge $g \in G$ (cf. Definition 4, *perceive*). The agent's knowledge of refers to the last retrieved resource, so the agent's knowledge is *local*.

4.2 Apply and Act

Algorithm 1–14, is the center of our attention: the agent's choice of an *appropriate* action after the plan fails. During each cycle, the agent checks if the plan's condition triple t is in g (cf. Definition 4, *apply*), which branches the agent's behavior to either continue with the plan by sending HTTP requests (cf. Definition 4, *act*), or searching for safe or unsafe affordances. We discuss the use of condition-action rules and of an LLM below.

Algorithm 1: *Agent loop*

Data: maze m, entry ε, goal γ

1 $\pi, \Gamma \leftarrow planner(m, \varepsilon, \gamma)$; /* receive plan, e.g. from BFS planner */
2 **forall** *plan steps* π_i *in* π **do**
   /* determine next operation, target uri, and condition triple */
3 $\quad \langle op_i, u_i, body_i \rangle \leftarrow act(\pi_i)$;
4 $\quad t_i \leftarrow apply(\pi_i)$;
5 $\quad$ **if** $op_i = GET$ **then**
6 $\quad\quad d' \leftarrow transfer(\langle GET, u_i, \emptyset \rangle)$; /* enact perception */
7 $\quad\quad g \leftarrow perceive(d', g)$; /* update knowledge with percept */
8 $\quad$ **else**
9 $\quad\quad update(\langle o_i, u_i, body_i \rangle)$; /* enact manipulation */
10 $\quad$ **end**
11 $\quad$ **if** $\Gamma \in g$ **then**
12 $\quad\quad$ *break*; /* goal reached, end execution */
13 $\quad$ **else if** $t_i \notin g$ **then**
   /* determine action suggested by environment */
14 $\quad\quad \langle op_a, u_a, body_a \rangle \leftarrow act(apply(g))$;
15 $\quad\quad insertAfter(\langle op_a, u_a, body_a \rangle, \pi_i)$; /* insert action into plan */
16 **end**

Condition-Action Rules. We use condition-action rules with Notation3 (N3) [4] to define the agent's behavior. N3 rules have the form $\{b_1...b_n\}$`log:implies`$\{h\}$., with `log:implies`[6] expressing an implication. The antecedent of the rule, $b_1...b_n$, and the conclusion, h, are RDF triple patterns (s, p, o) as $(s, p, o) \in (U \cup B \cup V) \times (U \cup V) \times (U \cup B \cup L \cup V)$ (cf. Definition 1) with V as set of variables that may replace constants to give patterns, where h follows from $g \in G$, if $h \in g$ or $\{b_1...b_n\} \subseteq g$. These rules represent the agent's *apply* and *act* functions to influence its internal state and choose HTTP operations. Agents may alter the plan by following safe affordance triples and send safe requests (GET) to linked URIs and merge retrieved knowledge in g (cf. *perceive* in Definition 4). When the environment describes desired requests with unsafe affordances with `http` and `httpm`, e.g. POST, the agent may conclude to add a new operation to the plan, too, but does not necessarily expect new knowledge (getting new insight would be done during a subsequent perception operation). Afterwards, the agent applies the N3 rules again, as long as the goal is not reached or the agent fails for good as it recognizes no alternative actions.

Example 4 (N3 rules). To find an alternative route, A1 has a rule (see below) that states whenever a resource ?a has a `maze:green` relation to another resource ?b, the agent can send an HTTP GET request to ?b and thus proceed through the maze. To use `maze:green` as link, the environment has to state that `maze:green` is a "useful" relation and is related to a `maze:exit` resource. Our maze states these relations at the entry.

---

[6] `log`: http://www.w3.org/2000/10/swap/log#.

```
{
    ?a maze:green ?b .    # maze states maze:green as safe affordance
} => {
    [] http:mthd httpm:GET ;
       http:requestURI ?b .    } .
```

LLM. Instead of relying on rules, the agent may consult an LLM to find a suitable action. To give meaningful input, the agent transforms its available data to be understandable for the LLM that is to natural language. LLMs can reduce the inherent rigidity of condition-action rules, but come with their own challenge for configuration: as part of *apply*, the agent constructs a prompt P as input for the LLM with a description of the task, the current percept and formatting instructions for T_{out}. The agent uses the LLM's API with an HTTP POST whose body contains the prompt, where the response contains the generated output sequence $T_{out} = \arg\max_T \mathcal{L}_C(T|P)$. We structure the template as follows:

- Setting, task and description of the maze structure
- Plan and current step therein
- RDF dataset of triples describing the current percept
- Information deemed useful, e.g. relations of affordances
- Formatting instructions for HTTP method, request URI, and body

T_{out} is cleaned of fragments (e.g. trailing whitespaces), checked (e.g. for valid URI formats), and parsed to triples to g according to agent's formatting instructions to match an operation in O and added to the plan, to be executed during the next iteration of the agent's loop in *act*.

4.3 Using Affordances in the Environment

Environments may expose cues for safe and unsafe interaction [5], here via triples K that are used in an *affordance* (see Sect. 3). Agents can look for cues as alternatives to their planned actions, however, there are no guarantees that these lead to success or optimal behavior [17]. While the agent needs to detect affordances, the environment has to describe what action may be performed, the requirements, and the consequences. With the consequence of affordances, e.g. unlocking a cell, an agent can recognize if an action is useful, e.g. specified in temporal logic [12], safe, or unsafe.

Safe Affordance. The environment can give hints on how to reach a goal based on affordance links which require only safe requests. To use safe affordances, the environment needs to sustain action descriptions in its graph D, where such triples are part of the direction URIs N, here maze:green. Agents interpreting affordance links have to understand what affordance is tied to which goal (e.g. maze:green leads to maze:exit), which must defined at design time or suggested by the environment and discovered at run time via an annotation to the link. While using a safe affordance does not change the state of the environment, it changes the internal state of the agent with its knowledge g - with the next perception step, a new representation can lead to the discovery of new links.

Example 5 (Safe affordance links). A1 does not find t in g after perceiving /cells/5, but instead (/cells/5, maze:green, /cells/123), which matches the rule from Example 4. Thus, A1 appends an additional operation for HTTP GET to <cells/123> to the plan (informally: A1 follows "the trail").

Unsafe Affordance. In our scenario, we specify with K the requirements as combination of URI where the method shall be sent and necessary contained data. The agent has to check if either already $K \in g$, may be inferred from g, or π needs a suspension for a search. While the environment needs time to react, the success or failure of an affordance is visible via resource state changes. Here, we forgo access control measures to protect certain resources - if necessary, the environment has to define how the agent can obtain the required permissions (which can be an affordance on its own).

Example 6 (Unsafe affordance actions). /cells/5 defines dyn:Action with an HTTP POST request as operation and a body that contains a dyn:keyValue of a dyn:RedKey (see Example 2). A1 may try if the interaction with /cells/5 leads to a state of the cell that allows π to be continued, however, there is no guarantee. Here, A1 will temporarily suspend π and look for the dyn:RedKey.

5 Evaluation and Results

5.1 Evaluation Metrics

Execution Cost: We measure the execution costs [17] as a numeric value for an agent such that every HTTP request during plan execution and for using safe and unsafe affordances increases the costs by 1.

Plan Stability Distance: We measure the plan stability distance after [9] as $D(\pi', \pi)$ between an original plan π and a new plan π' as the number of actions that appear in π' but not in π, plus the number of actions that appear in π but not in π'. With a small plan stability distance, a new plan is close to π.

Execution Performance: We measure the wall-clock time for each run in seconds: the agent software runs on a Lenovo ThinkPad T14 with an i7-10510U CPU and 15 GB RAM. LLMs run on a industrial-grade HPE ProLiant DL380 Gen10 with an Intel Xeon Silver 4214R with 64 cores and 800 GB RAM.

5.2 Experiment Setup and Results

We implement our agents[7] after Definition 4 with the capabilities to store and manipulate RDF graphs with Node.js and N3.js[8]. The execution of HTTP operations uses TypeScript's *fetch*. We implement our maze with BOLD [19].

---
[7] https://github.com/wintechis/agents-llm-affordances
[8] https://github.com/rdfjs/n3.js

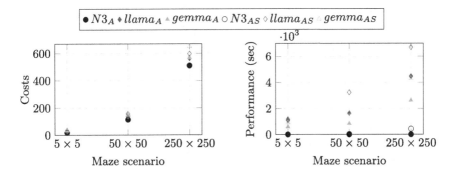

Fig. 2. Average costs and performance of all setups

As LLMs, we use Meta's Llama3 70b[9], Google's Gemma2 27b[10] and Alibaba's Qwen2 72b[11], as all reflect the state of the art at the time of writing and are open source. We set the LLMs' temperature to zero to reduce the probability of hallucinations and produce consistent answers. Table 1 shows our measured results for all variations averaged over 50 repetitions, visualized in Fig. 2. We give values for expected ideal costs and stability *in italics*: ideal costs show the minimal number of actions to traverse the maze with a plan, while stability shows the necessary adaptions for the plan, e.g. an ideal run with a cost of 14 and stability 4 is a plan with 14 steps where 2 steps were adapted (note that plan stability distance is symmetrical). Thus, a run with equal distance finds all necessary plan adaptions to be successful, while the cost shows the agent's effort to do so. We distinguish the following setup variations:

- number of cells as small (5 × 5), mid (50 × 50), and big (250 × 250), where more cells increase the plan's length and complexity
- setups with only unsafe affordances (A): we lock up to eight cells with hidden keys, but the agent may follow the original plan after using environmental cues for unlocking
- setups with unsafe *and* safe affordances (AS): additional to unsafe affordances, we remove connections from [2|6|10] cells as deviations, but do not provide means to manipulate the cells - the agent has to adapt the plan by following affordance links to reach its goal.

Agents using Qwen2 failed to complete all scenarios, so we could not carry out measurements. For successful agents, N3 adapts most efficiently, especially in complex scenarios. Gemma2 (G) shows consistent performance across different setups and sizes, while Llama3 (L) struggles more with adaptability in smaller, complex mazes but improves relatively in larger scenarios. Safe affordance increases the costs and decreases the performance for all agents:

---

[9] https://llama.meta.com/llama3/.
[10] https://huggingface.co/google/gemma-2-27b.
[11] https://github.com/QwenLM/Qwen2.

Table 1. Measured result - the suffix indicates environments with unsafe affordances (*A*) or unsafe+safe affordances (*AS*). Performance is given in seconds.

Agent	Small (5 × 5)			Mid (50 × 50)			Big(250 × 250)		
	Cost	Stability	Perf.	Cost	Stability	Perf.	Cost	Stability	Perf.
ideal$_A$	*14.0*	*0*	—	*110.0*	*0*	—	*499.0*	*0*	—
$N3_A$	24.0	0	9.1	125.0	0	50.9	534.0	0	543.9
$llama_A$	29.4	0	1096.9	137.0	0	1655.2	561.6	0	4461.6
$gemma_A$	32.6	0	585.4	147.6	0	858.9	568.8	0	2630.2
$qwen_A$	—	—	—	—	—	—	—	—	—
ideal$_{AS}$	*14.0*	*4*	—	*110.0*	*12*	—	*499.0*	*20*	—
$N3_{AS}$	26.0	4	8.8	131.2	12	53.8	543.0	20	671.6
$llama_{AS}$	34.3	4	1176.6	158.0	12	3237.9	597.5	20	6693.2
$gemma_{AS}$	37.4	4	927.3	157.0	12	1679.3	650.2	20	4301.2
$qwen_{AS}$	—	—	—	—	—	—	—	—	—

Unsafe affordances (A): All agents reach 0 stability, which indicates no need to change the original plan, as intended. Costs increase with maze size for all agents with a notable development, e.g. for N3 and Llama3 in setup *A* the difference decreases from 20% to only 5% more actions. When we compare the cost overhead to ideal behavior in both setups, we notice a generally high overhead in the small maze (N3: 71%, L&G >100%) as few extra actions already have a big impact. In larger mazes, the overhead is reduces e.g. for mid (N3: 13%, L: 24%, G: 33%) and big (N3: 7%, L: 12%, G: 14%) where Llama3 and Gemma2 perform similarly, with Gemma2 slightly worse in cost but better in performance.

Unsafe and safe affordances (AS): All agents match the ideal stability and adapt the plan successfully. Costs are generally higher compared to setup *A*, which is expected due to additional actions with detours of safe affordances, with an overhead in the small (N3: 85%, L&G >140%), mid (N3: 19%, L: 43%, G: 42%), and big maze (N3: 8%, L: 20%, G: 30%). Llama3 struggles more with small mazes (higher cost and lower performance) but becomes more competitive in larger mazes. Gemma2 performs well across all sizes, with better performance than Llama3. N3 performs best overall, with lower costs and better performance, especially in larger mazes, but again LLMs close in, up to a 10% difference.

6 Discussion

The Web as an environment is hard for agents as it is dynamic and wants interaction. With affordances, agents may use plans to be efficient but also adapt to dynamic environments. Take as example an online shop: you search for a product, add it to the cart, enter your credentials, and finally access the checkout. As the procedures are similar across shops, agents may use a "shopping" plan from a library instead of exploring every time anew. Adaptions to use unanticipated actions may be necessary ad hoc, e.g. to circumvent discount advertisements or broken links. LLMs as support for such agents are a starting point where rules are too rigid, but still, they have their limitations.

We see that in all scenarios rule-based and some LLM-based agents eventually created a successful plan. While N3 agents' success was mostly based on design time assumptions, e.g. which predicates signify affordances, LLMs received prompts together with processed information. As we use LLMs without pre-training, the ability to adapt is quite surprising when compared to rule-based agents. We find, however, that some precautions may mitigate hallucinations and prompt deviations [34], and speed up the performance. The combination of long-term and short-term data for appropriate actions reduces the search space, otherwise an agent run becomes a random search. Using contextual data, e.g. key entities, from the local percept of the graph as knowledge injection improved the selected actions [22]. Interestingly, when faced with the maximum context length of LLMs, we tried a sliding window for long-term data: a subset of the plan before and after the current step yields an improvement of output quality and speed, using only about 1, 200 input tokens across all setups.

Our agent relies on the action selection to adapt a given plan instead of failing for good. While a rule-based approach selects predictable, stable actions, an LLM might derail the plan with nonsensical answers. If the agent executes actions without evaluation, manipulations can leave the task unsolvable, which happened in our runs in one case. Furthermore, the prompt has to stress that the plan failed, otherwise, the LLM suggests executing the plan as is, resulting in a loop. Despite a successful adaption with most LLMs, we see caveats - obviously, the performance to solve the mazes, given a dedicated machine to run the models, is disappointing considering settings without reliance on commercial providers.

We agree with Kambhampati et al. [14] on the limits of LLM's planning capabilities; LLMs need a lot of aid to provide beneficial additions to plans. Especially for safe affordances, the prompt has to get quite direct that the option of an alternative is viable and exists, plus the agent has to check the answers to avoid executing actions already known to fail. With a carefully constructed prompt, however, erroneous agent behavior may be corrected: during testing, we found a (meanwhile repaired) bug where the LLM's suggested action was inserted in the plan with an offset. Despite the error, the LLM interpreted the shifted plan in subsequent interactions correctly and suggested correct actions.

All in all, we conclude that agents may use LLMs for affordances successfully, although only with precautions: while action affordances can be understood, the use does not come naturally. Affordance use depends on the provided data where we find that less data was needed than anticipated with a combination of long-term planning data and the task (with sliding windows) and short-term from the percept (with knowledge injection). We find that hallucinations are countered quite well with a hybrid memory approach and a low temperature [13,23]. Slow inference remains a nevertheless problem [34]. We see that with provided context-awareness, an LLM can be beneficial for adaptive agents.

Limitations. We use a custom, simplistic vocabulary to describe affordances to be interpreted in natural language (a "key" unlocks a "lock"). Engineering the prompt to stress important parts was a manual step, especially the emphasis on optional links and the relevance of certain triples which requires manual tuning and makes generalization across different tasks harder. Although some resilience towards processing nonsense input exists [7], it is questionable if an LLM suggests a reasonable action without further context, e.g. of `rdfs:label`.

7 Conclusion and Outlook

We combined a model-based agent with affordances for accurate decision-making in a maze environment, where actions for adaption are suggested by either fixed rules or an LLM. We compare the performance of N3 rules, Llama3, Gemma2, and Qwen2 and find that some LLMs can identify affordances and make use of it to adapt, while others cannot. With a combination of long-term planning data (with sliding windows) and short-term from the percept, less data was needed than anticipated. While the cost and stability to adapt is competitive for LLMs, the performance struggles. As beneficial side effects, we noticed the correction of offset errors in planning. Next, we want to compare how LLMs choose appropriate actions using a history of prompts and how a deliberately manipulated percept influences error correction.

References

1. Amundsen, M.: Building Hypermedia APIs with HTML5 and Node. O'Reilly (2012)
2. Aßfalg, N., Schneider, H., Käfer, T.: Integrated planning and execution on read-write linked data. In: Proceedings of the All the Agents Challenge (ATAC) at the 20th International Semantic Web Conference (ISWC), pp. 25–31 (2021)
3. Bandini, S., Manzoni, S., Vizzari, G.: Web sites as agents' environments: general framework and applications. In: Environments for Multi-Agent Systems II, pp. 235–250. Springer, Heidelberg (2006)
4. Berners-Lee, T.: Notation3 logic - an RDF language for the Semantic Web. https://www.w3.org/DesignIssues/Notation3.html
5. Boissier, O., et al.: Agents on the web (Dagstuhl Seminar 23081). Dagstuhl Rep. **13**(2), 71–162 (2023). https://doi.org/10.4230/DagRep.13.2.71
6. Charpenay, V., Käfer, T., Harth, A.: A unifying framework for agency in hypermedia environments. In: Engineering Multi-Agent Systems, pp. 42–61. Springer, Cham (2022)
7. Cherepanova, V., Zou, J.: Talking Nonsense: Probing Large Language Models' Understanding of Adversarial Gibberish Inputs (2024). https://arxiv.org/abs/2404.17120
8. Fang, R., et al.: Teams of LLM Agents can Exploit Zero-Day Vulnerabilities (2024). https://doi.org/10.48550/arXiv.2406.01637
9. Fox, M., et al.: Plan stability: replanning versus plan repair. In: Proceedings of the Sixteenth International Conference on International Conference on Automated Planning and Scheduling, ICAPS 2006, pp. 212–221. AAAI Press (2006)
10. Georgeff, M.P., Lansky, A.L.: Reactive reasoning and planning. In: Proceedings of the 6th National Conference on Artificial Intelligence, vol. 87, pp. 677–682 (1987)

11. Gleizes, M.P.: Self-adaptive complex systems. In: Multi-Agent Systems, pp. 114–128. Springer, Heidelberg (2012)
12. Harth, A., et al.: Towards representing processes and reasoning with process descriptions on the web. Trans. Graph Data Knowl. **2**(1), 1:1–1:32 (2024). https://doi.org/10.4230/TGDK.2.1.1
13. Huang, J., Chang, K.C.C.: Towards reasoning in large language models: a survey (2023). https://arxiv.org/abs/2212.10403
14. Kambhampati, S., et al.: LLMs Can't Plan, But Can Help Planning in LLM-Modulo Frameworks (2024). https://arxiv.org/abs/2402.01817
15. Kirrane, S.: Intelligent software web agents: a gap analysis. J. Web Semant. **71**, 100659 (2021). https://doi.org/10.1016/j.websem.2021.100659
16. Kirrane, S., Decker, S.: Intelligent agents: the vision revisited. In: Proceedings of the 2nd Workshop on Decentralizing the Semantic Web co-located with the 17th International Semantic Web Conference (ISWC 2018) (2018)
17. Koenig, S.: Agent-centered search. AI Mag. **22**(4), 109 (2001). https://doi.org/10.1609/aimag.v22i4.1596
18. Komenda, A., et al.: Domain-independent multi-agent plan repair. J. Netw. Comput. Appl. **37**, 76–88 (2014). https://doi.org/10.1016/j.jnca.2012.12.011
19. Kfer, T., Charpenay, V., Harth, A.: Bold: a benchmark for linked data user agents and a simulation framework for dynamic linked data environments (2023). https://doi.org/10.48550/arXiv.2307.09114
20. Lemée, J., et al.: Domain-expert configuration of hypermedia multi-agent systems in industrial use cases, vol. 2023-May, pp. 2499–2501 (2023)
21. Liu, X., et al.: AgentBench: Evaluating LLMs as Agents (2023). https://arxiv.org/abs/2308.03688
22. Martino, A., et al.: Knowledge injection to counter large language model (LLM) hallucination. In: The Semantic Web: ESWC 2023 Satellite Events, pp. 182–185. Springer, Cham (2023)
23. Mialon, G., et al.: Augmented language models: a survey (2023). https://arxiv.org/abs/2302.07842
24. Norman, D.A.: Affordance, conventions, and design. Interactions **6**(3), 38–43 (1999). https://doi.org/10.1145/301153.301168
25. Omicini, A., Ricci, A., Viroli, M.: Artifacts in the A&A meta-model for multi-agent systems. Auton. Agent. Multi-Agent Syst. **17**(3), 432–456 (2008). https://doi.org/10.1007/s10458-008-9053-x
26. Pan, S., et al.: Unifying large language models and knowledge graphs: a roadmap. IEEE Trans. Knowl. Data Eng. **36**(7), 3580–3599 (2024). https://doi.org/10.1109/TKDE.2024.3352100
27. Pearl, J.: Probabilistic Reasoning in Intelligent Systems: Networks of Plausible Inference. Morgan Kaufmann (1988)
28. Rao, A.S., Georgeff, M.P., et al.: BDI agents: from theory to practice. In: Proceedings of the 1st International Conference on Multi-Agent Systems, vol. 95, pp. 312–319 (1995)
29. Richardson, L., Ruby, S.: RESTful Web Services, 1st edn. O'Reilly Media (2007)
30. Russell, S., Norvig, P.: Artificial Intelligence: A Modern Approach, 3rd edn. Prentice Hall Press, USA (2009)
31. Schmid, S., Schraudner, D., Harth, A.: MOSAIK: an agent-based decentralized control system with stigmergy for a transportation scenario. In: The Semantic Web, pp. 697–714. Springer, Cham (2023)

32. de Silva, L., et al.: BDI agent architectures: a survey. In: Bessiere, C. (ed.) Proceedings of the Twenty-Ninth International Joint Conference on Artificial Intelligence, IJCAI 2020, pp. 4914–4921 (2020). https://doi.org/10.24963/ijcai.2020/684
33. Tamma, V., Payne, T.R.: Is a semantic web agent a knowledge-savvy agent? IEEE Intell. Syst. **23**(4), 82–85 (2008). https://doi.org/10.1109/MIS.2008.69
34. Wang, L., et al.: A survey on large language model based autonomous agents. Front. Comput. Sci. **18**(6) (2024). https://doi.org/10.1007/s11704-024-40231-1
35. Winikoff, M., Cranefield, S.: On the testability of BDI agent systems. J. Artif. Intell. Res. **51**, 71–131 (2014)
36. Wooldridge, M.: Intelligent agents: the key concepts. In: Multi-Agent Systems and Applications II, pp. 3–43. Springer, Heidelberg (2002)
37. Yao, S., et al.: Webshop: towards scalable real-world web interaction with grounded language agents (2023). https://doi.org/10.48550/arXiv.2207.01206

Enriching RDF Data with LLM Based Named Entity Recognition and Linking on Embedded Natural Language Annotations

Michael Freund[1](✉), Rene Dorsch[1], Sebastian Schmid[2], Thomas Wehr[2], and Andreas Harth[1,2]

[1] Fraunhofer Institute for Integrated Circuits IIS, Nürnberg, Germany
{michael.freund,rene.dorsch,andreas.harth}@iis.fraunhofer.de
[2] Friedrich-Alexander-Universität Erlangen-Nürnberg, Nürnberg, Germany

Abstract. In this paper, we present a processing pipeline for transforming natural language annotations in RDF graphs into machine-readable and interoperable semantic annotations. The pipeline uses Named Entity Recognition (NER) and Entity Linking (EL) techniques based on a foundational Large Language Model (LLM), combined with a Knowledge Graph (KG) based knowledge injection approach for entity disambiguation and self-verification. Through a running example in the paper, we demonstrate that the pipeline can increase the number of semantic annotations in an RDF graph derived from information contained in natural language annotations. The evaluation of the proposed pipeline shows that the LLM-based NER approach produces results comparable to those of fine-tuned NER models. Furthermore, we show that the pipeline using a chain-of-thought prompting style with factual information retrieved via link traversal from an external KG achieves better entity disambiguation and linking than both a pipeline without chain-of-thought prompting and an approach relying only on information within the LLM.

Keywords: Natural Language Processing · KG enhanced LLM · RDF

1 Introduction

Resource Description Framework (RDF) data is widely used in various domains where semantic information and knowledge needs to be stored in an interoperable and machine-readable format. For instance, in the Internet of Things (IoT) domain, through the World Wide Web Consortium's (W3C) Web of Things (WoT) Architecture [4], where RDF is used to describe the capabilities and APIs of sensors and actuators to create semantic interoperability [3]. Similarly, in archaeology [17], where RDF is used to describe discovered historical objects and their properties, such as where the objects were found and what materials the objects are made of, to create a comprehensive and interoperable knowledge base. RDF data typically consists of two parts: the main part, which is machine-readable information based on formal ontologies, and a smaller part, which includes natural language annotations such as free-form text found in labels and comments.

When RDF data is created by non-experts who are unfamiliar with relevant ontologies and established RDF data modeling techniques, or by experts who assume that the data will only be accessed by humans, they often annotate some relevant information using solely natural language comments or labels for simplicity. However, only ontology-based information is machine-readable and interoperable. The information in natural language annotations embedded within RDF data can still be highly relevant and valuable. But, the embedded information often remains largely inaccessible to machines, due to the challenges of processing natural language, and to humans when the annotations are in a language they do not understand.

Recent advances in natural language processing (NLP) techniques, particularly with foundational large language models (LLMs), have significantly progressed the field [10]. Unlike traditional Named Entity Recognition (NER) models, which rely on extensive labeled training data adapted to specific application domains, large foundational LLMs can perform effectively without the need for domain-specific labeled data or fine-tuning [16]. Even without fine-tuning, foundational LLMs have been shown to outperform fine-tuned domain-specific models, for instance in the medical field [14].

Therefore, a pipeline that uses a foundational LLM for NER to extract relevant entities from natural language annotations in RDF data, in combination with Entity Linking (EL) to map the recognized entities to concepts in existing Knowledge Graphs (KGs) or ontologies, could be flexible and capable of making natural language information machine-accessible and interoperable. With that approach RDF data could be made more descriptive without the need for domain-specific model fine-tuning.

Extracting relevant information from text in RDF data is challenging due to the typically short and variable nature of embedded natural language annotations, which often consist of single-word labels or short sentences in comments. The inherent ambiguity of natural language, where a concept can be described in multiple ways and the meaning of a term can change based on context, further complicates this task. Unlike ontology-based information, which follows a formal schema, natural language annotations are unstructured, making systematic parsing and interpretation difficult.

In order to extract information from natural language annotations in RDF graphs, we propose such an extraction pipeline that uses LLM-based NER techniques to identify relevant entities in embedded natural language annotations. After entity recognition, the pipeline links the identified entities to concepts in pre-selected KGs and ontologies using a retrieval augmented generation (RAG) [5] approach with semantic similarity search based on vector embeddings to retrieve the URIs of identified concepts. By dereferencing these URIs, RDF documents with additional factual information are retrieved, which are then used for entity disambiguation and self-verification through knowledge injection techniques [7], mitigating the risk of hallucination. Finally, the pipeline generates based on the identified entities RDF triples and integrates them into the original graph, converting the natural language information into ontology-based, machine-readable, and interoperable information.

The key contributions of our work are as follows:

- The introduction of a pipeline for transforming natural language annotations embedded in RDF graphs into additional semantic annotations using LLMs, NER, and EL.
- The introduction of a KG and knowledge injection-based method for entity disambiguation and self-verification.
- An evaluation of our LLM-based NER system compared to a traditional NER method, along with an assessment of the performance of the introduced pipeline compared to an LLM-only approach.

2 Running Example

To better illustrate the contributions of this work, we use the semantic description of an IoT Bluetooth Low Energy sensor following the Web of Things recommendation as a running example throughout this paper.

The RDF graph in turtle serialization describing the API of the sensor, is illustrated in Listing 1.1[1] and mainly uses the Thing Description ontology [1].

Listing 1.1. RDF graph describing the sensor introduced in the running example in Turtle serialization with relevant information contained in natural language text. Note that the contents of td:hasForm statements are omitted for simplicity.

```
1   [] a td:Thing ;
2     td:title "Flower"@en ;
3     td:description "Xiaomi Flower Care sensor in room 40."@en ;
4     td:hasPropertyAffordance [
5       td:name "temperature" ;
6       jsonschema:readOnly "True"^^xsd:boolean ;
7       td:description "In degrees Celsius."@en ;
8       td:hasForm [ ... ] ] ;
9     td:hasPropertyAffordance [
10      td:name "humidity" ;
11      jsonschema:readOnly "True"^^xsd:boolean ;
12      td:description "The humidity value in %."@en ;
13      td:hasForm [ ... ] ] .
```

---

[1] Well-known prefixes are omitted in all listings, but can be looked up on http://prefix.cc/.

From the semantic annotations, we can see that the sensor provides two property affordances: one named `temperature` and the other named `humidity`, and that both affordances are read-only. Additional information is contained in natural language via the three `td:description` (lines 3, 7, 12) and the two `td:name` annotations (lines 5, 10). Based on those annotations, we as humans can infer that the device is a Flower Care sensor manufactured by Xiaomi, deployed in room 40, and capable of measuring relative humidity in percent and temperature in degrees Celsius. But this information is currently only contained in natural language annotations and is therefore inaccessible to machines without the capability to process natural language and is therefore not interoperable.

In the following sections, we demonstrate how we extract the information using NER, associate and link the detected entities with selected vocabularies and KGs, disambiguate and verify the result, and generate triples to make the information machine-readable.

3 Related Work

Natural language processing is a difficult task that has challenged researchers for decades [12]. A subtask of the NLP domain is Named Entity Recognition, where the objective is to identify and classify relevant entities in text, where the meaning of entities may be context-dependent, requiring sequence labeling techniques [13]. Entity Linking complements NER by matching identified entities with existing concepts in ontologies or KGs. EL systems generate a list of potential candidates and perform entity disambiguation by considering the context in which the entity is used in the text. After successfully linking an entity to an associated concept, additional factual information can be used in further processing steps [20] and the entity can be integrated in the existing RDF graphs in the form of additional RDF triples.

With the recent increase in popularity of LLMs, such as the GPT series of models [6], and their ability to process natural language text [26], researchers have investigated the usability of LLMs for NER tasks. Wang et al. [23] explored the application of LLMs to NER tasks by introducing an approach called GPT-NER. The GPT-NER approach transforms the sequence labeling task into a generation task to exploit the strengths of LLMs. To address issues such as hallucination and overprediction when no entities are detected, a simple self-verification step is included, where the LLM is prompted again and asked to verify the given label. Monajatipoor et al. [11] used LLMs for a specialized NER task in the biomedical domain and found that providing relevant in-context examples via the input prompt is key for good performance.

Our work builds on the insights of the two LLM-based NER approaches and refines them by additionally using entity linking to identify the entities corresponding concepts in selected ontologies and KGs in order to leverage the factual knowledge through knowledge injection [7], where relevant information contained in a KG is provided as additional input to the LLM in order to improve the quality of the generated output. The introduced pipeline uses knowledge injection in an entity disambiguation and self-verification step.

To interact with LLMs, both approaches use so-called prompt engineering techniques, which involve natural language inputs that define what the LLM should do. A commonly used basic prompt format consists of an *instruction*, a *context*, and an *input text*. The instruction provides a general task description, the context offers few-shot examples, and the input text is the text that needs to be processed [15]. The quality of the prompt can in general impact the performance of the LLM in the desired task [22]. To solve more complex tasks, a more advanced chain-of-thought prompt [24] can be used, in which several intermediate steps of reasoning are used to reach the final goal.

The two previously introduced NER approaches use a simple prompt consisting of instruction, context, and input text. In contrast, we use chain-of-thought prompting in our pipeline.

To counter the tendency of LLMs to generate irrelevant or false outputs, known as hallucinations, a widely used approach is retrieval augmented generation [5]. The RAG method involves retrieving related information through dense vector similarity search and incorporating it as additional input during inference, i.e., injecting additional knowledge. Matsumoto et al. [8] implemented RAG in combination with KGs in a framework called KRAGEN. The authors embedded the entire KG, including all entities and relations, into a vector database to retrieve the factual information during inference.

In contrast, our approach focuses on embedding only the corresponding URIs of entities in the KG and using link traversal to retrieve the additional information, allowing for greater flexibility and precise linking to identified entities.

4 Enriching RDF Data Based on Natural Language Annotations

Our processing pipeline to enrich RDF data based on embedded natural language annotations is illustrated in Fig. 1. Enriching RDF data with additional semantic information extracted from natural language annotations helps to improve the understandability and interoperability of the data, making it more useful for various applications in research [19] and industry [2]. The proposed annotation pipeline consists of four sequential steps: Named Entity Recognition, Entity Linking, Disambiguation and Self-Verification, and Triple Generation. The steps are discussed in detail below.

Named Entity Recognition. The first step in the proposed processing pipeline is the recognition of relevant named entities. We start by retrieving the initial RDF data and extracting natural language annotations, such as labels, comments, or descriptions, using SPARQL queries. The extracted natural language annotations form our input set, which means that no duplicate entries are processed. Next, we use a foundational large language model to perform the NER task. The use of a foundational LLM-based NER approach allows us to bypass the need to fine-tune traditional NER models for entity recognition across various domains, which typically requires the manual generation of a significant number of labeled examples used as training data [18]. To interact with and instruct

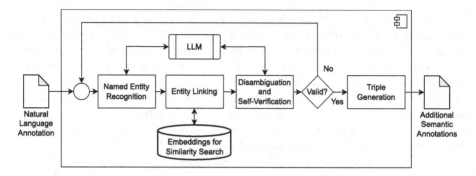

Fig. 1. The processing pipeline for transforming natural language annotations embedded in RDF data into semantic annotations. Key steps and data flows are illustrated.

the LLM, we create a template that implements a prompting strategy based on chain-of-thought prompting, where information and definitions about the NER task and the target domain are introduced, followed by a classical prompt containing a task description, a small set of few-shot examples, and the natural language annotation to be classified. In addition, we provide a clear specification of the expected response format, which improves the consistency of the LLM output and eases the overall parsing of the generated output, allowing the integration of the output information into further steps of the processing pipeline. For each group of named entities that the processing pipeline needs to recognize, we manually create a template that implements the chain-of-thought prompting for the specific domain. Groups of named entities include units of measurement in text (e.g., degrees Celsius) or symbolic form (°C), observable properties (e.g., temperature), or location-specific entities (e.g., room 40). Finally, after recognizing the named entities in a natural language annotation, we task the LLM to verify that the discovered entities have been assigned the correct label in the context of the natural language annotation, similar to the validation approach introduced by GPT-NER [23].

Example 1. In the context of the running example, the input set I containing all natural language annotations is defined as $I = \{$"Xiaomi Flower Care sensor in room 40.", "temperature", "In degrees Celsius.", "rel . humidity", "The rel . humidity value in %." $\}$. The NER step processes the input set and generates the output set R containing corresponding recognized entities where $R = \{$ "Xiaomi", "room 40", "temperature", "degrees Celsius", "humidity", "%"$\}$

Entity Linking. The second step of the processing pipeline is entity linking. During the EL process, we use the named entities identified in the NER step to construct entity embeddings using a pre-trained deep learning model. The embeddings are low-dimensional, dense vector representations that encapsulate the semantic and syntactic properties of the named entities [21]. By calculating semantic similarity scores between the embeddings of the recognized named entities and those of pre-selected domain-specific concepts from KGs or ontologies, we are able to generate a list of potential candidates to which a named entity might be related. The candidate list may have zero entries if no similar concept exists in the pre-selected KGs or ontologies, or only one entry if only one similar concept is found within a given distance. The next task is to disambiguate and select the most appropriate candidate from the list, taking into account the context provided by the original natural language annotation. This disambiguation is done in the next step of the pipeline.

Example 2. The entity linking step uses the set R as input and transforms R into a set E, which contains, for each detected named entity, a candidate list of potential corresponding ontology or KG entries. In the running example, we recognize entries from Wikidata[2], QUDT[3], and a local floor plan KG[4].

Only the named entity for humidity produced a candidate list containing more than one entry, since there exist entries for *relative humidity* and the absolute humidity simply called *humidity*. Note, that if the candidate list contains only one entry, that entry is directly added to the set E instead of the list. Therefore, the set E is given by E = {"wd:Q1636958", "ex:room40", "wd:Q11466", "qudt:DEG_C", ["wd:Q2499617", "wd:Q180600"], "qudt:PERCENT"}.

Disambiguation and Self-verification. After performing the linking step, the next stage of the processing pipeline disambiguates candidates when more than one potential match is found. If only one candidate is present in the list, the pipeline verifies the result to ensure the linked entity is meaningful.

If more than one possible candidate is discovered during the entity linking step, the pipeline first retrieves relevant RDF documents containing definitions and information about the potential candidates using link traversal starting from the URIs identified in the EL step. The factual information contained in the retrieved RDF documents can then be provided as additional input to the LLM. Specifically, we inject the additional factual knowledge from the RDF data into the prompt, along with the recognized entity and the original natural language sentence. The task of the LLM is to verify that the identified concept, based on the corresponding definition in the RDF document, is correct for the recognized entity in the context of the sentence and to explain the reasoning steps that lead to the decision. In a second step, the original task and the generated reasoning output are fed back to the LLM, which is tasked with generating a final decision if the concept is correct and outputting either **yes** or **no**. This approach allows the LLM to disambiguate entries in the candidate list using external information that may change over time and was not available during training using a chain-of-thought prompting style.

The disambiguation approach can also be seen as self-verification, where additional external knowledge is added as context for the LLM to determine if the

---

[2] prefix wd: https://www.wikidata.org/wiki/.
[3] prefix qudt: https://qudt.org/vocab/unit/.
[4] prefix ex: https://example.com/.

entity is being used correctly. Therefore, when the candidate list contains only one entry, we apply the same steps as in the disambiguation approach to verify that the candidate is used in the correct context.

If the results are not as expected, or if the candidate list cannot be disambiguated or verified, we restart the whole pipeline, repeating the previous entity recognition and linking steps, and providing the error information as additional input in the prompt.

Example 3. The Disambiguation and Self-Verification step uses the set E as input and produces the disambiguated and verified set E'. In the running example, only one ambiguity occurred, in the case of humidity. The original natural language annotation is The humidity value in %. and the recognized entity is humidity. The candidate list contains wd:Q180600 for humidity and wd:Q2499617 for relative humidity with the following natural language definitions extracted from the RDF data:

1. humidity: amount of water vapor in the air
2. relative humidity: ratio of the partial pressure of water vapor in humid air to the equilibrium vapor pressure of water at a given temperature

For each candidate, the original natural language annotation, the recognized entity, and the RDF data containing, among other things, the natural language definition presented above are then fed into the LLM, which is tasked with deciding if the definition is the correct one in the context of the natural language annotation.

The LLM determines that in this context, relative humidity is the correct definition because relative humidity is a ratio, and ratios are measured in percent. The self-verification process did not produce any errors for all other entities.

Therefore, the resulting set E' is given as E' = {"wd:Q1636958", "ex:room40", "wd:Q11466", "qudt:DEG_C", "wd:Q2499617", "qudt:PERCENT"}.

Listing 1.2. RDF graph describing the sensor, with additional triples generated based on the embedded natural language annotations.

```
1   @prefix ex: <https://example.com/> .
2
3   [] a td:Thing ;
4     td:title "Flower"@en ;
5     td:description "Xiaomi Flower Care sensor in room 40."@en ;
6     schema:manufacturer wd:Q1636958 ;
7     schema:location ex:room40 ;
8     td:hasPropertyAffordance [
9       td:name "temperature" ;
10      jsonschema:readOnly "True"^^xsd:boolean ;
11      sosa:observes wd:Q11466 ;
12      td:description "In degrees Celsius."@en ;
13      qudt:unit qudt:DEG_C ;
14      td:hasForm [ ... ] ] ;
```

```
15      td:hasPropertyAffordance [
16        td:name "humidity" ;
17        jsonschema:readOnly "True"^^xsd:boolean ;
18        sosa:observes wd:Q2499617 ;
19        td:description "The humidity value in %."@en ;
20        qudt:unit qudt:PERCENT ;
21        td:hasForm [ ... ] ] .
```

Triple Generation. The final step in the processing pipeline is to generate RDF triples consisting of a *subject*, *predicate*, and *object* for insertion into the existing RDF graph, thereby making the information extracted from natural language annotations machine-readable and interoperable. The object position of the new triple contains the result of the NER and EL processes described in the previous steps, where named entities were identified and linked to corresponding entities in existing ontologies and KGs.

To find an appropriate predicate for the identified object, we use a pre-defined hashmap that maps different types of identified entities to their corresponding predicates. The use of a hashmap ensures that the relationships between entities are correct and stored in a scalable way. Additionally, the mapping can be modified and adjusted to include other ontological terms without the need to retrain a statistical model. For the subject position of the new triple, we reuse the same subject as in the original natural language annotation, preserving the context of the information.

Example 4. The Triple Generation step uses E' as input and generates the set of all output triples T. For instance, consider the first entry in E' where the NER step detected Xiaomi with the label manufacturer. The linking step linked the entity to wd:Q1636958, and the disambiguation and self-verification step confirmed the correctness. Therefore, we know that a manufacturer has been recognized, and we look up the predicate associated with the label manufacturer in the hashmap, which returns the property schema:manufacturer. The original natural language annotation was associated with a blank node of type td:Thing, so the new triple will also be associated with the same blank node. The generated triple $t_1 \in T$ is therefore given as t_1 = ([], schema:manufacturer, wd:Q1636958).

After all additional triples have been generated, all elements of the set T are added to the RDF graph, as shown in Listing 1.2. The information previously contained only in natural language annotations is now available as machine-readable and interoperable semantic annotations (lines 6, 7, 11, 13, 18, 20), resulting in a much more expressive RDF graph overall.

5 Empirical Evaluation

We evaluated the introduced pipeline using the Gemma2 27B language model which is part of Googles Gemma family of language models [9] and we link found entities against the Wikidata KG, and classes in the QUDT ontology.

In the empirical evaluation, we aim to answer the following two research questions:

- **R1:** How accurate is the foundational LLM-based NER approach for extracting entities from RDF graphs with natural language annotations compared to the classical approach of fine-tuning a domain-specific NER model?
- **R2:** How accurate is the pipeline using chain-of-though prompting in identifying and linking correct entities compared to the pipeline without chain-of-though prompting and to an approach using only the LLM?

All results and scripts to reproduce the evaluation can be found on GitHub[5].

5.1 R1: Comparison of LLM-Based NER with Existing Approaches

In our comparative analysis, we focus on the recognition of units of measure in textual form (e.g., degrees Celsius) and symbolic form (e.g., °C). In addition, we recognize observable properties (e.g., temperature). We evaluated our proposed pipeline against a fine-tuned English language NER model[6]. The model was fine-tuned using the spaCy[7] framework on a synthetic data corpus specific to the IoT and WoT domains, generated using the llama3 70B language model based on a small hand-crafted dataset [25]. The corpus includes three SI base units with corresponding observed properties (time, length, and mass) and three derived SI units with corresponding observed properties (temperature, frequency, and acceleration), as well as percent as a pseudo-unit for dimensionless ratios such as the observed property relative humidity. Our corpus contains a total of 5,769 entries. A training script and the data corpus are available on GitHub[8].

For the comparison with the fine-tuned NER model and the pipeline, we use a dataset from the IoT/WoT domain. The dataset consists of a total of 69 RDF graphs describing APIs of IoT devices. Of these, 68 RDF graphs come from the W3C WoT TD implementation report[9] and are based on APIs of working systems created by contributing organizations such as Siemens AG, Intel, and Oracle. In addition, one RDF graph was manually created describing the interface of a Xiaomi Flower Care sensor used in the running example. The 69 RDF graphs contain a total of 620 natural language annotations in the form of `td:name` or `td:description` of which 44 contain references to units or observed properties.

The results of the evaluation in Table 1 show that the foundational LLM-based NER approach is able to correctly detect 40 unit and observed property references in the annotations and reaches a similar F1 score as the classical approach of fine-tuning a NER model. However, the results also show that the

---

[5] https://github.com/FreuMi/ner_pipeline.
[6] https://spacy.io/models/en#en_core_web_lg.
[7] https://github.com/explosion/spaCy.
[8] https://github.com/FreuMi/NER_Training.
[9] https://w3c.github.io/wot-thing-description/testing/report.html; available as a single RDF file at https://www.vcharpenay.link/talks/td-sem-interop.html.

Table 1. Comparison of classical fine-tuning and LLM-based NER approaches. The average runtime per annotation is reported in seconds.

	Classical Approach	LLM-based Approach
True Positives	33	40
False Positives	2	9
False Negatives	11	4
F1 Score	0.84	0.86
Avg. Runtime [s]	0.009	27.9

LLM-based NER approach produces more than four times as many false positives as the classical approach, but is able to detect more true positives. The difference in true positives is likely explained by a lack of training data for the classical approach. Adding even more labeled training data to the corpus could allow the model to generalize better and recognize more entities in complex scenarios, such as those involving different contexts or rare entity types.

Overall, the LLM-based NER approach achieves comparable results to a classical fine-tuned NER system, but requires additional processing of the detected entities to reduce the number of false positives. Additionally the runtime of the LLM-based approach is much higher than the classical approach due to higher computational requirements.

5.2 R2: Pipeline Evaluation

To evaluate the effectiveness of our proposed pipeline, which integrates NER, RAG-based entity linking, and LLM-based disambiguation with self-verification through chain-of-thought prompting, we conducted a comparison with two alternative approaches. The first comparison is with a pipeline implementation that uses similar techniques except for the chain-of-thought prompting style. The second comparison is with an implementation that uses prompting methods that rely solely on the LLM's internal data and information to identify and validate detected entities.

For the evaluation, we used a dataset consisting of 50 manually generated example natural language annotations focusing on units of measurement with ambiguities, inspired by sensor data sheets. The task is to identify the name of the unit of measurement entity based on the symbolic unit representation in the context of the sentence. The dataset is publicly available on GitHub[10].

An example from the dataset where disambiguation is needed is the natural language annotation *Time measured in S*. Here, the unit stands for seconds, even though the letter S is capitalized. The S does not stand for the unit Siemens, which also uses the letter S to describe electrical conductance, because that interpretation does not make sense in the context of the natural language annotation.

---

[10] https://github.com/FreuMi/ner_pipeline/tree/main/evaluation/dataset.

Another example where self-verification is required, is the natural language annotation *The length l is 10 m*. The NER system might detect l, which typically stands for liters, and m, which typically stands for meters, as symbolic units. However, in the context of this annotation, only the unit m is relevant, since the letter l is simply the name for a certain length and there is no reference to the unit liters.

Table 2. Comparison of the pipeline with chain-of-thought (CoT), the pipeline without CoT, and an approach using an LLM-only method. The average runtime for each approach per annotation is reported in seconds.

	Pipeline w/ CoT	Pipeline w/o CoT	LLM only
True Positives	39	40	41
False Positives	3	12	25
False Negatives	2	1	0
F1 Score	0.94	0.86	0.77
Avg. Runtime [s]	326.8	42.7	21.3

The results in Table 2 show that the LLM-only approach achieves an F1 score of 0.77 and successfully detects all possible entities, i.e. the 41 true positives. However, it fails to filter out entities that do not fit the context of the sentence, resulting in a high number of false positives.

The pipeline approach without chain-of-thought prompting achieves an F1 score of 0.86, with approximately half the number of false positives compared to the LLM-only approach, while maintaining an almost similar number of 40 true positives. These results suggest that the inclusion of external factual information in the form of RDF data helps the LLM to make more accurate decisions about whether the correct entity has been identified.

Finally, the pipeline approach with chain-of-thought prompting achieves the highest F1 score of 0.94, providing an additional improvement of approximately 0.08 over the pipeline approach without chain-of-thought prompting. This approach further reduces the number of false positives, but detects almost the same number of 39 true positives compared to the 40 and 41 true positives of the previous approaches. These results suggest that the combination of external knowledge and the chain-of-thought reasoning capabilities of the LLM improves its ability to determine whether an entity is correctly used in a sentence.

Overall, the chain-of-though based processing pipeline achieves the best results in our evaluation, but also has the highest average runtime, which is about 7 times slower than the pipeline without chain-of-thought prompting and about 15 times slower than the LLM-only approach.

6 Conclusion and Future Work

In this paper, we presented a processing pipeline for improving the machine-readability and interoperability of RDF data by extracting relevant information contained only in natural language annotations to generate semantic annotations. The pipeline employs LLM-based named entity recognition and semantic similarity-based entity linking. To disambiguate when multiple linkable candidates are found during the EL step, the pipeline uses the context, i.e., the natural language annotation, the candidate, and the RDF definition of the candidate retrieved from the ontology or KG using link traversal as input to the LLM to evaluate if the entity makes sense in the given context using chain-of-though prompting. If only one entity is found, the pipeline uses the same approach as for disambiguation by injecting the detected factual information as input to the LLM to verify the results. In our evaluation, we found that the foundational LLM-based NER approach performs on par with the traditional approach of fine-tuning an NER model for a specific domain. Additionally, we demonstrated the effectiveness of the pipeline using entity linking and the chain-of-though based disambiguation and self-verification steps, compared to a pipeline implementation without chain-of-though prompting and to an implementation relying only on the LLM without external information.

Future work will focus on further optimization of the pipeline, such as exploring the performance of other LLMs, especially smaller models with reduced computational requirements, to reduce the processing time and also enable potential deployment of the pipeline at the network edge in an IoT context.

Acknowledgement. This work was partially funded by the German Federal Ministry for Economic Affairs and Climate Action (BMWK) through the Antrieb 4.0 project (Grant No. 13IK015B) and the MANDAT project (Grant No. 16DTM107A).

References

1. Charpenay, V., Käbisch, S.: On modeling the physical world as a collection of things: the W3C thing description ontology. In: European Semantic Web Conference, pp. 599–615. Springer (2020)
2. Freund, M., Rott, J., Dorsch, R., et al.: FAIR Internet of Things data: enabling process optimization at Munich airport. In: European Semantic Web Conference. Springer (2024)
3. Kaebisch, S., McCool, M., Korkan, E., Kamiya, T., Charpenay, V., Kovatsch, M.: Web of Things (WoT) Thing Description 1.1 (2023). https://www.w3.org/TR/wot-thing-description/
4. Lagally, M., Matsukura, R., McCool, M., et al.: Web of Things (WoT) Architecture 1.1 (2023). https://www.w3.org/TR/wot-architecture/
5. Lewis, P., et al.: Retrieval-augmented generation for knowledge-intensive NLP tasks. Adv. Neural. Inf. Process. Syst. **33**, 9459–9474 (2020)
6. Mann, B., et al.: Language models are few-shot learners. arXiv preprint arXiv:2005.14165, vol. 1 (2020)

7. Martino, A., Iannelli, M., Truong, C.: Knowledge injection to counter large language model (LLM) hallucination. In: European Semantic Web Conference, pp. 182–185. Springer (2023)
8. Matsumoto, N., et al.: Kragen: a knowledge graph-enhanced rag framework for biomedical problem solving using large language models. Bioinformatics **40**(6) (2024)
9. Mesnard, T., Hardin, C., Dadashi, R., et al.: Gemma: open models based on Gemini research and technology. arXiv preprint arXiv:2403.08295 (2024)
10. Min, B., et al.: Recent advances in natural language processing via large pre-trained language models: a survey. ACM Comput. Surv. **56**(2), 1–40 (2023)
11. Monajatipoor, M., et al.: LLMs in biomedicine: a study on clinical named entity recognition. arXiv preprint arXiv:2404.07376 (2024)
12. Nadkarni, P.M., Ohno-Machado, L., Chapman, W.W.: Natural language processing: an introduction. J. Am. Med. Inf. Assoc. **18**(5), 544–551 (2011)
13. Nasar, Z., Jaffry, S.W., Malik, M.K.: Named entity recognition and relation extraction: state-of-the-art. ACM Comput. Surv. (CSUR) **54**(1), 1–39 (2021)
14. Nori, H., et al.: Can generalist foundation models outcompete special-purpose tuning? Case study in medicine. arXiv preprint arXiv:2311.16452 (2023)
15. Pan, S., Luo, L., Wang, Y., Chen, C., Wang, J., Wu, X.: Unifying large language models and knowledge graphs: a roadmap. IEEE Trans. Knowl. Data Eng. (2024)
16. Qin, C., Zhang, A., Zhang, Z., et al.: Is chatgpt a general-purpose natural language processing task solver? arXiv preprint arXiv:2302.06476 (2023)
17. Rantala, H., Ikkala, E., Rohiola, V., et al.: Findsampo: a linked data based portal and data service for analyzing and disseminating archaeological object finds. In: European Semantic Web Conference, pp. 478–494. Springer (2022)
18. Satheesh, K., Jahnavi, A., Iswarya, L., Ayesha, K., Bhanusekhar, G., Hanisha, K.: Resume ranking based on job description using SpaCy NER model. Int. Res. J. Eng. Technol. **7**(05), 74–77 (2020)
19. Scheffler, M., Aeschlimann, M., Albrecht, M., et al.: FAIR data enabling new horizons for materials research. Nature **604**(7907), 635–642 (2022)
20. Sevgili, Ö., Shelmanov, A., Arkhipov, M., et al.: Neural entity linking: a survey of models based on deep learning. Semant. Web **13**(3), 527–570 (2022)
21. Shen, W., Li, Y., Liu, Y., et al.: Entity linking meets deep learning: techniques and solutions. IEEE Trans. Knowl. Data Eng. **35**(3), 2556–2578 (2021)
22. Wang, S., Zhao, Z., Ouyang, X., Wang, Q., Shen, D.: Chatcad: interactive computer-aided diagnosis on medical image using large language models. arXiv preprint arXiv:2302.07257 (2023)
23. Wang, S., Sun, X., Li, X., et al.: GPT-NER: named entity recognition via large language models. arXiv preprint arXiv:2304.10428 (2023)
24. Wei, J., et al.: Chain-of-thought prompting elicits reasoning in large language models. Adv. Neural. Inf. Process. Syst. **35**, 24824–24837 (2022)
25. Whitehouse, C., Choudhury, M., Aji, A.F.: LLM-powered data augmentation for enhanced cross-lingual performance. arXiv preprint arXiv:2305.14288 (2023)
26. Yang, J., et al.: Harnessing the power of LLMs in practice: a survey on chatgpt and beyond. ACM Trans. Knowl. Discov. Data **18**(6), 1–32 (2024)

TEDME-KG Metrics Framework: A Metrics Framework for TEmporal Data Modelling Evaluation in Knowledge Graphs

Sepideh Hooshafza[1(✉)], Beyza Yaman[1], Alex Randles[1], Mark Little[1,2,2], and Gaye Stephens[1]

[1] ADAPT Research Centre, School of Computer Science and Statistics, Trinity College Dublin, Dublin, Ireland
Sepideh.hooshafza@gmail.com
[2] Trinity Health Kidney Centre, Trinity College Dublin, Dublin, Ireland

Abstract. Temporal data modelling in Knowledge Graphs (KGs) refers to the practice of modelling time-varying information in a structured manner. Representing temporal data, enhances KGs' capabilities by providing comprehensive time-dependent information, making it important for knowledge representation and analysis in various real-world applications. Organisations can enhance data management, user experience, and decision-making processes in their KG applications by selecting the most suitable temporal data modelling approach. Many approaches have been proposed to model temporal data in KGs but they have not been fully evaluated with different metrics. Evaluating different temporal data modelling approaches in KGs using a metrics framework assists objective comparison, informed decision-making, and the advancement of the field. This paper proposes TEDME-KG Metrics Framework, a Metrics Framework for TEmporal Data Modelling Evaluation in Knowledge Graphs. The framework was developed through a comprehensive literature review and the application of the Goal Question Metrics (GQM) method during the design and evaluation of KGs for selected temporal data modelling approaches.

Keywords: Knowledge Graph · Evaluation · Metrics Framework · Temporal data modelling

1 Introduction

In recent years, Knowledge Graphs (KGs) have drawn great attention from academia and industry as a method for modelling data [1, 2]. KGs are defined as entities (nodes) and relationships (edges) in a semantic network [3]. They are used in a variety of applications, including natural language processing, question answering machines, and recommendation systems [4]. KGs are widely used for large-scale data integration from disparate sources and are critical to many businesses [5]. Companies such as Facebook, Amazon, IBM, LinkedIn, Uber, and Airbnb utilise KGs in their data infrastructure [6]. The

Resource Description Framework (RDF) is one of the most common types of KG models.[7] RDF is a standard language for data representation and interchange on the Web. RDF-based KGs are widely used in practice and adhere to World Wide Web Consortium (W3C) standards [8, 9]. One of the challenges in designing KGs, is representing "time" because RDF itself does not provide a standardised way of representing temporal information. RDF lacks a built-in method of representing triple time as a concept because it is primarily a framework for subject-predicate-object triple representations of knowledge [10]. Temporal data modelling in KGs refers to the practice of modelling time-varying information in a structured manner. Temporal data models ensure time consistency in a data model, and assist researchers in analysing the historical data and ultimately improve accuracy and decision making [11, 12]. For example, a study by Zuo et al., showed that incorporating temporal information improved the accuracy of drug-target interaction prediction in a KG [13]. There is a gap in the existing literature regarding the evaluation of temporal data modelling approaches. There is no comprehensive list of metrics based on different aspects for evaluating temporal data modelling approaches in KGs. Furthermore, the absence of a systematic methodology for developing and evaluating these metrics makes it difficult to assess the comprehensiveness and robustness of evaluation results. Current temporal data modelling approaches yield different performance outcomes in terms of querying efficiency and complexity, necessitating further evaluation and comparison.

Therefore, the aim of this study is to design and evaluate a metrics framework for TEmporal Data Modelling Evaluation in KGs (TEDME-KG Metrics Framework). According to the ISO definition (ISO/TR 13054:2012)[1], a framework is a logical structure for classifying and organising complex information. In this context, a metrics framework consists of a set of metrics organised in a logical structure. Designing such a framework could help knowledge graph engineers, ontologies, and semantic web researchers consider different aspects in a structured way when evaluating temporal data modelling approaches in KGs.

2 Case Study - FAIRVASC Project

FAIRVASC is a research initiative by the European Vasculitis Society and the RITA European Reference Network, involving ten European partners focused on patient care for ANCA vasculitis [14, 15]. These partners include several national registries such as RKD[2] in Ireland, UKIVAS in the UK, and others from France, Czech Republic, Poland, Germany/Austria/Switzerland, and Sweden. The RKD registry, established in 2012, is dedicated to researching rare kidney diseases in Ireland and is a key contributor to this paper. The FAIRVASC ontology[3] standardises terminology across seven registries to manage ANCA vasculitis data effectively. Feedback from each registry helps refine these terms, which are incorporated into the ontology, covering aspects such as Patient, Diagnosis, and Clinical Outcomes.

---

[1] https://www.iso.org/obp/ui/#iso:std:iso:tr:13054:ed-1:v1:en.
[2] https://www.tcd.ie/medicine/thkc/our-research/rita-ireland-vasculitis-riv-registry-and-bio bank/.
[3] https://ontologies.adaptcentre.ie/fairvasc/.

Patients with ANCA vasculitis can experience relapse, which refers to the recurrence or worsening of disease symptoms after a period of improvements or remission. Understanding the specific medications used by patients during these critical periods is essential for designing more effective treatment plans and managing patient conditions more accurately. Without this temporal information, it is challenging to correlate medication use with patient outcomes, particularly in relation to relapse events [16].

This paper focused on a subset of the RKD registry, containing patient details and medication records for 600 patients, stored in a relational database. The data included patient class which is linked to patient-related information like patient ID, gender, year of birth, and patient overview class. Patient overview class which is linked to relapse date and treatment class, and treatment class which contains details regarding the medications utilised for a patient as a treatment plan. The subset of ontology based on the dataset without including medication start date and stop date was considered as a base model. (Fig. 1).

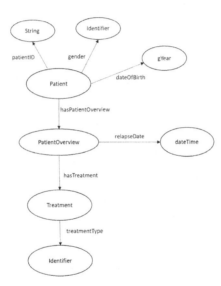

Fig. 1. Base model (ontology without including medication time)

The base model lacks the capability to present medication start and stop dates. This limitation is significant because each patient may use one or more medications during their disease progression, both before and after relapses. Modelling medication start and stop dates presents a significant challenge. Each medication has two associated dates: a start date and a stop date. The complexity increases when a patient uses the same medication at different intervals, resulting in multiple start and stop dates for a single medication. This situation complicates the modelling process, as it requires accurately capturing and representing these temporal relationships. To address this gap, Sect. 3.3 will extend the base model to include temporal classes that represent medication start and stop dates and the proposed metrics framework will be refined during the KG design and evaluation process.

3 TEDME-KG Metrics Framework

3.1 Methodology

This study was conducted in two phases. In the first phase, a literature review was performed to identify and extract metrics related to the evaluation of data modelling approaches in KGs. Following this, a thematic analysis was conducted, leading to the development of the first iteration of the metrics framework. In the second phase, an experiment was conducted. The metrics framework was refined during the design of two knowledge graphs based on two selected temporal data modelling approaches using the FAIRVASC project case study. The purpose of the second phase is outlined below:

Verify Theoretical Assumptions: To confirm that the theoretical underpinnings of the metrics framework align with actual data modelling practices and outcomes in RDF-based KGs.

Test Framework Applicability: To assess how well the metrics framework can be applied to various temporal data modelling approaches and its effectiveness in providing meaningful evaluations.

Refine and Improve: To identify any shortcomings or areas for improvement in the metrics framework based on real-world testing, leading to further refinements.

During the KG design process, the first iteration of the metrics framework was reviewed using the Goal Question Metrics (GQM) method [17]. The GQM method is a systematic approach for defining and interpreting metrics and involves three levels:

Goal: At the highest level, for each metrics category, a specific goal was established to provide a clear direction for the evaluation process.

Question: For each goal and its corresponding metrics subcategories, a set of questions was formulated to assess whether the goals are being met. These questions help in translating the abstract goals into concrete queries that can guide the evaluation process.

Metric: For each question, metrics were identified or updated based on the requirements that emerged during the KG design process. These metrics provide the necessary data to answer the questions effectively.

Following this systematic process, the second iteration of the metrics framework was designed.

3.2 Preliminary Metrics Selection

In this literature review, various metrics for evaluating data modelling approaches in KGs were identified. Initially, the search was focused on identifying metrics specifically for evaluating temporal data modelling approaches. However, a few papers were found that directly addressed the evaluation of temporal data modelling. Consequently, the scope of the search was expanded to include general metrics for evaluating data modelling in RDF-based KGs.

The literature review was conducted using the CIMO (Context, Intervention, Mechanism, and Outcome) question framework for developing the literature review question as follows, Context: RDF, Intervention: Assessment, Mechanisms: metrics, and Outcome: modelling. The question was, what are the evaluation metrics related to the data modelling approaches in RDF?

The Scopus[4] and IEEE Xplore[5] databases were searched for the literature review. For each concept based on the CIMO framework, relevant keywords were considered. The search query used for both databases was:

"(rdf OR "resource description framework") AND (evaluation OR assessment OR assess OR evaluate) AND (checklist OR criteria OR metric OR measure OR framework OR process OR step OR stage) AND (representation OR assertion OR model OR design) ".

Retrieved literature was imported into Covidence[6], systematic review software, for screening. Firstly, titles and abstracts were reviewed for inclusion. The final selection of papers was determined by doing a full-text review of the screened literature.

Selection of Studies: Search queries returned 1361 papers, of which 134 duplicate papers were removed. Screening of titles and abstracts led to the exclusion of 1182 papers and retention of 45 papers for full-text review. After reading the full texts, 10 papers were included in the literature review. Based on hand searching and forward citation, four papers were also identified for inclusion. 14 papers were included in the review. Information from each included paper was entered into a data extraction form that was incorporated into Covidence. A single researcher thoroughly reviewed the publications before manually entering the data into the data extraction table. Records of the reference and metrics were made. In order to do analysis, the data extraction form was then exported into a spreadsheet as a table.

Data Synthesis: An inductive thematic analysis of the retrieved literature was conducted to categorise metrics related to the evaluation of data modelling approaches in RDF [18]. The thematic analysis was conducted in 4 steps:

Step 1: A metrics subcategory was assigned to each extracted item based on their concept. The metrics subcategories were not unique per item. For example, where 2 items referred to the same concept, the metrics subcategory was repeated for both.
Step 2: Similar metrics subcategories were grouped together.
Step 3: A metrics category was defined and assigned for each set of metrics subcategories.
Step 4: The metrics categories were reviewed and agreed by a second researcher.

Literature Review Results: 20 papers published between 2009 and 2021 were identified for inclusion. A total of 49 metrics were extracted from the literature. Removal of duplicates and matching the metric names, resulted in 16 unique metrics. The 16 identified metrics were categorised in three metrics categories including modelling, uplifting, and querying and three subcategories included implementation support and

---
[4] https://www.scopus.com/home.uri.
[5] https://ieeexplore.ieee.org/xplore/home.jsp.
[6] https://www.covidence.org/.

standardisations, complexity, and performance. Based on the identified metrics, metrics subcategories, and metrics categories the first iteration of the metrics framework was designed (Table 1).

Table 1. TEDME-KG Metrics Framework – 1st Iteration

Metrics category	Metrics subcategory	Metrics
Modelling	1. Implementation support and standardisation	Implementation support tool[57, 58]
		Adherence to W3C standards[57, 58]
	2. Complexity	Additional triple generation[59]
		Number of nodes[59]
		Number of triples[59]
		Resource redundancy[59]
Uplifting	1. Implementation support and standardisation	Implementation support tool[57, 58]
		Adherence to W3C standards [57, 58]
	2. Complexity	Number of triple maps[60]
		Number of term maps[60]
	3. Performance	Storage space
Querying	1. Implementation support and standardisation	SPARQL query language support [58, 61]
		Triple store support tool [58]
	2. Performance	Query length requirement to execute a particular task [59, 62]
		Data load time [63–66]
		Query response time [57, 59, 61–65, 67–70]

3.3 Experiment

In this section, an experiment was conducted to evaluate two different temporal data modelling approaches and to update the TEDEME-KG metrics framework. The use case is based on the FAIRVASC project, which was discussed in Sect. 2.

The temporal data modelling approaches used in this experiment, were "Standard Reification" and "Named Graph" [29, 33]. The Standard Reification approach enables the expression of the temporal dimension of RDF triples using RDF through two main steps. First, a triple is represented by a Statement instance, with the subject, predicate, and object indicated separately in three distinct triples. Second, assertions are made about that instance as if it were a statement. Named Graphs facilitate the assignment of a URI to a collection of triples, allowing for statements to be made about the entire set.

Firstly, two ontologies were designed based on these two approaches, incorporating medication start time and stop date into the base model (Fig. 2).

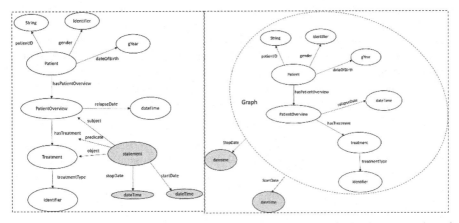

Fig. 2. Adding medication time to the base model based on Standard Reification approach (Left side) and Named Graph approach (Right side)

Once the ontologies were created, R2RML codes were developed considering two approaches (Fig. 4). The data were uplifted using the R2RML-F[7] uplift engine and RDF data were generated for each approach. As for the experiment's setup, the two RDF datasets were loaded on a GraphDB[8] triple store running on a single pc with a 4-cores CPU and 32 GB of main memory.

A temporal competency question was used to test the metrics in "Querying" category. The question was, "Which medications used by a patient with ID = 12, one month before and one month after his/her relapse?".

Two SPARQL queries were created based on each approach. The queries were executed, and the KG modelling approaches were evaluated based on the defined metrics. The SPARQL queries, and uplift codes (extended parts) are available on GitHub repository.[9] It should be noted that given the confidentiality of medical data, the dataset was not uploaded in the repository.

During the KG design process, a goal was defined at each step considering the metrics categories. Based on these goals, relevant questions were formulated and assigned to the appropriate metric subcategories. The metrics were then reviewed, and the following changes were applied.

Framework Refinement: Five changes were made to the framework's first iteration. Change 1: In the modelling category, "Implementation support tool" was changed to "Ontology editor support." This adjustment made the metric more explicit, as the ontology editor tool was the only implementation support tool identified at this design stage.

---

[7] https://github.com/chrdebru/r2rml.
[8] https://graphdb.ontotext.com/.
[9] https://github.com/sepidha/PhD/tree/KGSWC-2024.

Change 2: In the modelling category, "Adherence to W3C" was changed to "Ontology adherence to W3C." This adjustment made the metric more explicit, as ontology adherence to W3C was the only form of adherence identified at this design stage. Change 3: In the Uplifting category, "Implementation support tool" was changed to "Uplift engine support." This adjustment made the metric more explicit, as the uplift engine support was the only implementation support tool identified at this design stage. Change 4: In the Uplifting category, "Adherence to W3C" was changed to "Uplift adherence to W3C." This adjustment made the metric more explicit, as uplift adherence to W3C was the only form of adherence identified at this design stage. Change 5: In the Uplifting category, "storage space" was revised to "RDF data storage space." This adjustment clarified the metric and made it more explicit.

Based on the specified changes and updates, the second version of the TEDME-KG metrics framework was developed (Table 2). In this version, goals and questions were incorporated to enhance clarity and aid in the evaluation of temporal data modelling approaches. The modifications in the metrics are highlighted in bold.

Table 2. TEDME-KG Metrics framework -2nd iteration

Goal	Metrics Category	Metrics Subcategory	Question	Metric
Evaluating temporal data modelling approaches in terms of modelling aspect	Modelling	Implementation support and Standardisation	Are the approaches supported by an ontology editor?	**Ontology editor support**
			Do the approaches use W3C recommendations?	**Ontology adherence to W3C**
		Complexity	What is the number of nodes?	Number of nodes
			What is the number of triples?	Number of triples
			What is the number of additional generated triples?	Additional triple generation
			What is the number of redundant resources?	Resource redundancy

(*continued*)

Table 2. (*continued*)

Goal	Metrics Category	Metrics Subcategory	Question	Metric
Evaluating temporal data modelling approaches in terms of uplifting aspect	Uplifting	Implementation support and Standardisation	Is there any uplift engine available for uplifting data based on the modelling approach?	**Uplift engine support**
			Does the uplifting use W3C recommendation?	**Uplift adherence to W3C**
		Complexity	What is the number of triple maps?	Number of triple maps
			What is the number of term maps?	Number of term maps
		Performance	What is the storage space after uplifting data?	**RDF Data storage space**
Evaluating temporal data modelling approaches in terms of querying aspect	Querying	Implementation support and Standardisation	Are the approaches supported by a W3C query language?	W3C Query language support
			Are these approaches supported by a triple store?	Triple store support
		Performance	What is the data load time in the triple store?	Data load time
			What is the query length?	Query length
			What is the query response time?	Query response time

Experiment Results: This section presents the results of evaluating Standard Reification and Named Graph. Using the TEDME-KG metrics framework, the evaluation focused on three main aspects: modelling, uplifting, and querying. Below are the detailed findings for each category.

Modelling: Neither the standard reification approach nor the named graph approach supports ontology editor tools, indicating a potential gap in user-friendly tools for ontology design. However, both approaches adhere to W3C standards, ensuring compatibility with established semantic web technologies. In terms of complexity, the standard reification approach has more nodes and triples, indicating a more granular or fragmented representation of data and more detailed relationships, which adds complexity. It also generates a higher number of additional triples and shows higher resource redundancy, suggesting more duplicated resources and potential inefficiency.

Uplifting: Both approaches support uplift engines, facilitating data transformation, and adhere to W3C standards during data uplifting, ensuring consistency with web standards. The standard reification approach requires more triple maps and term maps, indicating a more complex mapping process and higher complexity in the data transformation rules.

Querying: Both approaches support W3C query languages, ensuring compatibility with standard querying practices, and are supported by triple stores, facilitating their practical use in data management systems. In terms of performance, the named graph approach has a significantly faster data load time and query response time, indicating better performance in data ingestion and query execution. The named graph approach has a slightly shorter query length, indicating marginally more concise query formulation.

Experiment Summary: The standard reification approach is more complex, with more nodes, triples, additional triples, triple maps, and term maps. The named graph approach outperforms the standard reification approach in terms of storage space, data load time, and query response time. Named graph shows better efficiency with less resource redundancy and lower storage requirements. Both approaches adhere to W3C standards and support necessary tools for uplift and querying. If detailed and rich data representation is needed, the standard reification approach might be suitable despite its complexity. However, if efficiency and performance are prioritised, the named graph approach is the better choice.

The results for each metric for the two approaches, Standard Reification and Named Graph, are provided in Table 3.

Table 3. Evaluation Results of Named Graph and Standard Reification based on TEDME-KG Metrics Framework

Metrics Category	Metrics Subcategory	Metric	Approach 1: Standard Reification	Approach 2: Named Graph
Modelling	Implementation support and Standardisation	Ontology editor support	No	No
		Ontology adherence to W3C	Yes	Yes
	Complexity	Number of nodes	11636	9345
		Number of triples	43348	28618
		Additional triple generation	34414	19684
		Resource redundancy	42134	25113

(*continued*)

Table 3. (continued)

Metrics Category	Metrics Subcategory	Metric	Approach 1: Standard Reification	Approach 2: Named Graph
Uplifting	Implementation support and Standardisation	Uplift engine support	Yes	Yes
		Uplift adherence to W3C	Yes	Yes
	Complexity	Number of triple maps	5	3
		Number of term maps	17	12
	Performance	RDF Data storage space	5125	3387
Querying	Implementation support and Standardisation	W3C Query language support	Yes	Yes
		Triple store support	Yes	Yes
	Performance	Data load time	2000	1000
		Query length	20	18
		Query response time	400	100

3.4 Detailed Description of the Metrics

This section provides an in-depth overview of the 16 metrics utilised in the TEDME-KG Metrics Framework (Table 4). Each metric is detailed with the following structure:

Definition: An explanation of the metric and its method of measurement, if applicable.

Range/Answer: The possible values or answers for the metric.

Unit: The unit used for measurement, if applicable.

Table 4. Metrics details used in TEDME-KG Metrics framework.

No	Metric	Definition	Range/Answer	Unit
1	Ontology editor support	This metric shows whether an approach is supported by a tool for ontology development	Yes/No	Not Applicable

(continued)

Table 4. (*continued*)

No	Metric	Definition	Range/Answer	Unit
2	Ontology adherence to W3C	This metric shows whether an approach uses the W3C recommendations	Yes/No	Not Applicable
3	Number of nodes	This metric shows the number of nodes used for modelling	0-n	Node
4	Number of triples	This metric shows the number of triples used for modelling	0-n	Triple
5	Additional triple generation	This metric shows the number of additional triples generated when adding time to the model. (Number of triples after adding time aspect – Number of triples before adding time aspect)	0-n	Triple
6	Resource redundancy	This metric shows whether the existing approach generates redundant resources or not. (Number of triples and nodes after adding time aspect) - (number of triples and nodes before adding time aspect)	0-n	Triple and Node
7	Uplift engine support	This metric shows whether an approach is supported by an uplifting engine	Yes/No	Not Applicable
8	Uplift adherence to W3C	This metrics shows whether the uplifting uses a W3C recommendation	Yes/No	Not Applicable
9	Number of triple maps	This metric shows the number of triple maps used in the uplift code. A triple map defines the logical table, which may be a table or view from a relational database	0-n	Triple maps

(*continued*)

Table 4. (*continued*)

No	Metric	Definition	Range/Answer	Unit
10	Number of term maps	This metric shows the number of term maps in the uplift code. The subject map, the predicate map and the object map are called term maps	0-n	Term maps
11	RDF Data storage space	This metric shows the size of the RDF dataset generated by the approach after uplifting data	0-n	KB
12	W3C Query language support	This metric shows whether an approach is supported by a W3C query language recommendation	Yes/No	Not Applicable
13	Triple store support	This metric shows whether an approach is supported by a triple store	Yes/No	Not Applicable
14	Data load time	This metric shows the time in which the RDF data is being uploaded in the triple store	0-n	millisecond
15	Query length	This metric shows the query length required to retrieve information to execute a particular task	0-n	Line
16	Query response time	This metric shows the time in which the query is being executed	0-n	millisecond

4 Related Work

Previous research in the area of KGs has primarily focused on data modelling without considering 'time' or temporal data [34]; however, there has been some research on modelling temporal data using KGs, and a number of RDF-based approaches to modelling temporal data in KGs exist, including RDF-Star, Standard Reification, Named Graph [29, 33, 35]. These approaches propose changing in RDF structure to represent temporal dimension [10]. In the context of temporal data modelling evaluation in KGs, a few research exist that has explored temporal data modelling approaches in knowledge graphs and evaluated the approaches mostly based on performance metrics [19, 21, 23]. In a research published by Frey, J et al. in 2017, different approaches to meta data modelling in semantic web were summarised [23]. The identified approaches were evaluated solely on their querying performance, with no consideration given to other aspects. In research conducted by Sen et al. in 2020, approaches for meta-knowledge assertion in the web of data based on 4 dimensions including time, fuzzy, provenance, and trust were explored [21]. The identified approaches were evaluated based on query performance metrics and a few modelling metrics proposed by the author. However, the methodology for selecting these metrics was not discussed.

The novelty of our proposed framework lies in its systematic development process, which is conducted in two steps: a comprehensive literature review and a practical experiment. This structured approach ensures a robust foundation and practical applicability. Additionally, the TEDME-KG metrics framework covers various aspects of temporal data modelling in a categorised manner, providing a thorough and organised evaluation method. This comprehensive and systematic approach differentiates our framework from existing methodologies, offering a more holistic and detailed evaluation of temporal data modelling in KGs.

5 Conclusion and Future Work

In this study, the TEDME-KG Metrics Framework was developed, encompassing three main metrics categories: Modelling, Uplifting, and Querying, and three metrics subcategories: Implementation Support and Standardisation, Complexity, and Performance. The framework includes a total of 16 metrics, with specific goals defined for each category and a corresponding list of questions for each subcategory to facilitate temporal data modelling evaluation. The TEDME-KG Metrics Framework offers a comprehensive and structured approach to evaluating temporal data modelling in Knowledge Graphs (KGs). By incorporating a systematic methodology based on the Goal Question Metrics (GQM) method and informed by an extensive literature review, the framework addresses the critical need for standardised evaluation metrics in this domain. The framework's design was iteratively refined through an experimental phase involving the application of two temporal data modelling approaches, ensuring its theoretical foundations aligned with practical outcomes. This process highlighted the framework's applicability in providing meaningful evaluations, while also identifying areas for further refinement.

The benefits of the TEDME-KG Metrics Framework are outlined below:

Objective Comparison: By providing a standardised set of metrics, the framework enables objective comparison between different temporal data modelling approaches, facilitating informed decision-making.

Comprehensive Evaluation: The framework covers various aspects of temporal data modelling, including modelling, uplifting, and querying, ensuring a thorough evaluation.

Support for Best Practices: By aligning with W3C standards, the framework promotes the adoption of best practices in the field of Knowledge Graphs.

Facilitating Research and Development: The TEDME-KG Metrics Framework provides a foundational tool that researchers and practitioners can use to advance the field of temporal data modelling, encouraging further innovation and development.

This framework not only bridges a significant gap in the current literature but also sets a foundation for future developments in the systematic assessment of temporal data in KGs. As future work, expert feedback from semantic web researchers will be incorporated to further validate and refine the framework's design, appropriateness, and usability. These insights will ensure that the TEDME-KG Metrics Framework remains robust and practical for real-world applications.

Acknowledgement. This work was conducted as part of the fulfilment of a PhD. The PhD scholar (S. Hooshafza) is funded by the ADAPT Centre for Digital Content Technology under the SFI Research Centres Programme (Grant 13/RC/2106_P2). The FAIRVASC project has received funding from the European Union's Horizon 2020 research and innovation programme under the EJP RD COFUND-EJP N° 825575.

References

1. Zou, X.: A survey on application of knowledge graph. J. Phys. Conf. Ser. **1487**, 012016 (2020)
2. Yan, J., et al.: A retrospective of knowledge graphs. Front. Comp. Sci. **12**(1), 55–74 (2018)
3. Wang, J., et al.: 3DRTE: 3D rotation embedding in temporal knowledge graph. IEEE Access **8**, 207515–207523 (2020)
4. Chen, Z., et al.: Knowledge graph completion: a review. IEEE Access **8**, 192435–192456 (2020)
5. Fensel, D., et al.: Introduction: what is a knowledge graph? In: Fensel, D., et al. (eds.) Knowledge Graphs: Methodology, Tools and Selected Use Cases, pp. 1–10. Springer, Cham (2020)
6. Hogan, A., et al.: Knowledge graphs. Synth. Lect. Data Semant. Knowl. **12**(2), 1–257 (2021)
7. Daniele, D.A., et al.: RSP-QL semantics: a unifying query model to explain heterogeneity of RDF stream processing systems. Int. J. Semant. Web Inf. Syst/ (IJSWIS) **10**(4), 17–44 (2014)
8. Zou, L., Özsu, M.T.: Graph-based RDF data management. Data Sci. Eng. **2**(1), 56–70 (2017)
9. Kim, Y., Kim, B., Lim, H.: The index organizations for RDF and RDF schema. In: 2006 8th International Conference Advanced Communication Technology (2006)
10. Wang, H.-T., Tansel, A.: Temporal extensions to RDF. J. Web Eng. **18**, 125–168 (2019)
11. Ni, R., et al.: Specific time embedding for temporal knowledge graph completion (2020)
12. Zhang, X., Qian, J., Wu, Q.: Research on the solution to redundancy of temporal data. In: 2010 International Conference on Computer Application and System Modeling (ICCASM 2010) (2010)
13. Li, Y., et al.: Drug-target interaction predication via multi-channel graph neural networks. Brief Bioinformation **23**(1) (2022)
14. Yaman, B., et al.: Towards a rare disease registry standard: semantic mapping of common data elements between FAIRVASC and the European joint programme for rare disease (2022)
15. McGlinn, K., et al., FAIRVASC: a semantic web approach to rare disease registry integration. Comput. Biol. Med. **145** (2022)
16. Salama, A.D.: Relapse in anti-neutrophil cytoplasm antibody (ANCA)–associated vasculitis. Kidney Int. Rep. **5**(1), 7–12 (2020)
17. Berander, P., Jönsson, P.: A goal question metric based approach for efficient measurement framework definition. In: Proceedings of the 2006 ACM/IEEE International Symposium on Empirical Software Engineering, pp. 316–325. Association for Computing Machinery: Rio de Janeiro (2006)
18. Braun, V., Clarke, V.: Thematic analysis, in APA handbook of research methods in psychology, vol 2: Research designs: Quantitative, qualitative, neuropsychological, and biological, pp. 57–71. American Psychological Association: Washington, DC, US (2012)

19. Orlandi, F., Graux, D., Sullivan, D.O.: Benchmarking RDF Metadata Representations: Reification, Singleton Property and RDF* (2021)
20. Ermolayev, V., et al.: Ontologies of time: review and trends. Int. J. Comput. Sci. Appl. **11**, 57–115 (2014)
21. Sen, S., et al.: State-of-the-art approaches for meta-knowledge assertion in the web of data. IETE Tech. Rev. **38**(6), 672–709 (2021)
22. Junior, A.C., Debattista, J., O'Sullivan, D.: Assessing the quality of R2RML mappings. In: SEM4TRA-AMAR@ SEMANTiCS (2019)
23. Frey, J., et al.: Evaluation of metadata representations in RDF stores. Semant. Web **10**(2), 205–229 (2019)
24. Sen, S., et al.: RDFM: an alternative approach for representing, storing, and maintaining meta-knowledge in web of data. Expert Syst. Appl. **179**, 115043 (2021)
25. Bellini, P., Nesi, P., Pantaleo, G.: Benchmarking RDF stores for smart city services (2015)
26. Chawla, T., et al.: Storage, partitioning, indexing and retrieval in big rdf frameworks: a survey. Comput. Sci. Rev. **38**, 100309 (2020)
27. Haase, P., Mathäß, T., Ziller, M.: An evaluation of approaches to federated query processing over linked data. In: Proceedings of the 6th International Conference on Semantic Systems, p. Article 5. Association for Computing Machinery, Graz (2010)
28. Saleem, M., et al.: How representative is a SPARQL benchmark? an analysis of RDF triplestore benchmarks (2019)
29. Hernández, D., Hogan, A., Krötzsch, M.: Reifying RDF: what works well with wikidata? (2015)
30. Kim, J., et al.: Query performance evaluation of OWL storage model (2011)
31. Kilintzis, V., Beredimas, N., Chouvarda, I.: Evaluation of the performance of open-source RDBMS and triplestores for storing medical data over a web service (2014)
32. Schmidt, M., et al.: SP^2Bench: a SPARQL performance benchmark (2009)
33. Carroll, J., et al.: Named graphs, provenance and trust. In: Proceedings of the 14th international conference on World Wide Web WWW 05, vol. 14, pp. 613–622 (2004)
34. Käfer, T., Wins, A., Acosta, M.: Modelling and analysing dynamic linked data using RDF and SPARQL. In: PROFILES@ISWC (2017)
35. Hartig, O.: Foundations of RDhÆ and SPARQnÆ (An Alternative Approach to Statement-Level Metadata in RDF). In: AMW (2017)

SFARDE: A Knowledge-Centric Semantic Strategic Framework for Heritage Artifact Recommendation Integrating Generative AI and Differential Enrichment of Ontologies

Archit Chadalawada[1] and Gerard Deepak[2(✉)]

[1] Manipal Institute of Technology Bengaluru, Manipal Academy of Higher Education, Manipal, India
[2] Department of Computer Science Engineering, BMS Institute of Technology and Management, Bengaluru, India
gerard.deepak.christuni@gmail.com

Abstract. Strategic models for heritage artifact recommendation that integrates generative AI and ontologies is the need of the hour. This paper proposes a knowledge-centric semantic model for heritage artifact recommendation which outgrows knowledge from the user perspective and the perspective of the dataset. The user query is enhanced with generated metadata from the Web 3.0 and classified using a strong Bi-LSTM model, and the artifacts are outgrown through generative AI using Gemini Pro as well as ontologies through standard knowledge repositories. This framework computes enhancing differential semantics through Second Order Co-occurrence Pointwise Mutual Information and CoSimRank used at several stages in the pipeline. The Dolphins Echolocation Algorithm is encompassed for optimization of the initial solution set to transform it into a more optimal solution set, using CoSimRank as an objective function. The overlap of generation of tags and the populated knowledge graphs in the dataset perspective to yield a matching instance reprioritization vector is the underlying strategy of the model, which uses both knowledge which is generated from the user perspective through preprocessed user queries, as well as the dataset perspective. An overall precision of 96.07%, recall of 97.07%, and F-measure of 96.57%, with a low False Discovery Rate of 0.04 makes the proposed SFARDE framework a best-in-class model for heritage artifact recommendation.

Keywords: Artifact Recommendation · Bi-LSTM Classifier · Culture and Heritage · Gemini Pro · Generative AI · Knowledge Graphs · Ontologies · Semantic AI · Web 3.0

1 Introduction

Heritage artifact recommendation refers to the suggestion of relevant documents from a dataset of culture and heritage related artifacts, based on some user query. It represents a very important and culturally significant domain for knowledge-centric recommendations of artifacts. Knowledge-centric recommendations use the power of knowledge graphs and ontologies to yield highly relevant facets based on user queries and hence play a crucial role in the knowledge-centric frameworks utilizing the structural metadata of the Web 3.0. The use of generative AI models in recommender systems enables the personalization of recommended facets based on user preferences, as well as categorization of unstructured data from a database of artifacts. The recent years have been marked by an exponential growth and expansion semantic AI and Web 3.0 technologies. Hence, in the era of the Web 3.0, knowledge-centric recommendation systems which encompass the power of generative AI and semantics-based computations for the differential enrichment of ontologies are in critical demand, particularly in the domain of culture and heritage, which is a topic of deep connection and relevance to most of the world's population.

Motivation: A strategic framework for heritage artifact recommendation in the era of the Web 3.0 that encompasses knowledge graphs and ontologies is needed due to its societal significance and cultural impact. In the era of the semantic web, semantically driven frameworks which are sandwiched with generative AI and strong optimization models are required, on account of the strength and similarity in information density of the contents of the Web 3.0.

Contribution: The proposed framework has the following major contributions. The generation of metadata is done in the perspective of the user's query words. User perspective-based metadata generation and its classification using a powerful deep learning Bi-LSTM framework is one of the contributions of this paper. Generating tags and captions through the use of generative AI with Gemini Pro from the dataset of artifacts and subsequently, ontology generation and its enrichment through Google Knowledge Graph API, YAGO and NELL to yield populated knowledge graphs is another novel contribution of the proposed framework. Differential semantics-based filtering of entities through Second Order Co-occurrence Pointwise Mutual Information (SOC-PMI) and CoSimRank at different junctures within the pipeline of the proposed framework, as well as the inclusion of the Dolphins Echolocation Algorithm [16] for achieving the most optimal solution set by transforming the initial solution set is also a contribution of this paper. Most importantly, this framework sandwiches semantic artificial intelligence through strong semantic similarity measures with differential thresholds and strategically combines a generative AI and a deep learning model at different stages of the architectural pipeline. Optimization algorithms with subsequent incremental outgrown additive knowledge through Onto-Collab and the Google Knowledge Graph API and enrichment using knowledge store repositories is a distinguished contribution of the proposed framework.

Organization: This paper contains 6 Sections. Section 2 explains the "Related Works", and the "Proposed System Architecture" is discussed in Sect. 3. Sections 4 and 5 explain the "Implementation" as well as the "Performance Evaluation and Results" of the proposed framework respectively. Lastly, Sect. 6 presents the "Conclusion" of the paper.

2 Related Work

In his paper, Newman [1] has described the essential components of knowledge flows in great detail, through a perspective of agents, artifacts, as well as transformations. The paper has an emphasis on domain neutrality and scalability, using ontologies built from a wide variety of artifacts, utilizing a pragmatic approach for knowledge-based systems design. Tarus et al. [2] have presented a review of recommender systems, with the domain of choice being e-learning, enriched through the use of ontologies. They have described the prominent current methods and representation languages used and conducted an analysis of future trends. It was noted that hybrid recommendation frameworks improved the overall effectiveness of the recommenders. Xinyu et al. [3] have proposed a knowledge-centric recommendation framework, encompassing semantic AI techniques for smarter solution designs in engineering problems. Most importantly, the proposed framework incorporated an aspect of context-awareness as well as enrichment of knowledge diversity within its pipeline, thus improving its overall effectiveness.

A document recommendation system for annotations of keywords with a focus on using online social tagging services, rather than user-based ratings was put forth by Ziyu et al. [4]. The model recommends user-relevant research papers and web artifacts through representation learning, graphs, and semantic AI. Weng et al. [5] have described a recommendation system for research documents through the use of ontologies and analysis of community networks to take into account user preferences, with an overall precision of 93%. Furthermore, Rei et al. [6] have proposed a strategic method for metadata generation based on multimodal data consisting of textual, tabular, and image-based data. The dataset used was specific to cultural heritage as a domain of choice and was classified using a variety of multitask classifiers such as convolutional neural networks, transformers and gradient tree boosting for different data types.

Belhi et al. [7] have developed an approach using machine learning for the multimodal classification and annotation of cultural heritage assets based on image data classified by means of a convolutional neural network, in tandem with metadata incorporated into a traditional neural network architecture. The approach relies on transfer learning as a means of knowledge transfer between the joint model. Michalakis et al. [8] have presented a survey on the current context-awareness methodologies and applications in the field of cultural heritage. The paper analyzed various existing models on the basis of their context incorporation mechanisms as well as systems integrations methods. Additionally, Gardner et al. [9] have proposed a semantics-oriented strategy for dealing with competency problems in documented dataset artifacts in training data for natural language processing models. The strategy incorporates statistical models to identify and remove dataset artifacts that lead to subtle biases and spurious correlations between input features and output labels.

The work by Sülün et al. [10] puts forth a novel approach which recommends reviewers for a variety of software artifacts in the project development pipeline. The model uses the traceability graphs of the various software artifacts to suggest the most suitable reviewer based on their knowledge of different artifacts. Habibi et al. [11] have presented a method for keyword extraction from the entities yielded by an automatic speech recognition algorithm. Following this, a clustering-based approach was utilized by the proposed framework for the recommendation of conversation documents. Nava et al. [12]

have developed a novel framework for document recommendation through the generation of ontologies for semantic similarity between the documents. It includes semantic search, knowledge retrieval and filtering of documents using semantic similarity measures for modeling document similarity within the ontology.

A system heritage and cultural item recommendation system using an edge intelligence approach run on mobile devices has been proposed by Su et al. [13]. The framework utilizes the power of edge computation for semantic search as well as machine learning computation to provide user-centric cultural item recommendations, along with a cloud-based analysis of heterogenous cultural heritage artifacts. Pauwels et al. [14] have developed a system for analyzing large volume datasets of virtual heritage artifacts containing multimodal data in the form of text records, image data, videos, point cloud and 3D models. The paper describes a context-based metadata generation scheme for ontology generation in the domain of heritage artifacts. Maree et al. [15] have put forth a strategic framework for the recommendation of Palestinian cultural heritage artifacts by means of the Holy-Land Ontology with additional knowledge outgrowth using the Art & Architecture Thesaurus with DBPedia knowledge store repositories. It recommends artifacts that are highly relevant to the user query through semantic similarity computations with dataset artifacts and keyword extraction as well as past user preferences. Additionally, user location is also integrated into the system to provide more context to the system.

3 Proposed System Architecture

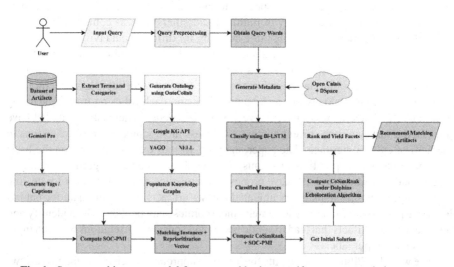

Fig. 1. System architecture model for proposed heritage artifact recommendation system.

Depicted in Fig. 1 is the proposed system architecture of the SFARDE framework, which is a knowledge-centric framework for heritage artifact recommendation encompassing the integration of generative AI and ontology enrichment. From Fig. 1, it is indicated that the framework is driven by user query. The query input by the user for searching heritage-based artifacts is subjected to preprocessing. The preprocessing involves tokenization, lemmatization, stop word removal and Named Entry Recognition (NER). The individual query words, Q_W are yielded after preprocessing of the input query. The individual query words are subjected to generation of metadata. Metadata is directly generated from the open web spaces and the Web 3.0 using DSpace and Open Calais as tools of choice. The metadata generated is extremely large to handle in terms of its scale, volume and heterogeneity. Henceforth, it needs to be classified before it is handled and is subjected to classification using a Bi-LSTM as a classifier of choice. The classified instances which are outputted from Bi-LSTM classifier are further used.

A Bi-LSTM or Bidirectional LSTM classifier is commonly used for strong deep learning classification in sequence-based data in natural language processing applications. It consists of 2 layers, each comprising of a unidirectional LSTM model. One layer considers past context from previous data, while the other considers future contextual data. This enables the model to provide a more holistic prediction, based on a richer context frame, incorporating both previous and subsequent data contexts. This presence of both forwards and backwards processing of sequential data enables Bi-LSTM frameworks to better model the relationships between words, as compared to the traditional unidirectional LSTM framework, in which contextual processing of sequential data only occurs in one direction.

Subsequently, the dataset of heritage artifacts is subjected to extraction of terms and categories. These terms and categories are directly obtained from the dataset, which is further subjected to generation of tags and captions via the Gemini Pro model of the Generative AI.

Generative AI refers to deep learning systems with the capability of data generation in text, image or other formats, in response to user prompts. They achieve this through the unsupervised learning of a foundation model, using vast amounts of unlabeled data, and possibly model fine-tuning using some labeled data as well. Models trained in this manner are referred to as Large Language Models. One such popular model is the Google Gemini Pro 1.0, with multimodal capabilities in image, video, audio and textual data. It is a highly performant model, boasting a quality index of 62, with a token generating capacity of 87 output tokens per second. It yields a 71.8% accuracy in the Massive Multitask Language Understanding (MMLU) benchmark. Henceforth, the Gemini Pro 1.0 being a lightweight model optimized for speed and efficiency, and thus working well in high-volume tasks, is a suitable choice for the proposed SFARDE framework.

Also, in order to generate the ontologies relevant to that of the domain of the dataset, the terms and categories obtained from the dataset of artifacts are used for the generation of ontologies using the OntoCollab tool. OntoCollab is used for detailed ontology generation but only up to 16 levels, inclusive of 1 level of individuals. The reason for restricting the ontology generation to 16 levels is to avoid deviance from the actual content of the dataset. Once the ontology is generated using OntoCollab, it is further passed into the Google Knowledge Graph API, YAGO and NELL knowledge basis and knowledge store repositories. The knowledge graphs are yielded by the Google Knowledge Graph API and the sub-graphs are generated through the structural framework of the Web 3.0. YAGO and NELL are queried through agents via SPARQL and their respective APIs to yield large instances of knowledge. Then Google Knowledge Graph API and the entities yielded from YAGO and NELL are further aggregated together by creating at least a single individual link between them by computing their terminal Shannon's entropies and establishing at least one link to generate a single, large populated knowledge graph.

The single, large populated knowledge graph and the generated tags and captions from the Gemini Pro API are subjected to computation of SOC-PMI with a threshold of 0.60. The reasons for empirically minimizing the threshold to 0.60 is due to the strength of SOC-PMI and also to yield a large number of matching instances. These matching instances are then assembled on the basis of their Shannon's entropy as a reprioritization vector. The reprioritization vector is used to compute SOC-PMI and CoSimRank with a threshold of 0.75 for SOC-PMI and a step deviance of 0.15 for CoSimRank with the classified instances outputted from the Bi-LSTM classifier. The reasons for choosing these threshold and step deviance measures are again on the basis of strength of the SOC-PMI measure and CoSimRank, and also on account of the voluminous amount of instances that born from the Bi-LSTM classifier. This successfully yields the initial solution.

$$f^{pmi}(t_i, w) = log_2 \frac{f^b(t_i, w) \times m}{f^t(t_i) f^t(w)} \qquad (1)$$

$$f(w_1, w_2, \beta) = \sum_{i=1}^{\beta} \left(f^{pmi}(X_i^{w_1}, w_2) \right)^{\gamma} \qquad (2)$$

$$SOCPMI(w_1, w_2) = f(w_1, w_2, \beta_1) + f(w_2, w_1, \beta_2) \qquad (3)$$

Equations (1), (2) and (3) depict the steps in the computation of SOC-PMI between two words, w_1 and w_2. It is a semantics similarity measure that quantifies the similarity and identifies semantic relationships between the tags and captions yielded by the Gemini Pro model, and the populated knowledge graph from the database of artifacts. Equation (1) computes the pointwise mutual information between w_1 and w_2, which quantifies how strongly associated they are based on their relative frequencies of occurrence. The set of neighboring words are denoted as X for w_1 and w_2. The aggregate pointwise mutual information between both of the words is then calculated in Eq. (2), considering β common neighbors and an association parameter of γ, which adjust the weights of each of the values in the summation. Lastly, the SOC-PMI between the words calculated as per Eq. (3), is the sum of aggregate positive pointwise mutual information values calculated using separate values of β. A larger value of SOC PMI between 2 words indicates larger semantic similarity and association between them.

$$S(a, b) = \frac{1}{C(a,b)} \sum_{x \in N(a)} \sum_{y \in N(b)} S(x, y) \qquad (4)$$

The calculation of the CoSimRank between 2 entities, a and b, is shown in Eq. (4). CoSimRank is a semantic similarity measure used to compute the similarity between two nodes in a graph based on recursively evaluating similarities between their neighbors, where N(a) and N(b) describe the sets of neighbor nodes of entities a and b respectively, with C(a,b) being a normalization constant. Hence, for each set of entities a and b, their similarity is evaluated based on the extent of similarity of their neighbor nodes respectively. A larger value of CoSimRank indicates high degree of similarity between

2 entities and vice versa. It is encompassed for determining similarities of the classified instances born out of the Bi-LSTM pipeline with the matching instances yielded from the dataset perspective. It is also used as a criteria function for the Dolphins Echolocation metaheuristic optimization algorithm to transform the initial solution set into an optimal solution set that yields the best facets for heritage artifact recommendation.

On account of the large and voluminous nature of the initial solution set, it requires further computation in order to yield the best-in-class matching facets for which the CoSimRank is further computed under a multi-agent setup using the Dolphins Echolocation Algorithm. The Dolphins Echolocation Algorithm is a metaheuristic optimization algorithm. The Dolphins Echolocation Algorithm as metaheuristic optimization model is a bio-inspired algorithm, mimicking the mechanism used by dolphins to locate their prey through echolocation. This algorithm outperforms many existing methods, with much lower computational cost, on account of a lower number of optimization parameters. The hunting process of dolphins was modeled as a constrained optimization problem and mapped onto a metaheuristic optimization algorithm. The predefined rules based on which search agents explore the feasible solution set mimics the process of using echolocation to locate prey. Searching of the solution space is done in a similar manner to which dolphins restrict their search space to a more focused area upon reaching near its prey, by means of increasing their click rate. Hence, the Dolphins Echolocation Algorithm as a metaheuristic optimization model achieves a faster convergence rate as compared to other algorithms and enables a computationally efficient transformation of the initial solution set into a much more optimal solution set within the proposed framework.

The CoSimRank is computed under the Dolphins Echolocation Algorithm by running it until it reaches consensus wherein the initial solutions are transformed into the feasible and optimal solution set. The CoSimRank at this juncture is set to a step deviance of 0.12 due to the strength of the CoSimRank measure as well as the Dolphins Echolocation Algorithm. This yields the final optimal solution set, which is subjected to ranking and followed by the recommendation of matching facets in the increasing order of its CoSimRank, and these matching facets are initially recommended. Based on a user click on the facets, the artifacts from the dataset are yielded, by matching these facets with the terms and categories of the dataset. The search halts based on user satisfaction. If the user is unsatisfied, this continues recursively until there are no more recorded clicks by the user.

4 Implementation

The proposed framework was implemented in Python within the Google Colaboratory IDE. The metadata was generated from the preprocessed query words using Open Calais and DSpace as tools of choice, which are popular semantic web tools for the linking of rich semantic metadata in the form of entities, facts and events onto documents. The preprocessing of query words was achieved using tokenization, lemmatization, stop word removal and NER via the Python NLTK library and ontology generation was conducted using OntoCollab. The Google Knowledge Graph API, YAGO and NELL knowledge bases and knowledge store repositories were queried through a multi-agent setup by means of SPARQL query. The Gemini Pro API was used for the generation of tags and

Table 1. Algorithm

Algorithm - Knowledge-Centric Semantic Heritage Artifact Recommendation Integrating Generative AI and Differential Enrichment of Ontologies
Input: User query, dataset of heritage artifacts, YAGO and NELL knowledge base and knowledge store repositories
Output: Highly relevant heritage artifacts matching the user query, ontology and knowledge graph of heritage domain from the dataset perspective

Begin
Step 1: Load the set of heritage artifacts A_h from the dataset
Step 2: Load user query by assigning Query ← getUserQuery()
Step 3: //Obtain the query words by preprocessing user queries:
    tokenizedQ_w ← Query.tokenize()
    lemmatizedQ_w ← tokenizedQ_w.lemmatize()
    stopwordremovedQ_w ← lemmatizedQ_w.removeStopWords()
    Q_w ← stopwordremovedQ_w.namedEntryRecognition()
Step 4: //Generate metadata using Open Calais and DSpace:
    GeneratedMetadata ← Q_w.generateMetadata(OpenCalais,DSpace)
Step 5: //Classify generated metadata using a Bi-LSTM model:
    ClassifiedInstances ← BiLSTM.classify(GeneratedMetadata)
Step 6: //Generate the tags and captions using Gemini Pro:
    TagsAndCaptions ← GeminiPro.generateTagsAndCaptions()
Step 7: Extract Terms ← A_h.extractTerms()
Step 8: Extract Categories ← A_h.extractCategories()
Step 9: //Generate ontology using OntoCollab on the extracted terms and categories:
    HeritageOntology ← OntoCollab.generateOntology(Terms, Categories)
Step 10: //Outgrow auxiliary additive knowledge into the ontology to generate populated knowledge graphs:
    PopulatedKnowledgeGraph ← GoogleKG.populate(HeritageOntology,YAGO,NELL)
Step 11: Initialize MatchingInstances ← NULL
Step 12: //Determine the matching instances:
    for each entity K_i in PopulatedKnowledgeGraph:
      for each entity T_j in TagsAndCaptions:
        SOC_PMI ← computeSOCPMI(K_i,T_j)
        if SOC_PMI >= 0.60:
          MatchingInstances.append(K_i)
        end if
      end for
    end for
Step 13: //Compute the reprioritization vector:
    ReprioritizationVector ← assembleInstances(MatchingInstances, ShannonsEntropy)
Step 14: Initialize the InitialSolutionSet ← NULL
Step 15: for each instance I_s in the InitialSolutionSet:
    SOC_PMI ← computeSOCPMI(I_s,ReprioritizationVector)
    if SOC_PMI >= 0.75:
      COSIMRANK ← computeCoSimRank(I_s,ReprioritiztionVector)
      if COSIMRANK < 0.15:
        InitialSolutionSet.append(I_s)
      end if
    end if
    end for
Step 16: //Execute the **Dolphins Echolocation Algorithm** under CoSimRank with a step deviance of 0.12 to yield the optimized solution set:
    OptimizedSolutionSet ← InitialSolutionSet.applyDolphinsEcholationAlgorithm()
Step 17: //Rank the optimized solution set in increasing order of CoSimRank:
    OptimizedSolutionSet.orderByCoSimRank()
Step 18: Set RecommendedFacets ← NULL
Step 19: Set Satisfied ← False
    while not Satisfied:
      RecommendedFacets ← recommendFacets()
      RecommendedFacets.display()
      SelectedFacet ← userClick()
      RecommendedArtifacts ← SelectedFacet.yieldArtifact()
      Satisfied ← userFeedback()
    end while
End

captions with generative AI. The SOC-PMI and CoSimRank were computed using agents which were implemented with AgentSpeak. Furthermore, the Dolphins Echolocation Algorithm was also sandwiched through an agent setup, which was designed using JADE.

The dataset used was a hybrid, customized dataset, which was curated using the "Artifact Database, Huntington County N.D." dataset by the Veterans Curation Program (2017), the "Digital Artefacts of Rural Tourism: The Case Study of Poland" dataset by Karol Krol (2022), and the "Antarctic Artefacts Database" dataset by Lazer et al. (2006). The single, large dataset for heritage artifact recommendation was formalized by annotating each of these datasets through custom annotators and reprioritization of the records based on the annotations. Furthermore, to improve the artifact density, Web crawling was also enabled to incorporate existing Web documents and website artifacts, which were also annotated along the same lines, to generate the enhanced final dataset onto which the experimentations were conducted by the proposed model and all other baseline models.

Table 1 describes the algorithm used by the proposed SFARDE framework, whereby user queries are loaded and preprocessed through tokenization, lemmatization, stop word removal, and NER. Following this, metadata is generated using Open Calais and DSpace and is subsequently classified by encompassing a Bi-LSTM classifier. Independently, the dataset of artifacts is loaded and subjected separately to generation of tags and caption using Gemini Pro as well as generation of the ontology via OntoCollab with incremental additive knowledge outgrowth through Google Knowledge Graph API with YAGO and NELL as knowledge base and knowledge store repositories. The matching instances determined through computation of SOC-PMI with a threshold of 0.60 are then used to determine the reprioritization vector from the dataset perspective on the basis of Shannon's entropy.

Further, the classified instances along with the reprioritization are subjected to computation SOC-PMI with a threshold of 0.75 and CoSimRank with a step deviance of 0.15 to furnish the initial solution set. This initial solution set is then optimized using the agent-based Dolphins Echolocation Algorithm under CoSimRank, with a step deviance of 0.12. Lastly, CoSimRank is used for the ranking of matching facets and are recommended to the user, which recursively recommends heritage artifacts upon user clicks, until the user is satisfied.

5 Performance Evaluation and Results

The proposed SFARDE model, which is a strategic framework for knowledge-centric heritage artifact recommendation with the integration of generative AI and differential enrichment of ontologies, utilized precision, recall, accuracy and F-measure percentages as the primary metrics for performance evaluation, as well as the False Discovery Rate (FDR) as an auxiliary metric. These performance metrics quantify the relevance of the results and FDR is a measure of false positives rate in the model, thereby indicating the error rate in the framework. This clearly completes both the positive evaluation of results for the optimal framework, as well as the evaluation of results with a negative bias approach. Table 2 indicates that the proposed SFARDE framework has yielded the highest average precision, recall, accuracy and F-measure percentages of 96.07%, 97.07%, 96.57% and 96.57% respectively, with an FDR of 0.04, which is lower than all of the baseline models.

Table 2. Performance comparison between the proposed SFARDE and other baseline models

Model	Average Precision %	Average Recall %	Average Accuracy %	Average F-Measure %	FDR
RSRT [17]	92.18	93.74	92.96	92.95	0.08
Hipikat [18]	92.45	93.84	93.15	93.14	0.08
NRKA [19]	93.07	94.07	93.57	93.57	0.07
Proposed SFARDE	**96.07**	**97.07**	**96.57**	**96.57**	**0.04**

The Eqs. (5), (6), (7), (8) and (9) were used for computing the average precision, recall, accuracy, F-measure and FDR respectively.

$$Precision = \frac{Retrieved \cap Relevant}{Retrieved} \quad (5)$$

$$Recall = \frac{Retrieved \cap Relevant}{Relevant} \quad (6)$$

$$Accuracy = \frac{Proportion\ correct\ of\ each\ query\ passed\ ground\ truth\ test}{Total\ No.\ of\ Queries} \quad (7)$$

$$F - measure = \frac{2(Precision * Recall)}{(Precision + Recall)} \quad (8)$$

$$False\ Discovery\ Rate = 1 - Positive\ Predicted\ Value \quad (9)$$

The performance of the proposed SFARDE model is compared by baselining it with against 3 distinct models named the RSRT, Hipikat and NRKA frameworks. The RSRT yields an average precision of 92.18%, recall of 93.74%, accuracy of 92.96% and F-measure of 92.95%, along with the highest FDR of 0.08. The Hipikat framework has produced an overall precision, recall, accuracy, and F-measure of 92.45%, 93.84%, 93.15% and 93.14%, as well as an FDR of 0.08. Additionally, the NRKA framework has achieved a 93.07% overall precision, 94.07% recall, 93.57% accuracy, and 93.57% F-measure, with an FDR of 0.07. From Table 2, it is clearly significant that the proposed SFARDE has outperformed all other baseline models and has yielded the highest average precision, accuracy, recall and F-measure percentages, with the lowest value of FDR. Figure 2 depicts the graph of precision vs number of recommendations for all of the models.

The main reason that the proposed SFARDE framework outperforms all baseline models is that it is a knowledge-centric model for heritage artifact recommendation, which uses the metadata which is generated from the structural space of the Web 3.0 to enrich user query words and the metadata is further classified and handled using a Bi-LSTM model to make it more atomic in nature. Apart from this, the dataset of artifacts is subjected to generation of auxiliary knowledge using OntoCollab as an ontology, which is further enriched by knowledge bases and knowledge store repositories like Google Knowledge Graph API, YAGO and NELL to yield the populated knowledge graph. Additionally, the Gemini Pro generative AI API generates tags and categories from the dataset perspective. Hence, the dataset perspective is incrementally modelled with auxiliary knowledge which is highly differential due to the reasons of making it grow from an ontology perspective, and an enrichment using Google Knowledge Graph, YAGO and NELL knowledge store repositories.

Furthermore, another perspective of generative AI is used to improve its diversity, which ensures that diversification of artifacts from a dataset perspective is achieved, which also differentiates the proposed framework in terms of its knowledge possessing capability. Most importantly, matching of instances to yield the reprioritization vector happens on computation of SOC-PMI from the populated knowledge graph and the tags and captions in which co-incidence is considered. Crucially, CoSimRank and SOC-PMI with differential step deviance measures and thresholds respectively is further computed between the reprioritization vector instances and the classified instances from the query perspective, which is the classified metadata that comes out of the Bi-LSTM pipeline. Subsequently, the Dolphins Echolocation Algorithm under CoSimRank as a criteria function is run iteratively on the initial solution set, helping to furnish the most feasible and optimal solution set, which is much denser. This indicates that the artifacts recommended by a combination of semantic similarity measures, namely SOC-PMI and CoSimRank, with empirically decided thresholds and step deviance measures which further yield differential entities due to the differential nature of the entities selected through the pipeline, as well as optimization using the Dolphins Echolocation Algorithm, are highly relevant to the user query.

Furthermore, the presence of a strong deep learning Bi-LSTM framework to classify the metadata which is generated through the query perspective, along with Gemini Pro to generate the tags and captions through generative AI and OntoCollab for generation of ontologies and further enrichment of these ontologies through a knowledge graph using Google Knowledge Graph API, with additional population by means of YAGO and NELL, make this model quite efficient in terms of its auxiliary knowledge inclusion and incremental knowledge addition. Lastly, differential early selected knowledge instances which undergo further differential refinement through empirically decided differential semantic similarity and step deviance measures, and optimization through the Dolphins Echolocation Algorithm, makes this framework the most appropriate, when compared to the other baseline models.

Although the RSRT model recommends artifacts specific to the software domain from repository transactions and is domain-specific, its performance is lower than expected because it uses Latent Semantic Indexing, which definitely yields subsequent auxiliary knowledge and contributes some level of semantic inclusiveness, but the knowledge is quite shallow in nature and semantic similarity computation measures are absent. Furthermore, traditional machine learning models are used to recommend the artifacts and also association rule binding helps in further refining the artifacts. However, although

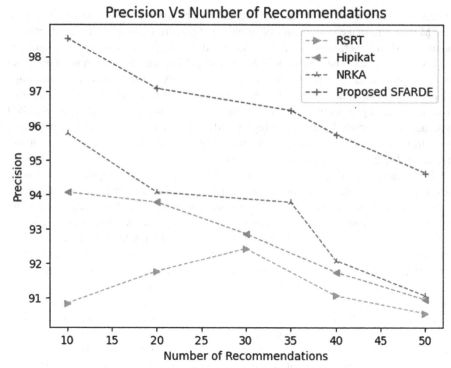

Fig. 2. Graph of precision vs number of recommendations of different models

these models are good, the amount of auxiliary knowledge generated through Latent Semantic Indexing is extensively shallow and machine learning models with association rule binding do not perform as expected. It can be further enhanced through the use of generative AI, deep learning frameworks. Even the incremental addition of auxiliary knowledge could have made it denser, which is missing in the RSRT model. Hence, due to the presence of shallow knowledge and an absence of semantic similarity measures, the RSRT model does not perform as well as the proposed SFARDE model.

The Hipikat framework recommends software artifacts which are domain-centric to software development. It is a tool that aggregates in-house software projects by means of syntactic analysis and document analysis. However, this model lacks any supporting semantic similarity computation mechanisms and incremental auxiliary knowledge addition. The use of generative AI with a deep learning pipeline as well as integration of the two different perspectives of the user query as well as the dataset is absent. Henceforth, the performance of the Hipikat model is lower than the proposed SFARDE model.

The reason for the NRKA model lagging is due to the use of a connectionist-based approach with a recurrent neural network along with Latent Semantic Indexing, where a content-based and collaborative filtering-based model is used. Again, the Latent Semantic Indexing generates highly shallow auxiliary knowledge. The recurrent neural network model works extensively well, however due to the shallow nature of auxiliary knowledge it does not perform as expected. Rather, it results in underfitting because of the deep learning model applied on the dataset, which generates a shallow amount of knowledge. The content-based model works but collaborative filtering requires ratings of the individual artifacts, however most artifacts available in the existing web cannot be rated. Even if it is rated based on user ratings, the associations can be made based on favoritism, rather than the actual presence of the contextual information. The query perspective is not formalized which would not work out in large-scale artifacts. Crucially, strong knowledge addition models, outgrowing incremental knowledge from the dataset and query perspectives, semantic similarity models for differential knowledge domain semantics are absent, and henceforth the NRKA also lags as compared to the proposed framework.

6 Conclusion

This paper has proposed a novel framework for heritage artifact recommendation in the era of Web 3.0 which strategically encompasses a user perspective-based generation of metadata and its classification using a strong deep learning Bi-LSTM model. This enhances the outcome of the recommendations by increasing the diversity of Web search as metadata obtained from the structure of the Web 3.0 reduces the cognitive gap between the knowledge in the existing infrastructure of the Web 3.0, and the knowledge that is made inclusive in the proposed framework's architectural pipeline. Secondly, the presence of Gemini Pro to generate the tags and captions in the dataset perspective and also incrementally generating ontologies which are highly relevant to the domain of the dataset, which is heritage and culture, along with further enhancement by using Google Knowledge Graph API, YAGO and NELL as knowledge bases and knowledge store repositories increases the auxiliary knowledge density from the dataset perspective. Most importantly, the information density and knowledge density are made extensively high and hence the cognitive gap is bridged between the entities in the framework's architectural pipeline as well as the knowledge in the World Wide Web. Information density refers to how tightly packed the information present within the data is, whereas knowledge density describes the depth of understanding and contextual richness of data within the pipeline. The model also includes strategic semantic similarity and relevance computation measures based on step deviance, namely using SOC-PMI and CoSimRank. Furthermore, the Dolphins Echolocation Algorithm for optimization makes the proposed framework much distinct and this is the reason why it yields the best-in-class matching recommendation facets for which the artifacts are mapped with the dataset and yielded. The clear demarcation between user query driven enrichment of knowledge through metadata and dataset-based enrichment of incremental knowledge through generative AI, ontologies, Google Knowledge Graph API and knowledge store repositories clearly balances the amount of dataset-generated the YAGO and NELL knowledge and user-generated knowledge which creates a perfect amalgam of the knowledge in two different perspectives, whereby the most relevant facets are yielded to the user. The proposed SFARDE framework has achieved an overall precision of 96.07%, recall of 97.07%, accuracy of 96.57%, F-measure of 96.57%, and a low False Discovery Rate of 0.04, hence making it a best-in-class heritage artifact recommendation framework.

References

1. Newman, B.: Agents, artifacts, and transformations: the foundations of knowledge flows. In: Handbook on Knowledge Management 1: Knowledge Matters, pp. 301–316 (2004)
2. Tarus, J.K., Niu, Z., Mustafa, G.: Knowledge-based recommendation: a review of ontology-based recommender systems for e-learning. Artif. Intell. Rev. **50**, 21–48 (2018)
3. Li, X., Chen, C.H., Zheng, P., Jiang, Z., Wang, L.: A context-aware diversity-oriented knowledge recommendation approach for smart engineering solution design. Knowl.-Based Syst. **215**, 106739 (2021)
4. Guan, Z., et al.: Document recommendation in social tagging services. In: Proceedings of the 19th International Conference on World Wide Web, pp. 391–400 (2010)
5. Weng, S.S., Chang, H.L.: Using ontology network analysis for research document recommendation. Expert Syst. Appl. **34**(3), 1857–1869 (2008)
6. Rei, L., et al.: Multimodal metadata assignment for cultural heritage artifacts. Multimedia Syst. **29**(2), 847–869 (2023)
7. Belhi, A., Bouras, A., Foufou, S.: Leveraging known data for missing label prediction in cultural heritage context. Appl. Sci. **8**(10), 1768 (2018)
8. Michalakis, K., Caridakis, G.: Context awareness in cultural heritage applications: a survey. ACM J. Comput. Cult. Herit. (JOCCH) **15**(2), 1–31 (2022)
9. Gardner, M., et al.: Competency problems: on finding and removing artifacts in language data. arXiv preprint arXiv:2104.08646 (2021)
10. Sülün, E., Tüzün, E., Doğrusöz, U.: Rstrace+: reviewer suggestion using software artifact traceability graphs. Inf. Softw. Technol. **130**, 106455 (2021)
11. Habibi, M., Popescu-Belis, A.: Keyword extraction and clustering for document recommendation in conversations. IEEE/ACM Trans. Audio Speech Lang. Process. **23**(4), 746–759 (2015)
12. Nava, R.V., Dominguez, V.H.M., Montalvo, J.G.: A document recommendation system using a document-similarity ontology. IEEE Lat. Am. Trans. **14**(7), 3329–3334 (2016)
13. Su, X., Sperlì, G., Moscato, V., Picariello, A., Esposito, C., Choi, C.: An edge intelligence empowered recommender system enabling cultural heritage applications. IEEE Trans. Ind. Inf. **15**(7), 4266–4275 (2019)
14. Pauwels, P., Di Mascio, D.: Interpreting metadata and ontologies of virtual heritage artefacts. Int. J. Herit. Dig. era **3**(3), 531–555 (2014)
15. Maree, M., Rattrout, A., Altawil, M., Belkhatir, M.: Multi-modality search and recommendation on Palestinian cultural heritage based on the holy-land ontology and extrinsic semantic resources. J. Comput. Cult. Herit. (JOCCH) **14**(3), 1–23 (2021)
16. Kaveh, A., Farhoudi, N.: A new optimization method: Dolphin echolocation. Adv. Eng. Softw. **59**, 53–70 (2013)
17. David, J.: Recommending software artifacts from repository transactions. In: New Frontiers in Applied Artificial Intelligence: 21st International Conference on Industrial, Engineering and Other Applications of Applied Intelligent Systems, IEA/AIE 2008 Wrocław, Poland, June 18–20, 2008 Proceedings 21, pp. 189–198. Springer, Heidelberg (2008)
18. Cubranic, D., Murphy, G. C.: Hipikat: recommending pertinent software development artifacts. In: 25th International Conference on Software Engineering, 2003. Proceedings, pp. 408–418. IEEE (2003)
19. David, J.: Navigation recommendation on knowledge artifacts (2008)

Leveraging Graph Models for Comprehensive Visual Analytics of Equine Heritage

Abdelkader Ouared[1]([✉]), Noureddine Belarbi[1], Abdelhafid Chadli[1], and Kebbal Seddik[2]

[1] Faculty of Mathematics and Computer Science, University of Tiaret, Tiaret, Algeria
{abdelkader.ouared,abdelhafith.chadli}@univ-tiaret.dz,
vetindex14@gmail.com
[2] Blida 1 University, Laboratory of LBRA (Biotechnology Research Laboratory Related to Animal Reproduction), Blida, Algeria
kebbal_seddik@univ-blida.dz

Abstract. Horse breeding and heritage involve intricate relationships spanning ancestors, descendants, bloodlines, and interdependent connections among horses. These interactions are naturally depicted through reproduction, lineage, crossbreeding to improve traits, genetic enhancement, and various events, forming a complex world of interactions. Effective preservation and management of equine heritage ($E\mathcal{H}$) necessitate advanced data modeling techniques to capture more complex spatiotemporal dependencies between horses. In this line of research, we propose a system called Kyle-$\mathcal{KG}$ (Knowledge Graph for Equine Heritage), a solution based on graph data modeling principles. To achieve this, we: (i) use a relational model for data sources from the Information Management System for $E\mathcal{H}$, (ii) employ a tracker component responsible for extracting and persisting data by mapping relational data to a graph database, and (iii) integrate a visual analytics component designed for decision-makers (e.g., farm managers, breeders). A case study with real-world equine heritage data demonstrates the effectiveness of our approach by providing comprehensive visual analytics through the value of relationships.

Keywords: Equine Heritage · Graph Data Model · Data Modeling · Graph database · Visual Analytics

1 Introduction

Horse breeding, deeply connected with society, stands out in cultural heritage preservation for its ancestral significance. Horses embody centuries-old traditions and our collective heritage [17]. The management of Equine Heritage ($E\mathcal{H}$) data presents unique challenges due to the complex relationships between horses, including ancestors, descendants, bloodlines, and events (e.g. [1,12,13]).

Indeed, the preservation and management of $E\mathcal{H}$ demand advanced data modeling approaches capable of capturing the intricate and diverse relationships inherent in this domain. The traditional use of relational databases for modeling $E\mathcal{H}$ is limited by three key factors when analyzing $E\mathcal{H}$ data. *First*, complex relationships: $E\mathcal{H}$ data often involves intricate relationships between horses, breeders, lineages, events, and other entities. Relational databases, with their focus on normalized tables, can struggle to represent these effectively, leading to a proliferation of tables and cumbersome joins to extract meaningful insights. *Second*, recursive queries: The *friend of a friend* (FOAF) problem aptly illustrates the difficulty. Tracing breeding lines or competition histories often requires recursive queries, which are not natively supported by most Relational Database Management Systems (RDBMS). *Third*, limited flexibility Adding new data types or relationships often necessitates schema modifications, potentially disrupting existing applications and requiring data migration.

Despite capturing the daily life of the farm among the horses, this information must be represented comprehensively to ensure it is both understandable and interpretable. To address this need, we propose a system called Kyle-$\mathcal{KG}$ (Knowledge Graph for Equine Heritage; "Kyle" means "Horse" in Arabic), which is a graph data modeling solution designed to enable decision-makers to effectively monitor and analyze $E\mathcal{H}$. To support these needs, Kyle-$\mathcal{KG}$ is built upon:

(i) An existing relational model for data sources used in the Information Management System (IMS) for $E\mathcal{H}$.
(ii) A tracker component responsible for extracting and persisting data by mapping relational data to a graph database (GraphDB).
(iii) A visual analytics platform tailored for decision-makers, such as farm managers and breeders.

Our system is driven by an intuitive graph data model specifically designed for the dynamic management of $E\mathcal{H}$ information. This model is meticulously crafted to meet the critical data needs of $E\mathcal{H}$ management, emphasizing the various relationships within community structures among horses. By constructing multiple graph types, such as interaction and similarity graphs, we enable comprehensive visual analytics of Equine Heritage.

Experimental validation using real-world $E\mathcal{H}$ data illustrates the effectiveness of our approach in enabling comprehensive visual analytics in this field.

We start by a preliminary and motivating scenario in Sect. 2 to highlight the usefulness of comprehensive visual analytics. In Sect. 3, we introduce our system. Section 4 details our tool support and application scenarios. We then present related work in Sect. 5. Finally, we conclude the paper in Sect. 6.

2 Preliminary and Motivating Example

Consider a scenario where a breeder, *Jane*, needs to trace the ancestral lineage of her prized mare, Najma, to make informed breeding decisions. In a traditional tabular database model, *Jane* starts by querying Najma's basic information from a horse table that includes fields like name, birth date, and registration number. However, to understand Najma's lineage, *Jane* must manually cross-reference this data with another table listing ancestors, where each row indicates parentage.

Table 1. Example of historical reproduction and lineage of IMS of $E\mathcal{H}$

Parent (Horse)	Reproduction Date & Time	Foal ID Number	Desc.	G-P (Horse)	Sire/Dam (Parent)	Great- G-P	Distant Relation
Najma	10/05/2020 14:00	ID-001	Foal X	Barq	Lu'lu'ah	Raihana	Cousin (Foal Y)
Najma	20/03/2021 10:30	ID-002	Foal Y	Raihana	Asif	Zahra	Cousin (Foal X)
Lu'lu'ah	05/09/2019 11:15	ID-003	Foal Z	Zahra	Faris	Noor	Nephew (Foal W)
Barq	12/11/2020 09:45	ID-004	Foal W	Najma	Lu'lu'ah	Raihana	Niece (Foal Z)
Raihana	25/06/2018 08:00	ID-005	Foal V	Asif	Zahra	Najma	Cousin (Foal U)
Asif	15/02/2017 16:20	ID-006	Foal U	Noor	Aisha	Najma	Cousin (Foal T)

Jane faces significant challenges in managing `Najma`'s $E\mathcal{H}$ due to the limitations of the current tabular data model. Firstly, tracing `Najma`'s ancestral lineage is complicated by fragmented data spread across various tables and spreadsheets, making it difficult to compile a comprehensive and accurate lineage. Additionally, the tabular model lacks explicit relationship representations, forcing *Jane* to manually infer connections based on shared identifiers or data patterns, which increases the risk of errors and omissions. Secondly, when analyzing `Najma`'s participation in events, *Jane* must filter event tables by her name or other identifiers. This process overlooks the broader context and limits her ability to correlate event results with `Najma`'s reputation and breeding potential without labor-intensive data aggregation and referencing external documentation. Thirdly, the tabular model's inability to integrate diverse data sources exacerbates these issues. Critical insights about `Najma`'s lineage, event participation, and ownership history remain fragmented, hindering a holistic understanding and strategic decision-making. *Jane*'s significant time and effort spent on manually assembling and validating data from disparate sources further complicates her ability to respond swiftly to breeding opportunities or challenges, leading to potential errors and delays.

Table 1 depicts the historical reproduction and lineage of horses, illustrating complex familial relationships beyond direct parentage. It includes columns for each horse's *parent, reproduction date* and *time, foal ID number, descendant foal, grandparent horse, parent horse* (sire/dam), *great-grandparent horse*, and *distant relations* (e.g., cousin, nephew, niece) with other foals. However, the main limitation of the above model is that it can traditionally only capture the characteristics of horses and does not work for data with complex correlations. In addition, most existing relational DBMS ignore multiple correlations between different entities and are not well-suited for 'Friend of a Friend (FOAF)' queries. For instance, consider the following query: *Find all horses that are indirectly related to Mare* `Najma`*'s offspring, considering up to two generations of separation* ?. This task parallels finding ascendants and descendants of a given horse, akin to finding a FOAF in social networking, which necessitates multiple joins and complex data traversal.

To overcome the limitations of traditional storage and better capture complex spatiotemporal dependencies, we need to translate user requirements into a given executable data model, whether from RDBMSs or recent NoSQL systems such as Neo4J. We should leverage an advanced database model to represent relationships between horses, ancestors, events, and owners as interconnected nodes and edges. This allows us to retrieve complete ancestral lineage in a single traversal and analyze event participation with contextual details.

3 Our System Kyle-$\mathcal{KG}$

To address the aforementioned issues, we propose a comprehensive process flow encompassing data extraction, persistence, analysis, and reporting (See Fig. 1). The $\mathcal{EH}$ management system has two main components: (i) extraction & persistence and (ii) analytics. The extraction & persistence component collects data on horse reproduction, lineage, and significant events, initially storing it in a relational database and subsequently converting it to a GraphDB. The analytics component uses Cypher queries and Machine Learning (ML), Graph Algorithms and Analytics on the GraphDB for complex analyses such as lineage tracing and pattern recognition. Results are displayed on a comprehensive and visual analytics to effectively manage and analyze $\mathcal{EH}$. The results of graph analysis can be communicated to the horse breeding manager to generate actionable reports for breeders regarding pollination, reproduction, vaccination, and feeding plans. The graph information system provides analysts with the ability to query the database based on data elements and their relationships. Based on the analysis, the manager has the necessary information to take corrective action on $\mathcal{EH}$.

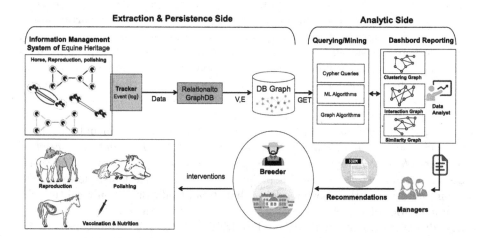

Fig. 1. Comprehensive Overview of Kyle-$\mathcal{KG}$: Data Extraction and Persistence on the Right, Graph Model Utilization on the Left

3.1 Graph Model Requirements

To effectively manage and explore the intricate relationships within $\mathcal{EH}$, we propose designing a specialized graph model. Our review of state-of-the-art literature on equine heritage information systems has identified the following critical requirements (RQs) that our graph model should address:

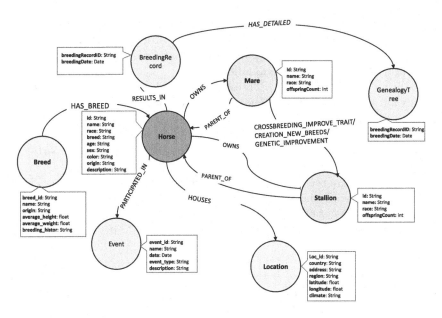

Fig. 2. Our Graph Model for $E\mathcal{H}$.

RQ.1:	Retrieve the ancestral lineage of a specific horse, including all predecessors.
RQ.2:	Find horses with similar characteristics to a specific horse.
RQ.3:	Find clusters of horses based on specific criteria.
RQ.4:	Identify various relationships between community structures among horses.
RQ.5:	Identify trends in $E\mathcal{H}$ management.

3.2 Our Graph Data Model

Based on the requirements discussed above, we will now design a graph model for *knowledge management of equine heritage* that can meet the data needs of important tasks: data integration, relationship mapping, and query efficiency. For equine heritage management, the following real-world entities are relevant: *horses, owners, lineages, events, locations,* and *breeds*.

We regard the $E\mathcal{H}$ information as a directed graph $G = (V, E)$ where V is a set of nodes representing entities and E a set of edges representing relationships between these entities. In our graph model, nodes represent various entities and have specific labels and properties.

- **Horses** $h \in H$ are represented by nodes with the label *Horse*. For each horse, we store properties such as name, birth date, and registration number.
- **Breeds** $b \in B$ are represented by nodes with the label *Breed*. For each breed, we store properties such as breed name and characteristics.

- **Breeding Records.** $br \in BR$ are represented by nodes with the label *BreedingRecord*. For each BreedingRecord, we store properties such as date of breeding, resulting offspring, and additional details like breeding method (natural cover, artificial insemination).
- **Stallions** $s \in S$ (optional): Represented by nodes with the label *Stallion* if you want to differentiate stallions from horses for breeding purposes. Properties can include stud fee.
- **Mares** $m \in M$ (optional): Represented by nodes with the label *Mare* if you want to differentiate mares from horses for breeding purposes. Properties can include offspring count.
- **Events** $e \in E$ are represented by nodes with the label *Event*. For each event, we store properties such as event name, date, and type (e.g., race, exhibition).
- **Locations** $loc \in LOC$ are represented by nodes with the label *Location*. For each location, we store properties such as name, latitude, longitude, and address.

The relationships (edges) between these nodes capture the interactions and associations within the $E\mathcal{H}$ domain:

- **Breedship:** Each horse node is connected to its BreedingRecord node through an edge labeled *result in*. This relationship captures the Genealogy Tree details.
  Breed: Each horse node is connected to a breed node through an edge labeled *belongs to*. This relationship identifies the horse's breed.
  Event Participation (optional): Each horse node can be connected to event nodes through edges labeled *participated in*. This relationship captures the horse's participation in various events.
  Event Location: Each event node is connected to a location node through an edge labeled *held at*. This relationship indicates where the event took place.
  Parentage (optional): If differentiating stallions and mares, a directed edge labeled *sire* can connect a horse node to a stallion node, and another edge labeled *dam* can connect a horse node to a mare node. This captures the parentage information. In addition, Mares and Stallions interact using: *Crossbreeding to improve traits*, *Creation of new breeds*, and *Genetic improvement*.

Additionally, we can store various properties for these relationships to capture more detailed information.

3.3 Exploiting the Graph Model

Using our intuitive graph model illustrated in Fig. 2, we can leverage it through various methods: Cypher Queries, ML (Machine Learning) algorithms, and Visual Analytics. In addition, Graph Algorithms ($\mathcal{G}$) can be applied to build inter-community graphs, where the links between communities reflect the spatial relationships. More specifically, the nodes in the graph are the communities, and the edges represent relationships between communities. We also encode weights on the edges as the relationship strength between communities can be different. Moreover, since there may be various relationships between communities that can help our prediction, we construct multiple graphs: interaction graph ($\mathcal{G}_I$) and similarity graph ($\mathcal{G}_S$).

Figure 3 Illustrates the interdependent relationship of different community structures among horses with geographic information or origin.

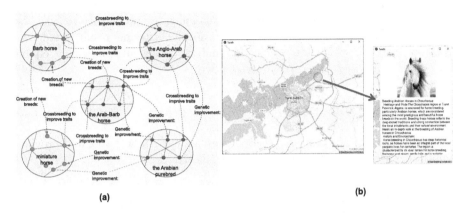

Fig. 3. Illustration of the interdependent relationship: (a) different community structures among horses, (b) Geographic information.

Horse Interaction Graph: To identify and emphasize the communities with higher interactions through historical horse trading, movements, and other equine-related activities, we analyze historical horse flows to construct interaction graphs among these communities. As depicted in Fig. 3(a), an interaction graph $\mathcal{G}I = (\mathcal{V}, \mathcal{E}, W_I)$ is formed to indicate frequent interactions between communities based on historical records of horse flows. Here, Ii, j denotes the number of horse flow records between communities i and j, with $W_I = I_{i,j}$ reflecting the weight between two communities, thereby highlighting those with more frequent horse flows.

Horse Similarity Graph: To uncover similarities in various aspects such as population composition, topographical distribution, building distribution, and traffic patterns related to equine activities, we analyze the correlations between communities. The similarity between communities extends beyond geographical distance to include these diverse aspects (See Fig. 3(b)). We construct a similarity graph $\mathcal{G}_S = (\mathcal{V}, \mathcal{E}, W_C)$ as a weighted undirected graph, where the weight $W_C = pi, j$ is determined by the Pearson correlation coefficient $p_{i,j}$ between communities i and j. This approach facilitates the identification of communities with similar characteristics and trends in $E\mathcal{H}$ management.

In a generic way, the analysis pipeline employs path representation, clustering, and graph algorithms. In this equine genetic analysis scenario, the input is a graph $G = (V, E)$ where V represents horses as nodes and E denotes genetic relationships as edges. Firstly, we defines path P from ancestors to descendants as sequences of genetic relationships: $P = R_{1j} \bowtie R_{2j} \ldots \bowtie R_{kj}$, $k \in [1, n]$, $j \in [1, J]$, with R_{1j} being the first ancestor $1j$ to descend from a given horse. And then, we apply K-means clustering to group horses based on genetic similarity into clusters C, $C = \{C_1, C_2, \ldots, C_k\}$, where each C_i represents a cluster of horses. Secondly, we applied graph algorithms ($\mathcal{G}$) using multiple graphs: $\mathcal{G}_I, \mathcal{G}_S$ to identify and analyze community structures among horses, revealing groups with common genetic ancestry. The output includes comprehensive reports and visualizations summarizing path representations, clustering results, and evolutionary insights, which are crucial for understanding and managing equine genetic heritage.

4 Proof of Concept

To emphasize our approach, this section presents our support tool and provides application scenarios to highlight its importance.

4.1 Tool Support

We have developed Kyle-$\mathcal{KG}$, an efficient knowledge management tool for horse breeding (see Fig. 4). This tool centralizes breeding information, tracks ancestry through a tree view, and logs events and solutions for future reference. Our implementation leverages Neo4j[1] to effectively manage and explore the rich interconnected data of the $\mathcal{EH}$ equine heritage at the Chaouchoua horse breeding center in Tiaret, Algeria. Our proposed solution consists of the following modules:

(i) *About the Horse*: This module describes the horse's characteristics, such as origin, physical traits, and genealogical lineage.
(ii) *Historical Events*: This module explores all events related to equine reproduction, breeding, and pollination.
(iii) *Visual Analytics*: This module enables users to explore complex genetic information through detailed summaries, making it easily understandable.
(iv) *Breeding Advice*: This module prioritizes genetic health and performance in breeding, prevents inbreeding, and preserves heritage. It educates breeders, optimizes pairings, and monitors outcomes to ensure a diverse and healthy horse population.

4.2 Application Scenarios

In our implementation, a decision maker's intent is translated into a Cypher query that interacts with the underlying Neo4j graph and presents the results through Visual Analytics. The following usage scenarios are explored based on

---
[1] https://neo4j.com/.

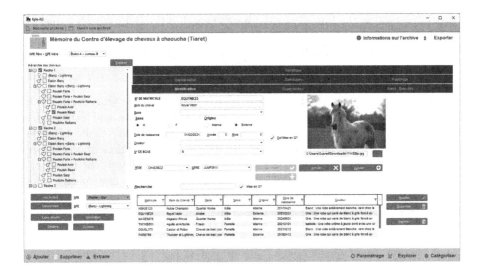

Fig. 4. Tool Support Overview: Kyle-$\mathcal{KG}$

our previous research questions: (**i**) Visualizing and identifying horses with similar characteristics. (**ii**) Exploring the ancestry and descent of a given horse. (**iii**) Clustering existing horses based on specific criteria. (**iv**) Conducting interactive cluster analysis. (**v**) Providing recommendations for equine heritage ($E\mathcal{H}$) based on genetic cluster ($\mathcal{G}_C$) analysis.

Visualizing and Identifying Horses with Similar Characteristics. In equine heritage management, breeders often need to identify horses with similar characteristics to inform breeding decisions. This process involves understanding the genetic traits, performance history, and lineage of horses. By utilizing a GraphDB and implementing the KNN algorithm via Cypher queries, breeders can conduct similarity analyses based on comprehensive, interconnected data.

```
MATCH (h:Horse {name: 'Najma'})-[:BELONGS_TO]->(b:Breed) ... MATCH
(h)-[:DESCENDED_FROM*1..3]->(ancestor:Horse) ... MATCH
(h)-[:PARTICIPATED_IN]->(event:Event) ... MATCH
(h)-[:OWNED_BY]->(owner:Owner) ... RETURN h.name,
apoc.algo.knn.cosineSimilarity(events, {weight: 0.8}).score,
    apoc.algo.knn.cosineSimilarity(ancestors, {weight: 0.6}).score,
    apoc.algo.knn.cosineSimilarity(owners, {weight: 0.4}).score ORDER BY ... LIMIT 5;
```

Listing 1.1: Cypher Query: finding similar horses using KNN algorithm

The Cypher query of Listing 1.1 starts by retrieving information related to Najma, including her breed, ancestral lineage (up to three generations), events she has participated in with their locations, and her current and past owners. The query returns the top 5 horses most similar to Najma based on the calculated similarity scores, prioritizing horses with higher event, ancestor, and owner similarity. The graph model captures rich relationships between horses, events, lineages, and owners, providing contextual data essential for accurate similarity analysis. This use case demonstrates how leveraging graph models and advanced algorithms, such as the k-nearest neighbors algorithm (k-NN) in Cypher queries, can unlock the full potential of $\mathcal{EH}$ data. It highlights the importance of relationships in enhancing knowledge management and decision support systems in this domain.

Tracing the Ancestry and Descent of a Given Horse. Our graph model enables us to trace the evolution of $\mathcal{EH}$. Users can navigate and explore $\mathcal{EH}$ through a graph-based visualization, allowing horse evolution to be retrieved via different graph paths. Our GraphDB contains information about horses, including names, breeds, and parent-child relationships. Below is the result of a Cypher query from Fig. 5, showing the ancestors/descendants of a specific horse named "Nadjma". This query demonstrates how Cypher can navigate relationships within a horse heritage knowledge graph and retrieve valuable information about ancestry and descent based on a simple description.

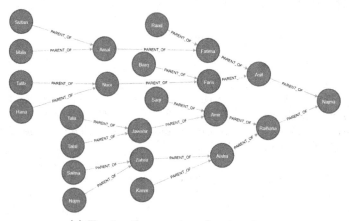

(a) Tracing the ancestry of a given horse

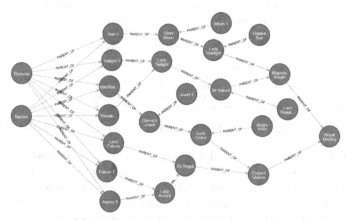

(b) Tracing the descent of a given horse

Fig. 5. Tracing the ancestry and descent of a given horse.

Note that this exploration is complex, as previously discussed in Sect. 2. We believe that this tool will significantly enhance our understanding of horse trends and their evolutionary characteristics over time. Additionally, tracing horse evolution helps in assessing a horse's impact within the community, improving research insights into the $E\mathcal{H}$ network, and facilitating data-driven research decisions, such as those related to reproduction and breeding.

Clustering the Existing Horses Based on Specific Criteria. Clusterization of existing horses is very beneficial for visualizing and understanding their relationships. We believe this material is of significant interest and demand within the animal husbandry community. Our proposed process for $E\mathcal{H}$ clustering is based on Neo4j.

In large graphs, the clusters or sub-graphs formed are also referred to as modules. Identifying these communities aids in comprehending various aspects of the graph, such as locating nodes with similar behaviors and identifying strongly connected groups. Community detection has become a crucial part of managing $E\mathcal{H}$, as it helps to explore intricate relationships among ancestors, descendants, and bloodlines.

The query in Listing 1.2 identifies clusters of horses based on their participation in events, breeds, and lineage connections. This information can be utilized for various purposes, such as enhancing predictions through machine learning techniques.

```
MATCH (h:Horse)-[:PARTICIPATED_IN]->(e:Event),
      (h)-[:BELONGS_TO]->(r:Race),
      (h)-[:DESCENDED_FROM]->(l:Lineage)
RETURN r.name AS Race,
       collect(DISTINCT h.name) AS Horses,
       collect(DISTINCT e.name) AS Events,
       collect(DISTINCT l.name) AS Lineages
ORDER BY Race
```

Listing 1.2: Clustering based on horse participation in events, races, and lineage

This query efficiently gathers relevant data for clustering, leveraging the flexibility and interpretability of the GraphDB model through comprehensive visual analytics (Fig. 6). By examining the clusters formed based on events, breeds, and lineages, we can gain deeper insights into the relationships and patterns inherent in $E\mathcal{H}$ data.

Indeed, a more in-depth analysis of the cluster data can be conducted using machine learning (ML) techniques. It is important to note that graph models significantly enhance the capabilities of ML, providing a robust framework for more accurate and insightful predictions and analyses.

Interactive Cluster Analysis Using $\mathcal{GI}$. To identify and highlight communities with significant interactions through historical horse trading and other equine-related activities, we analyze historical horse flows to construct interaction graphs among these communities. These $E\mathcal{H}$ graphs provide detailed information on interdependent relationships and various community structures among horses, such as *crossbreeding to improve traits, creation of new breeds,*

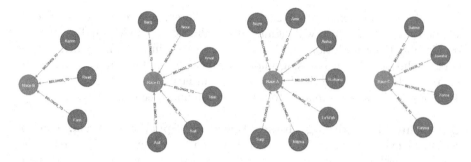

Fig. 6. Excerpt of clustering horses based on horse participation in events and races.

and *genetic improvement*, all meticulously extracted from IMS operational data. The exploration connects horses to their clusters and eventually to their interdependent relationships with other clusters. As depicted in Fig. 7, an interaction graph $\mathcal{GI}$ is formed to indicate frequent interactions between communities based on historical records of horse flows. This clustering of horses based on their participation in events and races is achieved using a Neo4j Cypher query to analyze their interdependent relationships. The horses are clustered into four distinct groups (see Fig. 7), with each cluster comprising horses from potentially diverse areas but sharing similar characteristics. For instance, the cluster labeled "Race B" demonstrates the preservation of breed purity. This analysis aims to reveal the similarities and differences between the user's $E\mathcal{H}$ and other datasets already stored in the GraphDB, all within the same context of the model under analysis.

Recommendations for $E\mathcal{H}$ Based on $\mathcal{G}_S$ Analysis. Leveraging the relationships facilitated by our graph model, we can envision a variety of further uses through $\mathcal{G}_S$ analysis. This includes prioritizing genetic health and performance in breeding, preventing inbreeding, and preserving heritage. For example, $\mathcal{G}_S$ analysis can be specifically applied in upcoming tests to evaluate our model's effectiveness in equine breeding management, with a particular focus on genetic health and performance. We constructed a $\mathcal{G}_S$ between both communities of horses. This analysis facilitates the identification of communities with similar characteristics and trends in $E\mathcal{H}$ management as recommendations for different purpose such as educating breeders, optimizing pairings, and monitoring outcomes to ensure diverse and healthy horses.

5 Related Work

GraphDB has proven instrumental in heritage management by facilitating efficient storage and retrieval of interconnected data. In the realm of cultural heritage, graph models have been employed to represent relationships between artifacts, historical figures, and events (e.g. [1,6,10,12,13]) for organizing and preserving historical narratives, showcasing how graph structures enhance data accessibility and contextual understanding. Graph data modeling in $E\mathcal{H}$ provides

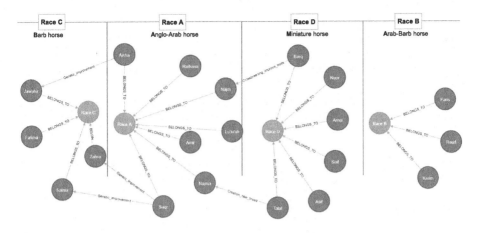

Fig. 7. Clustering of Horses Based on Event and Race Participation: An Analysis of Interdependent Relationships Using Neo4j Cypher Queries.

unique advantages for managing relationships among horses, genetic lineages, events, and ownership histories. Studies have used GraphDB to map genetic lineages and trace horses' historical participation in events, facilitating comprehensive data integration and enhancing interpretability (e.g. [8,13]). Authors in [5,7] presented extensive ancient DNA analysis from horses, uncovering two extinct lineages and highlighting significant genetic impacts. In another line of research, the authors of [18] addressed the analysis of genomic variants across diverse horse breeds, revealing historical genetic similarities, increased recent inbreeding, and new insights into the genetics of horses. In a similar trend, the authors in [18] provided a database for the epigenetic changes after horse exercise. HExDB[2] a database for epigenetic changes occurring after horse exercise. Similarly, [19] revealed demographic patterns of horse ownership and travel, highlighting urban horse keeping, vaccination practices, and the potential for rapid disease spread due to event attendance and international travel with untraceable horses. Other studies explored Knowledge Representation for lineages of $E\mathcal{H}$ using ontologies and semantic web technologies (e.g. [9,11,20]). At the same time, several studies proposed solutions to use ML techniques to analyze rider's effects on horse gait using on-body inertial sensors (e.g. [2–4]).

A similar study has been conducted to capitalize on the scientific effort of testing data for reproduction purposes using graph-based database models [14–16].

6 Conclusion

Our approach paves the way for improved preservation practices, fostering a deeper appreciation for equine heritage and its rich history. The proposed graph

---

[2] http://www.primate.or.kr/hexdb.

data model not only facilitates efficient data integration and relationship mapping but also ensures query efficiency. It serves as a robust framework for advanced queries and analyses, enabling stakeholders to leverage the full value of equine heritage data for informed decision-making and comprehensive management. Two pivotal avenues for future research are worth considering. Firstly, we aim to undertake exploratory statistical analyses, including correlational analysis, regression, or factor analysis. These approaches will deepen our understanding of horse behaviors and trends. Secondly, we intend to employ machine learning techniques to develop predictive models. These models will not only identify and predict horse behaviors but also assist in crafting specific dietary and training regimes tailored to optimize horse performance within varying system constraints.

References

1. Bogdanova, G., Todorov, T., Noev, N.: Using graph databases to represent knowledge base in the field of cultural heritage. Digit. Present. Preserv. Cult. Sci. Heritage **6**, 199–206 (2016)
2. Chavan, P., et al.: Horse race prediction using machine learning algorithms. In: International Conference on Computer & Communication Technologies, pp. 329–336. Springer (2023)
3. Cihan, P.: Horse surgery and survival prediction with artificial intelligence models: performance comparison of original, imputed, balanced, and feature-selected datasets. KAFKAS ÜNİVERSİTESİ VETERİNER FAKÜLTESİ DERGİSİ **30**(2) (2024)
4. Darbandi, H., Havinga, P.: A machine learning approach to analyze rider's effects on horse gait using on-body inertial sensors. In: 2022 IEEE International Conference on Pervasive Computing and Communications Workshops and other Affiliated Events (PerCom Workshops), pp. 248–253. IEEE (2022)
5. Detry, C., et al.: Tracking five millennia of horse management with extensive ancient genome time series (2019)
6. Dou, J., Qin, J., Jin, Z., Li, Z.: Knowledge graph based on domain ontology and natural language processing technology for Chinese intangible cultural heritage. J. Vis. Lang. Comput. **48**, 19–28 (2018)
7. Fages, A., et al.: Tracking five millennia of horse management with extensive ancient genome time series. Cell **177**(6), 1419–1435 (2019)
8. Fan, T., Wang, H., Hodel, T.: CICHMKG: a large-scale and comprehensive Chinese intangible cultural heritage multimodal knowledge graph. Heritage Sci. **11**(1), 115 (2023)
9. Gao, H., et al.: An equine disease diagnosis expert system based on improved reasoning of evidence credibility. Inf. Process. Agric. **6**(3), 414–423 (2019)
10. Gu, X., Xiao, Y., Hua, Z., Jin, H., Wang, B.: A novel approach for constructing intangible cultural heritage knowledge graphs. In: 2022 8th International Symposium on System Security, Safety, and Reliability (ISSSR), pp. 117–126. IEEE (2022)
11. Hreiðarsdóttir, G.E., Árnason, Þ., Svansson, V., Hallsson, J.H.: Analysis of the history and population structure of the Icelandic horse using pedigree data and DNA analyses (2014)

12. Huang, Y.Y., et al.: Using knowledge graphs and deep learning algorithms to enhance digital cultural heritage management. Heritage Sci. **11**(1), 204 (2023)
13. Origlia, A., et al.: Multiple-source data collection and processing into a graph database supporting cultural heritage applications. J. Comput. Cult. Heritage (JOCCH) **14**(4), 1–27 (2021)
14. Ouared, A.: Towards an explicitation and a conceptualization of cost models in database systems. In: Model and Data Engineering: 7th International Conference, MEDI 2017, Barcelona, Spain, 4–6 October 2017, Proceedings 7, pp. 223–231. Springer (2017)
15. Ouared, A., Kharroubi, F.Z.: Moving database cost models from darkness to light. In: Smart Applications and Data Analysis: Third International Conference, SADASC 2020, Marrakesh, Morocco, 25–26 June 2020, Proceedings 3, pp. 17–32. Springer (2020)
16. Ouared, A., Ouhammou, Y.: Capitalizing the database cost models process through a service-based pipeline. Concurr. Comput.: Pract. Exp. **35**(11), e6463 (2023)
17. Sigley, G.: The ancient tea horse road and the politics of cultural heritage in southwest China: regional identity in the context of a rising China. In: Cultural heritage politics in China, pp. 235–246. Springer (2013)
18. Todd, E.T., et al.: Imputed genomes of historical horses provide insights into modern breeding. Iscience **26**(7) (2023)
19. Uprichard, K., Boden, L., Marshall, J.: An online survey to characterise spending patterns of horse owners and to quantify the impact of equine lameness on a pleasure horse population. Equine Vet. J. **46**, 4–4 (2014)
20. Yinqiqige, B., et al.: The construction of mongolian lexical semantic net about horse culture and economy. In: 2023 IEEE 4th International Conference on Pattern Recognition and Machine Learning (PRML), pp. 444–448. IEEE (2023)

OWL2Vec4OA: Tailoring Knowledge Graph Embeddings for Ontology Alignment

Sevinj Teymurova[1], Ernesto Jiménez-Ruiz[1,2(✉)], Tillman Weyde[1], and Jiaoyan Chen[3]

[1] City St George's, University of London, London, UK
ernesto.jimenez-ruiz@city.ac.uk
[2] University of Oslo, Oslo, Norway
[3] The University of Manchester, Manchester, UK

Abstract. Ontology alignment is integral to achieving semantic interoperability as the number of available ontologies covering intersecting domains is increasing. This paper proposes OWL2Vec4OA, an extension of the ontology embedding system OWL2Vec*. While OWL2Vec* has emerged as a powerful technique for ontology embedding, it currently lacks a mechanism to tailor the embedding to the ontology alignment task. OWL2Vec4OA incorporates edge confidence values from seed mappings to guide the random walk strategy. We present the theoretical foundations, implementation details, and experimental evaluation of our proposed extension, demonstrating its potential effectiveness for ontology alignment tasks.

Keywords: ontology alignment · random walks · ontology embeddings · knowledge graph embeddings

1 Introduction

Knowledge graphs (and ontologies) are increasingly recognized as essential for successful AI implementations across various data science applications [14]. Ontology alignment isa crucial task to enable sntic interoperability and enhance the application of knowledge graphs. The ontology alignment process involves finding and harmonizing semantic connections between different ontologies. While current methods have advanced ontology alignment considerably, there remain obstacles to their widespread implementation [20,31].

The ontology matching community has contributed to the evolution of ontology alignment systems for the last twenty years with the organization of the annual Ontology Alignment Evaluation Initiative (OAEI) [35,36]. In recent years there has been a shift from traditional systems [11], using lexical and structural techniques, to systems using machine learning and (large) language models. Prominent examples include LogMap-ML [5], BertMap [21], DeepAlignment [29], VeeAlign [26], SORBETMatcher [15] and OLaLa [24]. The OAEI, with the new

Bio-ML track [22,23], has also evolved accordingly to attract and systematically evaluate such systems.

Knowledge Graph Embeddings (KGE) techniques [38,42] aim at capturing, in a low-dimensional continuous vector space, the structure and semantics of the graph. These low-dimensional representations enable the application of machine learning algorithms to graph-structured data in downstream tasks such as node classification, link prediction, or knowledge graph alignment [2]. Traditional KGE techniques commonly rely on one of the following: *(i)* geometric transformations, *(ii)* matrix factorization methods, and *(iii)* neural networks. Recent advancements in KGE have expanded to incorporate semantics beyond relational facts. These include encoding textual literals and integrating logical structures to capture richer semantic information within KG representations [6].

In this paper, we present OWL2Vec4OA, an extension of the ontology embedding system OWL2Vec* [4] tailored to the ontology alignment task. OWL2Vec* projects a given ontology into a graph, randomly walks over the graph to generate sequences of entities, and runs the language model Word2Vec [32] to generate embeddings of both entity URIs and words. OWL2Vec4OA, unlike OWL2Vec*, relies on potentially incomplete or inaccurate ontology alignments to bridge a given set of input ontologies. When projecting the ontologies and performing the random walks to create entity sequences, the confidence value of these seed mapping are used to bias the random walks (*i.e.*, edges with higher confidence values will have higher chances to be visited). Hence, OWL2Vec4OA allows for a tighter connection of the input ontologies given a set of seed mappings, while giving preference to edges with higher confidence.[1] OWL2Vec4OA currently relies on the ontology matching systems LogMap [27] and AML [13] to produce seed mappings.

Our experiments show that the embeddings computed by OWL2Vec4OA are more suitable to the ontology alignment tasks than the original OWL2Vec* vectors. OWL2Vec4OA embeddings also lead to promising ranking results in the OAEI's Bio-ML track by simply comparing the computed vectors between the relevant source and target entities.

The rest of the paper is organised as follows. Section 2 introduces the necessary notions behind OWL2Vec4OA. Section 3 presents relevant related work. OWL2Vec4OA is described in detail in Sect. 4. Section 5 provide experimental results of OWL2Vec4OA on the Bio-ML datasets. Finally, Sect. 6 concludes the paper and discusses potential lines for future work.

2 Preliminaries

Ontologies and Knowledge Graphs. Ontologies serve as structured, clearly defined representations of collectively agreed-upon concepts and relationships within a specific field or area of interest [17]. Widely applied in information retrieval, data integration, and knowledge-based systems, ontologies facilitate

---

[1] Edges projected from ontology axioms are given the highest confidence.

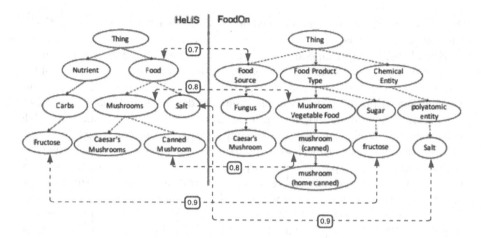

Fig. 1. Fragment of an alignment between HeLiS and FoodOn (adapted from [5]). The green dash arrow denotes mappings with confidence values ranging from [0,1]. Blue arrows represent the inverse of the predicate `rdfs:subClassOf`. (Color figure online)

semantic interoperability and reasoning across diverse applications. Knowledge graphs [25] have recently gained attention to represent entities and relationships within a graph-structured data model and have been very successful to improve search functionality, tailor user experiences, and inform strategic business choices [14]. Nonetheless, from the Semantic Web point of view, in essence, ontologies and knowledge graphs can be seen as equivalent notions (*i.e.*, OWL ontologies provide a formalization of the RDF graph data model [8], while knowledge graphs also imply the existence of such formalization). In this paper, we use knowledge graphs and ontologies interchangeably.

Ontology alignment is essential for data integration, semantic search, and cross-ontological reasoning. Ontology alignment can be defined as the process of identifying semantic relationships between elements (such as classes, attributes, and instances) of two or more ontologies. In this paper, we focus on atomic ontology matching where the goal is to establish equivalence or subsumption among atomic (*i.e.*, named) entities in the input ontologies [11,33]. Mappings are typically represented as a 4-tuple $\langle e, e', r, c \rangle$ where e and e' are entities from different ontologies; r is a semantic relation (*e.g.*, equivalence or subsumption); and c is a confidence value, usually, a real number within the interval $(0\ldots 1]$. For instance, Fig. 1 shows a fragment of an alignment between the ontologies HeLiS [10] and FoodOn [9]. The alignment indicates that the concept HeLiS:Fructose is similar to the concept obo:FOODON_03301305 (fructose) with a confidence value of 0.9. Confidence values are typically provided by ontology alignment systems and represent the degree of certainty associated with a correspondence between entities. Alignment systems often employ sophisticated algorithms that consider lexical similarities, structural relationships, and external knowledge sources. Methods such as cross-referencing, semantic similarity measures, and machine learning techniques can be applied to establish the mappings.

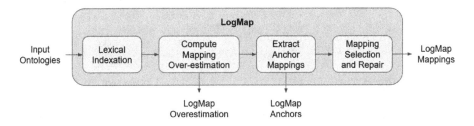

Fig. 2. General architecture of LogMap.

LogMap [27,28] is an efficient ontology alignment tool for large-scale ontologies which employs lexical indexation, logic-based reasoning, and semantic similarity computation in a multi-stage process. LogMap has demonstrated effectiveness in various challenges and applications, especially in biomedicine. As shown in Fig. 2, LogMap produces as output three different mappings sets: *(i)* LogMap overestimation are a large set of candidate mappings aiming for high recall while representing a manageable subset of all possible mappings; *(ii)* LogMap anchors are typicall a highly precise set of mappings; and *(iii)* LogMap mappings are the final computed mappings aiming at a balanced Precision and Recall. These set of mappings will be used in our experiments.

Random walks are key for embedding systems like RDF2Vec [37], OWL2Vec∗ and node2vec [16] to capture the structural and contextual information of a graph or network, so that the system can learn about the relationships between different nodes and their local neighborhoods. The PageRank algorithm [30] revolutionized web page ranking by employing a random walk model on the web's hyperlink structure. It simulates a "random surfer" traversing a graph of web pages, computing page importance based on the probability of the surfer landing on each page. This approach effectively captures the web's complex link topology to determine page significance, becoming a cornerstone in information retrieval and influencing various fields beyond web search. Random walks allow to process massive graphs without needing to consider all possible paths. Seminal works like DeepWalk [34] and node2vec [16] utilize random walks to generate node sequences for training embeddings, effectively capturing network topology. Wei [43] proposed an extension of the Metropolis-Hastings algorithm for sampling from large-scale networks, introducing strategies such as early rejection and biased sampling. RDF2Vec [37] introduced random walk-based embeddings for RDF knowledge graphs, later extended by Steenwinckel [39] with new walk extraction strategies. Cochez et al. [7] also introduced biased walk strategies to RDF2Vec with twelve different edge weighting functions.

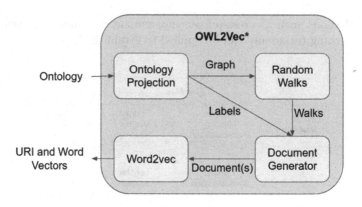

Fig. 3. General architecture of OWL2Vec*

OWL2Vec. Inspired by RDF2Vec [37], OWL2Vec* [4] was designed following similar principles but adapted to create embeddings for OWL ontologies and to take into account the lexical information of the ontologies (*e.g.*, literals in the form of labels and synonyms). Figure 3 depicts the architecture of OWL2Vec* for generating ontology embeddings. OWL2Vec* projects the input ontology into a graph and generates entity sequences via random walks over the ontology graph, then it generates different types of documents by substituting none, some, or all entity URIs by its lexical representation. Finally, the word embedding model Word2vec [32] is applied over the generated documents to compute embeddings for both URIs and words. Note that URIs are unique and thus their embeddings are contextual. OWL2Vec* has shown to outperform other approaches in intra-ontology subsumption and class membership prediction when both structural and lexical information was critical. OWL2Vec*, however, focuses on creating embedding for a single ontology. Although a set of ontologies can also be given as input, their graph representation will not be connected and thus the random walks will not generate sequences involving elements from different ontologies.

3 Related Work

The ongoing research in the ontology matching community evidences the need for more sophisticated techniques, as shown in the annual OAEI campaign [35] new methodologies and systems are developed to address this challenging problem in the Semantic Web. Otero-Cerdeira et al. [33] provides a comprehensive survey of ontology matching techniques.

These diverse approaches demonstrate the complexity of ontology alignment and the variety of techniques employed [12]. Nonetheless, ontology matching tools can now process more efficiently even the most complex ontologies, including those with hundreds of thousands of classes, encompassing billions of possible connections. Recent advancements in ontology matching have incorporated machine learning, including embedding-based techniques and (large) language models, showing promising results in improving alignment accuracy (*e.g.*, [5,21]). These approaches leverage vector representations of ontological elements to capture semantic relationships more effectively than traditional methods. By utilizing language models or domain-specific embedding algorithms, these techniques can identify nuanced similarities between concepts, potentially leading to more accurate and comprehensive ontology alignments. This new generation of systems can broadly be categorized into three categories. *(i) Direct embedding comparison:* Methods like ERSOM [44] and DeepAlignment [29] calculate distances between concept embeddings directly. *(ii) Supervised mapping classifiers:* VeeAlign [26], MEDTO [18], LogMap-ML [5], and SORBET [15] train classifiers using concept embeddings as input. This approach adds a layer of learning specific to the OM task but still relies on independent embeddings for the input ontologies. *(iii) Based on language models:* Recent methods such as BERTMap [21], BERTSubs [3] and OLaLa [24] rely on language models to implement task-specific models. For instance, BERTMap fine-tunes pre-trained language models (PLMs) using synonyms from the ontologies, while BERTSubs focuses on subsumption mapping prediction using context-based information that is transformed into text by templates.

Although systems based on language models have shown impressive results, they rely on pre-trained or large language models adding an important complexity layer for large matching tasks. Our approach OWL2Vec4OA leverages a simpler language model like Word2Vec to create tailored embeddings for the ontology matching task. Although Word2Vec does not create contextual embeddings (*i.e.*, same string with different meanings will get the same embedding), the sequences that OWL2Vec4OA creates include *things* (*i.e.*, entity URIs) in addition to strings, leading to contextual embeddings for the ontology entities as URIs are unique. The embeddings are tailored to the matching task as OWL2Vec4OA bridges the input ontologies with seed mappings computed by a (traditional) alignment system like LogMap [27] and AML [13] before performing the random walks to generate the URI and word sequences. Hence, unlike other approaches like LogMap-ML, the computed embeddings for both ontologies are in the same vector space and tightly related via the seed mappings.

Our evaluation uses the datasets of the OAEI's Bio-ML track [23], specialized benchmarks for evaluating machine learning-based OM systems. Currently, our experiments focus on direct embedding comparison, as the main purpose of this exercise was to evaluate the quality of the embeddings without any additional layer. The reported results are promising. In the near future we also plan to train a model with the computed embeddings similarly to LogMap-ML and the approach presented by Hao et al. [19].

4 Ontology Embeddings with OWL2Vec4OA

OWL2Vec4OA extends the OWL2Vec⋆ system with a mechanism to tailor the embeddings to the ontology alignment task involving two or more input ontolo-

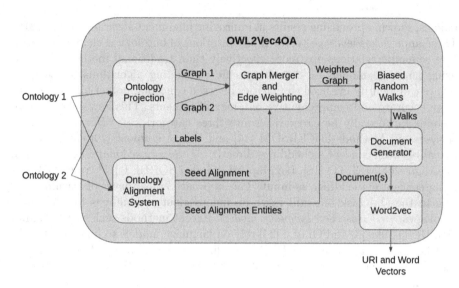

Fig. 4. General architecture of OWL2Vec4OA

gies. The main steps of our OWL2Vec4OA are depicted in Fig. 4,[2] and summarised as follows.

Ontology Projection. We use the same ontology projection rules as in OWL2Vec* [4]. The projection rules transform one or more ontology axioms into RDF triples (*i.e.*, labeled edges in the projected graph). Some axioms such as class subsumption (*e.g.*, obo:CHEBI_28757 (fructose) **rdfs:subClassOf** obo:FOODON_03420108 (sugar)) and annotations (*e.g.*, obo:FOODON_03420108 **rdfs:label** "sugar") have a one-to-one triple transformation; while more complex axioms require the application of projection rules. For example, the axiom obo:FOODON_03301391 (mushroom (canned)) **rdfs:subClassOf** RO_0001000 (derives from) **some** FOODON_03411261 (fungus) is transformed into the triple ⟨ obo:FOODON_03301391, RO_0001000, FOODON_03411261 ⟩. A directed labeled graph for each of the input ontologies is returned as the output of the projection.

Ontology Alignment. OWL2Vec4OA currently relies on the traditional ontology matching systems LogMap [27] and AML [13] to produce seed mappings. For example, Fig. 1 shows a subset of plausible mappings computed by an alignment system. These seed mappings are used to bridge the ontology graphs and thus enabling the execution of random walks over entities from different ontologies. Note that, seed mappings do not need to be accurate nor complete and their confidence values will be used to bias the random walks. As shown in Fig. 2,

---

[2] Source codes of OWL2Vec4OA are available here: https://github.com/Sevinjt/OWL2Vec4OA.

Input: Weighted graph $G = (V, E, U, W)$; seed entities S; walk depth wd; iterations $iter$
Output: Walks or entity sequences W
$W = \{\}$;
for k *in range(iter)* **do**
    // iterates over the seed entities
    for e *in* S **do**
        $current\_walk = [\,]$;
        Append $uri(G, e)$ to $current\_walk$;
        $current\_size = 1$;
        $focus = e$;
        while $current\_size < wd$ **do**
            Extract set of outer edges E_{focus} from $focus$ vertex;
            if $|E_{focus}| = 0$ **then**
                break;
            end
            // According to the probabilities of the edges in E_{focus} as in Equation 1
            Randomly select an outer edge l such that $l = (focus, v) \in E_{focus}$;
            Append $uri(G, l)$ to $current\_walk$; // URI of the link
            Append $uri(G, v)$ to $current\_walk$; // URI of the vertex
            $current\_size + +$;
            $focus = v$;
        end
        Add $current\_walk$ to W
    end
end
return W;   // Set of walks/sequences.

Algorithm 1: Biased Random Walks Algorithm

LogMap produces three sets of mappings of different quality that will be used as seed: *(i)* an overestimation of potential mappings (LogMap$_{over}$), *(ii)* highly precise mappings or anchors (LogMap$_{anch}$), and *(iii)* the regular output mappings (LogMap$_{out}$). In addition, LogMap$_{out}$ mappings are combined with AML mappings in our experiments (*i.e.,* LogMap$_{out}$ ∪ AML, and LogMap$_{out}$ ∩ AML).

Graph Merger and Edge Weighting. Unlike our predecessor OWL2Vec*, OWL2Vec4OA builds a single graph taking as input the graph projections of the ontologies to be aligned and a set of seed mappings. Mappings (*i.e.,* ⟨e, e', r, c⟩) have a direct graph representation as triples, for example, in Fig. 1 the following mapping (r = equivalence) was identified ⟨ HeLiS:Fructose, **owl:equivalentClass**, obo:FOODON_03301305 (fructose) ⟩ with confidence $c = 0.9$. OWL2Vec4OA assigns a weight to each edge or link as follows: *(i)* 1.0 if the edge was derived from ontology axioms; and *(ii)* c if the edge was derived from a mapping. The output is a labeled weighted graph $G = (V, E, U, W)$, where E is the set of edges built from the projected RDF triples, V is the set of vertices composed by the subjects and objects in these triples, U is the the set of URIs associated to the vertices and edges, and W is the set of weights associated to the edges. The function $weight(G, l)$ returns the weight for a given edge l, while the function $uri(G, e)$ returns the URI of a given entity e.

Biased Random Walks. OWL2Vec4OA, inspired by Cochez et al. [7], implements the biased random walker summarised in Algorithm 1. The algorithm takes as input a weighted labeled graph G and a set of seed entities S and performs (biased) random walks of depth wd starting from each of the seed entities. It optionally iterates over the seed entities more than once to allow for different walks for the same seed entity. In our setting, the seed entities represent the entities involved in the seed mappings computed in the alignment step. This way the walks are tailored to the alignment task, without the need of exploring the whole input ontologies. The bias in the random walk takes into account the weight assigned to each of the edges or links (l) in the graph G to assign a probability to each of the potential paths. Given $E_u = \{l_i = (u, v_i)\}$ with $i = 1..n\}$ and l_i an outer edge for u, the probability for each edge is computed as in Eq. 1.

$$Pr(l_j = (u, v_j)) = \frac{weight(G, l_j)}{\sum_{i=1}^{n} weight(G, l_i)} \quad (1)$$

For example, a walk of depth 3 starting from the seed entity HeLiS:Fructose (*i.e.*, an entity appearing in a mapping) could include the following sequence of URIs: HeLiS:Fructose, owl:equivalentClass, obo:FOODON_03301305 (fructose), rdfs:subClassOf, obo:FOODON_03420108 (sugar).

Document Generator and Word2Vec Embeddings. OWL2Vec4OA, as in OWL2Vec*, creates three types of documents from the generated walks W in the previous step: *(i)* structure document, *(ii)* lexical document, and *(iii)* combined document. The structure document is a direct representation of the walks as the sentences are composed by entity URIs. The lexical document replaces every URI occurrence in the walks by the respective lexical representation of the entity (*i.e.*, the occurrence of obo:FOODON_03420108 is replaced by its lexical representation "sugar", typically provided in the ontology via an rdfs:label annotation). Finally, the combined document randomly replaces in each walk some of the entity URI occurrences by its associated label. The three documents are merged and used to train a Word2Vec model with the skip-gram architecture. The trained Word2Vec model produces embeddings for both URI and word occurrences in the merged document. As introduced in Sect. 2, the URI embeddings can be seen as contextual embeddings as URIs are unique.

5 Evaluation

We have performed a preliminary evaluation of the suitability of the embeddings computed by OWL2Vec4OA in ontology alignment tasks. Particularly we have used the datasets provided by the OAEI's 2023 Bio-ML track[3]. The Bio-ML track included several tasks (*e.g.*, OMIM-ORDO, NCIT-DOID, SNOMED-NCIT-Pharm and SNOMED-NCI-Neoplas), involving biomedical ontologies with tens of thousands of classes, and reference alignments based on Mondo [41] and UMLS [1].

Table 1. OMIM-ORDO task with Walk depth 3, Walker iteration: iter = 1/iter = 5.

Seed Mappings	Hits@1	Hits@5	Hits@10	MRR
Train-Validation	0.01/0.01	0.02/0.02	0.05/0.05	0.04/0.04
LogMap$_{over}$	0.27/0.28	0.51/0.53	**0.60/0.62**	0.38/0.40
LogMap$_{anch}$	0.11/0.26	0.28/0.41	0.36/0.46	0.20/0.33
LogMap$_{out}$	0.26/0.27	0.41/0.45	0.47/0.53	0.33/0.36
LogMap$_{out}$ ∪ AML	0.30/0.31	**0.54/0.54**	**0.61/0.61**	**0.41/0.41**
LogMap$_{out}$ ∩ AML	**0.31/0.34**	0.49/0.50	0.54/0.54	0.40/0.41

Bio-ML presents two evaluation settings: global matching and local ranking. Global matching is evaluated with the traditional measures Precision and Recall, comparing a set of system-computed mappings with the reference set of mappings; while local matching evaluates the capacity of a system to rank a correct mapping given a pool of potential candidates. Bio-ML uses Mean Reciprocal Rank (MRR) and Hits@K (*i.e.*, cases where the correct mapping was ranked within the top-k) in the local matching setting.

Scoring Function and Settings. We have applied OWL2Vec4OA embeddings into the local matching tasks of Bio-ML. Mappings are scored and ranked according to the cosine similarity of the computed URI embeddings for the entities in the mapping. We have computed OWL2Vec4OA embeddings for different walk depths and iterations. We fixed the Word2Vec hyperparameters—the number of epochs and embedding dimension to 70 and 100, respectively. The experiments were conducted on a High-Performance Computing cluster with access to up to 48 CPUs, using the Slurm workload manager to ensure efficient resource allocation and job scheduling. Generated resources are available in Zenodo [40].

Impact of Seed Mappings and Number of Iterations. Table 1 shows the results over the OMIM-ORDO for different sets of seed mappings as introduced in

---

[3] Bio-ML Challenge [23]: https://krr-oxford.github.io/OAEI-Bio-ML/. Bio-ML 2023 Datasets: https://doi.org/10.5281/zenodo.8193375.

Sect. 4. We also used as seed the mappings provided as training and validation in Bio-ML, as shown in the first row, using training-validation mappings in isolation did not lead to promising results given their reduced size. The set of seed mappings leading to the best results was the union of LogMap$_{out}$ and AML mappings, with LogMap$_{out}$ ∩ AML and LogMap$_{over}$ leading to similar results. The results also show that OWL2Vec4OA is also able to handle noisy set of seed mappings like LogMap$_{over}$. The impact of additional iterations over the seed entities did not lead to a significantly increased performance.

Table 2. Results of OWL2Vec4OA and OWL2Vec* over four Bio-ML tasks, with different walk depths (wd).

Task	System	wd	MRR	Hits@1	Hits@5	Hits@10	Hits@20	Hits@30
OMIM-ORDO	OWL2Vec*	2	0.074	0.018	0.091	0.178	0.332	0.393
		3	0.073	0.018	0.090	0.170	0.318	0.381
		4	0.071	0.019	0.078	0.320	0.321	0.387
	OWL2Vec4OA	2	**0.586**	**0.533**	**0.637**	**0.657**	**0.672**	**0.693**
		3	0.402	0.306	0.512	0.587	0.650	0.685
		4	0.215	0.132	0.281	0.359	0.446	0.532
NCIT-DOID	OWL2Vec*	2	0.218	0.110	0.306	0.448	0.631	0.746
		3	0.175	0.074	0.251	0.377	0.561	0.690
		4	0.105	0.035	0.121	0.225	0.409	0.541
	OWL2Vec4OA	2	0.195	0.064	0.310	0.508	0.709	0.812
		3	0.358	0.181	0.573	0.741	0.872	0.924
		4	**0.609**	**0.442**	**0.840**	**0.928**	**0.970**	**0.984**
SNOMED-NCIT-N	OWL2Vec*	2	0.063	0.014	0.075	0.134	0.231	0.309
		3	0.068	0.017	0.079	0.142	0.238	0.308
		4	0.055	0.011	0.052	0.114	0.218	0.305
	OWL2Vec4OA	2	0.648	0.543	0.767	0.831	0.888	0.904
		3	0.605	0.484	0.746	0.813	0.872	0.899
		4	**0.805**	**0.747**	**0.872**	**0.888**	**0.902**	**0.910**
SNOMED-NCIT-P	OWL2Vec*	2	0.079	0.018	0.094	0.184	0.302	0.675
		3	0.078	0.018	0.092	0.181	0.292	0.667
		4	0.055	0.011	0.052	0.114	0.218	0.305
	OWL2Vec4OA	2	**0.436**	**0.342**	**0.534**	**0.583**	**0.609**	**0.967**
		3	0.311	0.190	0.435	0.502	0.558	0.933
		4	0.291	0.204	0.355	0.434	0.521	0.944

Comparison with OWL2Vec.* Following the results in Table 1, we set LogMap$_{out}$ ∪ AML as the seed mappings and the number of iterations to 1 in the subsequent experiments. We experimented with walk depths ranging from 2 to 4. We compared the performance of the embeddings computed with OWL2Vec4OA with those computed with the original OWL2Vec* version (using its multi-ontology setting). As expected, Table 2 shows that the ranking with OWL2Vec4OA embeddings considerably outperforms the ranking with OWL2Vec* embeddings in all tasks and for all evaluated walk depths, indicating that the OWL2Vec4OA embeddings are more suitable for ontology alignment tasks. The best results are obtained for the task SNOMED-NCIT-Neoplas with walk depth 4 where Hits@1 reach more than 80% of the cases. In other tasks, the results are also promising indicating that the embeddings computed by OWL2Vec4OA capture relevant features of the original entities that could be exploited by a subsequent machine learning model.

Impact of the Walk Depth. Increasing the walk depths has a positive impact in the tasks NCIT-DOID and SNOMED-NCIT-Neoplas; while for OMIM-ORDO and SNOMED-NCIT-Pharm longer paths seem to add noise to the embeddings. This is inline with the results obtained in the original OWL2Vec⋆ paper [4] where longer paths did not seem to lead to better results. It is worth mentioning that longer paths also increase the computation times. In the near future, we plan to perform an extended evaluation to better understand the impact of longer walks on different ontologies and matching tasks.

6 Conclusions and Future Work

We have presented OWL2Vec4OA, an extension of the ontology embedding system OWL2Vec⋆ [4]. OWL2Vec4OA has been tailored to the ontology alignment task by using a preliminary set of ontology alignments, possibly incomplete or inaccurate, to bridge a given set of input ontologies. These seed mappings and their confidence are key when performing biased random walks to create sequences of entities from both ontologies. The results section shows promising results where the OWL2Vec4OA embeddings lead to much better-ranking results than those computed by OWL2Vec⋆.

Currently, our experiments rely on direct embedding comparison which leads to good similarity results. However, predicting equivalent or subsumption mappings is a more complex task. In the near future, we aim at training machine learning models to better benefit from the features of the OWL2Vec4OA embeddings for an ontology alignment task. Prominent examples in the literature are LogMap-ML [5], which successfully applied a Siamese Neural Network; and Hao et al. [19], which explored the use of Graph Neural Networks (GNN). These approaches, however, created embeddings that were independent for each input ontology, unlike those computed by OWL2Vec4OA.

In addition, we also plan to conduct additional experiments to better understand the impact of the walk depth with different strategies to create entity sequences (*i.e.*, focusing on concepts and/or avoiding OWL constructs). Entity embedding can also be constructed using the word embedding associated to their labels, which may bring additional features with respect to the URI embeddings. Finally, once we have an end-to-end ontology alignment system in place, we aim to participate in the OAEI campaign and perform an extensive comparison with the state-of-the-art.

Acknowledgements. This research is funded by the Ministry of Education and Science of Azerbaijan Republic with support from City St George's, University of London. This work has also been partially supported by the Academy of Medical Sciences Network Grant (Neurosymbolic AI for Medicine, NGR1\1857), the project "XAI4SOC: Explainable Artificial Intelligence for Healthy Aging and Social Wellbeing" funded by the Agencia Estatal de Investigación (AEI), the Spanish Ministry of Science, Innovation and Universities and the European Social Funds (PID2021-123152OB-C22), the EPSRC project OntoEm (EP/Y017706/1), and the EU Projects: RE4DY (101058384, HORIZON-CL4-2021), Plooto (101092008, HORIZON-CL4-2022), SM4RTENANCE (101123490, DIGITAL-2022), and Tec4MaasEs (101138517, HORIZON-CL4-2023).

References

1. Bodenreider, O.: The Unified Medical Language System (UMLS): integrating biomedical terminology. Nucleic Acids Res. **32**(Database-Issue), 267–270 (2004). https://doi.org/10.1093/NAR/GKH061
2. Cai, H., Zheng, V.W., Chang, K.C.C.: A comprehensive survey of graph embedding: problems, techniques, and applications. IEEE Trans. Knowl. Data Eng. **30**(9), 1616–1637 (2018)
3. Chen, J., He, Y., Geng, Y., Jiménez-Ruiz, E., Dong, H., Horrocks, I.: Contextual semantic embeddings for ontology subsumption prediction. World Wide Web **26**(5), 2569–2591 (2023)
4. Chen, J., Hu, P., Jimenez-Ruiz, E., Holter, O.M., Antonyrajah, D., Horrocks, I.: OWL2vec*: embedding of OWL ontologies. Mach. Learn. **110**(7), 1813–1845 (2021)
5. Chen, J., Jiménez-Ruiz, E., Horrocks, I., Antonyrajah, D., Hadian, A., Lee, J.: Augmenting ontology alignment by semantic embedding and distant supervision. In: The Semantic Web: ESWC, pp. 392–408. Springer (2021)
6. Chen, J., Mashkova, O., Zhapa-Camacho, F., Hoehndorf, R., He, Y., Horrocks, I.: Ontology embedding: a survey of methods, applications and resources. arXiv preprint arXiv:2406.10964 (2024)
7. Cochez, M., Ristoski, P., Ponzetto, S.P., Paulheim, H.: Biased graph walks for RDF graph embeddings. In: Akerkar, R., Cuzzocrea, A., Cao, J., Hacid, M. (eds.) Proceedings of the 7th International Conference on Web Intelligence, Mining and Semantics, pp. 21:1–21:12. ACM (2017). https://doi.org/10.1145/3102254.3102279
8. Cuenca Grau, B., Horrocks, I., Motik, B., Parsia, B., Patel-Schneider, P.F., Sattler, U.: OWL 2: the next step for OWL. J. Web Semant. **6**(4), 309–322 (2008)
9. Dooley, D.M., et al.: FoodOn: a harmonized food ontology to increase global food traceability, quality control and data integration. NPJ Sci. Food **2**(1), 1–10 (2018)
10. Dragoni, M., Bailoni, T., Maimone, R., Eccher, C.: HeLis: an ontology for supporting healthy lifestyles. In: ISWC, pp. 53–69. Springer (2018)
11. Euzenat, J., Meilicke, C., Stuckenschmidt, H., Shvaiko, P., dos Santos, C.T.: Ontology alignment evaluation initiative: six years of experience. J. Data Semant. **15**, 158–192 (2011). https://doi.org/10.1007/978-3-642-22630-4_6
12. Faria, D., Pesquita, C., Mott, I., Martins, C., Couto, F.M., Cruz, I.F.: Tackling the challenges of matching biomedical ontologies. J. Biomed. Semant. **9**(1), 4:1–4:19 (2018). https://doi.org/10.1186/S13326-017-0170-9
13. Faria, D., Pesquita, C., Santos, E., Palmonari, M., Cruz, I.F., Couto, F.M.: The agreementmakerlight ontology matching system. In: On the Move to Meaningful Internet Systems. Lecture Notes in Computer Science, vol. 8185, pp. 527–541. Springer (2013). https://doi.org/10.1007/978-3-642-41030-7_38

14. Fensel, D.A., et al.: Knowledge Graphs: Methodology, Tools and Selected Use Cases. Springer (2020). https://api.semanticscholar.org/CorpusID:210975360
15. Gosselin, F., Zouaq, A.: SORBET: a Siamese network for ontology embeddings using a distance-based regression loss and BERT. In: International Semantic Web Conference, pp. 561–578. Springer (2023)
16. Grover, A., Leskovec, J.: node2vec: scalable feature learning for networks. In: Proceedings of the 22nd ACM SIGKDD International Conference on Knowledge Discovery and Data Mining, pp. 855–864 (2016)
17. Gruber, T.R.: Toward principles for the design of ontologies used for knowledge sharing? Int. J. Hum. Comput. Stud. **43**(5–6), 907–928 (1995)
18. Hao, J., et al.: Medto: medical data to ontology matching using hybrid graph neural networks. In: Proceedings of the 27th ACM SIGKDD Conference on Knowledge Discovery & Data Mining, pp. 2946–2954 (2021)
19. Hao, Z., Mayer, W., Xia, J., Li, G., Qin, L., Feng, Z.: Ontology alignment with semantic and structural embeddings. J. Web Semant. **78**, 100798 (2023). https://doi.org/10.1016/J.WEBSEM.2023.100798
20. Harrow, I., et al.: Ontology mapping for semantically enabled applications. Drug discovery today (2019)
21. He, Y., Chen, J., Antonyrajah, D., Horrocks, I.: BERTMap: a BERT-based ontology alignment system. In: Thirty-Sixth AAAI Conference on Artificial Intelligence (2022)
22. He, Y., Chen, J., Dong, H., Horrocks, I.: Exploring large language models for ontology alignment. In: Fundulaki, I., Kozaki, K., Garijo, D., Gómez-Pérez, J.M. (eds.) Proceedings of the ISWC 2023 Posters, Demos and Industry Tracks. CEUR Workshop Proceedings, vol. 3632. CEUR-WS.org (2023). https://ceur-ws.org/Vol-3632/ISWC2023_paper_427.pdf
23. He, Y., Chen, J., Dong, H., Jiménez-Ruiz, E., Hadian, A., Horrocks, I.: Machine learning-friendly biomedical datasets for equivalence and subsumption ontology matching. In: 21st International Semantic Web Conference. Lecture Notes in Computer Science, vol. 13489, pp. 575–591. Springer (2022). https://doi.org/10.1007/978-3-031-19433-7_33
24. Hertling, S., Paulheim, H.: OLaLa: ontology matching with large language models. In: Venable, K.B., Garijo, D., Jalaian, B. (eds.) Proceedings of the 12th Knowledge Capture Conference (K-CAP), pp. 131–139. ACM (2023). https://doi.org/10.1145/3587259.3627571
25. Hogan, A., et al.: Knowledge graphs. ACM Comput. Surv. (CSUR) **54**(4), 1–37 (2021)
26. Iyer, V., Agarwal, A., Kumar, H.: VeeAlign: multifaceted context representation using dual attention for ontology alignment. In: Proceedings of the 2021 Conference on Empirical Methods in Natural Language Processing (EMNLP), pp. 10780–10792. Association for Computational Linguistics (2021). https://doi.org/10.18653/V1/2021.EMNLP-MAIN.842
27. Jimenez-Ruiz, E., Cuenca Grau, B.: LogMap: logic-based and scalable ontology matching. The Semantic Web - ISWC, vol. 7031 (2011)
28. Jiménez-Ruiz, E., Cuenca Grau, B., Zhou, Y., Horrocks, I.: Large-scale interactive ontology matching: algorithms and implementation. In: Raedt, L.D., et al. (eds.) ECAI 2012 - 20th European Conference on Artificial Intelligence. Including Prestigious Applications of Artificial Intelligence (PAIS-2012) System Demonstrations Track, Montpellier, France, 27–31 August 2012. Frontiers in Artificial Intelligence and Applications, vol. 242, pp. 444–449. IOS Press (2012). https://doi.org/10.3233/978-1-61499-098-7-444

29. Kolyvakis, P., Kalousis, A., Kiritsis, D.: DeepAlignment: unsupervised ontology matching with refined word vectors. In: Proceedings of the 16th Annual Conference of the North American Chapter of the Association for Computational Linguistics: Human Language Technologies, 1–6 June 2018 (2018)
30. Lawrence Page, R.: The PageRank citation ranking: bringing order to the web. Technical report, Stanford Digital Library Technologies Project (1998)
31. Li, H., et al.: User validation in ontology alignment: functional assessment and impact. Knowl. Eng. Rev. **34** (2019)
32. Mikolov, T., Sutskever, I., Chen, K., Corrado, G.S., Dean, J.: Distributed representations of words and phrases and their compositionality. In: Advances in Neural Information Processing Systems, vol. 26 (2013)
33. Otero-Cerdeira, L., Rodríguez-Martínez, F.J., Gómez-Rodríguez, A.: Ontology matching: a literature review. Expert Syst. Appl. **42**(2), 949–971 (2015)
34. Perozzi, B., Al-Rfou, R., Skiena, S.: Deepwalk: online learning of social representations. In: Proceedings of the 20th ACM SIGKDD International Conference on Knowledge Discovery and Data Mining, pp. 701–710 (2014)
35. Pour, M.A.N., et al.: Results of the ontology alignment evaluation initiative 2023. In: Shvaiko, P., Euzenat, J., Jiménez-Ruiz, E., Hassanzadeh, O., Trojahn, C. (eds.) Proceedings of the 18th International Workshop on Ontology Matching (OM). CEUR Workshop Proceedings, vol. 3591, pp. 97–139. CEUR-WS.org (2023). https://ceur-ws.org/Vol-3591/oaei23_paper0.pdf
36. Pour, M.A.N., et al.: Results of the ontology alignment evaluation initiative 2022. In: Shvaiko, P., Euzenat, J., Jiménez-Ruiz, E., Hassanzadeh, O., Trojahn, C. (eds.) Proceedings of the 17th International Workshop on Ontology Matching (OM). CEUR Workshop Proceedings, vol. 3324, pp. 84–128. CEUR-WS.org (2022). https://ceur-ws.org/Vol-3324/oaei22_paper0.pdf
37. Ristoski, P., Rosati, J., Noia, T.D., Leone, R.D., Paulheim, H.: RDF2Vec: RDF graph embeddings and their applications, vol. 10, pp. 721–752 (2019). https://doi.org/10.3233/SW-180317
38. Rossi, A., Barbosa, D., Firmani, D., Matinata, A., Merialdo, P.: Knowledge graph embedding for link prediction: a comparative analysis. ACM Trans. Knowl. Discov. Data **15**(2), 14:1–14:49 (2021). https://doi.org/10.1145/3424672
39. Steenwinckel, B., Vandewiele, G., Agozzino, T., Ongenae, F.: pyRDF2Vec: a Python implementation and extension of RDF2Vec. In: European Semantic Web Conference, pp. 471–483. Springer (2023)
40. Teymurova, S.: OWL2VecOA Resources (2024). https://doi.org/10.5281/zenodo.13217801
41. Vasilevsky, N.A., et al.: Mondo: unifying diseases for the world, by the world. medRxiv (2022). https://doi.org/10.1101/2022.04.13.22273750. https://www.medrxiv.org/content/early/2022/05/03/2022.04.13.22273750
42. Wang, Q., Mao, Z., Wang, B., Guo, L.: Knowledge graph embedding: a survey of approaches and applications. IEEE Trans. Knowl. Data Eng. **29**(12), 2724–2743 (2017)
43. Wei, W., Erenrich, J., Selman, B.: Towards efficient sampling: exploiting random walk strategies. In: AAAI, vol. 4, pp. 670–676. Citeseer (2004)
44. Xiang, C., Jiang, T., Chang, B., Sui, Z.: ERSOM: a structural ontology matching approach using automatically learned entity representation. In: Proceedings of the 2015 Conference on Empirical Methods in Natural Language Processing, pp. 2419–2429 (2015)

Design of an Ontology-Driven Constraint Tester (ODCT) and Application to SAREF and Smart Energy Appliances

Tareq Md Rabiul Hossain Chy[✉], Henon Lamboro, Olivier Genest, Antonio Kung, Cécile Rabrait, Dune Sebilleau, and Amélie Gyrard

Trialog, Paris, France
{tareq.chy,henon.lamboro,olivier.genest,antonio.kung,cecile.rabrait,
dune.sebilleau,amelie.gyrard}@trialog.com

Abstract. The integration of smart energy appliances, such as thermostats, energy-efficient washing machines, and connected lighting, is essential for improving energy management in modern homes and power grids. To maximize their potential, these devices must communicate seamlessly and adhere to standardized protocols. This study introduces Ontology-Driven Constraint Tester (ODCT) designed to ensure interoperability among smart energy appliances. The tester focuses on validating datasets against the Smart Appliances REFerence (SAREF) ontology and its extension for energy systems, SAREF4ENER, both of which were established by ETSI SmartM2M to enhance interoperability in smart energy appliances. ODCT is applied to the Flexible Start use case from the Joint Research Centre's Code of Conduct for Energy Smart Appliances. Our methodology includes: 1) generating relevant datasets, 2) defining SHACL shape constraints, 3) designing a user-friendly web application, and 4) conducting compliance testing. Results show that ODCT effectively identifies deviations from the established ontologies, ensuring data accuracy and interoperability.

Keywords: Semantic Interoperability · Ontology Compliance Testing · SAREF Ontology · SAREF4ENER Ontology · Smart Energy Appliances · SHACL (Shapes Constraint Language) · Flexible Energy Management

1 Introduction

Global warming is accelerating, with the Earth's average temperature projected to rise by 1.5 °C above pre-industrial levels by 2030, primarily due to greenhouse gas emissions from human activities such as burning fossil fuels and deforestation [9]. This increase poses a significant threat, necessitating immediate and substantial reductions in emissions. In response, the European Union has set a goal of achieving carbon neutrality by 2050 as part of the European Green Deal, with an intermediate target of reducing net greenhouse gas emissions by 55% by

2030 compared to 1990 levels[1]. Energy-related activities, which accounted for 71.6% of global greenhouse gas emissions in 2022, are a critical focus for these efforts[2].

To reduce the impact of energy-related activities on global warming, some of the major levers for action are: transition to renewable energy sources, improvement of energy efficiency and development of demand-side flexibility [10]. Flexibility within power systems, defined as the ability to adjust electricity production, storage, or consumption in response to external signals, is crucial for grid stability and the integration of renewable energy [13]. The digitalization of the energy sector plays a vital role in achieving this flexibility, enabling the remote monitoring and management of energy-consuming devices like electric vehicles and smart energy appliances.

Interoperability is essential for effective energy management and the integration of diverse energy systems and devices. Existing standards often fall short in addressing the unique needs of smart energy appliances, necessitating the development of new solutions to ensure seamless communication and data exchange [6]. To address this gap, the Joint Research Centre (JRC) has developed a Code of Conduct (CoC) [7] for Smart Energy Appliances, setting new interoperability standards in Europe.

In response to these challenges, this study proposes the development of an Ontology-Driven Constraint Tester (ODCT) designed to verify whether datasets from smart appliances comply with a given set of ontologies. The current version of ODCT focuses on the Flexible Start use case from the JRC CoC, which is crucial for flexibility management and grid integrity.

ODCT focuses on ensuring ontology-based interoperability without mandating specific protocols or technologies, thereby allowing industries to implement solutions of their choice while ensuring they adhere to the defined ontologies. This flexibility makes ODCT both protocol and technology-agnostic, facilitating seamless integration and bridging between various competing solutions. By ensuring compatibility with the ontologies required by the JRC CoC, including the Smart Appliances REFerence (SAREF) ontology and its energy-specific extension, SAREF4ENER [4,5], ODCT aims to support the European Union's goals of achieving a connected, efficient, and sustainable energy ecosystem.

The paper is structured as follows: Sect. 2 reviews the current state of the art in ontologies within the energy sector, with a particular emphasis on challenges related to semantic interoperability. Section 3 proposes the Ontology-Driven Constraint Tester (ODCT) as a solution to address the challenges of interoperability, and also explains the Flexible Start use case. Section 4 outlines the research methodology, covering dataset generation, SHACL constraint definition, ODCT implementation, and compliance testing. Section 5 presents an in-depth analysis of ODCT's performance, highlighting key findings and offering suggestions for further enhancement of the ODCT's capabilities. Finally,

---

[1] https://commission.europa.eu/strategy-and-policy/priorities-2019-2024/european-green-deal_en, accessed on June 18, 2024.
[2] https://edgar.jrc.ec.europa.eu/report_2023, accessed on July 10, 2024.

Sect. 6 summarizes the research contributions, reflecting on the overall findings, and proposes avenues for future work in the area of ontology compliance and semantic interoperability in the energy sector.

2 Literature Review on Ontology-Based Semantic Interoperability in the Energy Sector

Nowadays, ontology-based solutions are crucial for enhancing semantic interoperability and preserving data meaning. This section reviews key concepts and the state of the art in this field.

2.1 Introduction to Ontologies in the Energy Sector

Ontologies are vital in the energy sector for managing the vast data generated by diverse systems and stakeholders, enabling seamless communication and integration across platforms. An ontology, as defined by Tom Gruber[3], is a formal specification of a shared conceptualization that organizes concepts, properties, and relationships into machine-readable models [11]. Among the ontologies developed for the energy sector, SAREF (Smart Appliances REFerence) [4] and its extension, SAREF for Energy (SAREF4ENER) [5], stand out. Developed under the ETSI SmartM2M initiative and endorsed by the European Union, these ontologies are specifically designed to enhance interoperability in smart appliances and energy systems. They play a critical role in standardizing information exchange, thereby improving grid stability, supporting renewable energy integration, and enhancing overall energy efficiency [1].

In contrast, the World Wide Web Consortium's (W3C) Semantic Sensor Network (SSN) [3] and Sensor, Observation, Sample, and Actuator (SOSA) [8] ontologies offer broader applicability across the Internet of Things (IoT) but require significant customization for effective use in the energy sector. SAREF and SAREF4ENER, with their targeted focus on energy management, offer a more direct and specialized solution for the sector's unique challenges. Table 1 provides a comparison of these ontologies, highlighting key features relevant to the development of an Ontology-Driven Constraint Tester (ODCT).

Based on Table 1, SAREF and SAREF4ENER are the most suitable ontologies for the initial version of our Ontology-Driven Constraint Tester (ODCT). Their focus on smart appliances and energy management, combined with high interoperability and low integration complexity, makes them ideal for this application. While SSN and SOSA are valuable for broader IoT uses, SAREF and SAREF4ENER provide more targeted and efficient solutions for the energy sector, particularly in managing domain-specific data and integrating seamlessly with existing infrastructures. These strengths are essential for advancing interoperability and data exchange in the energy sector. To fully understand the foundation of these ontologies, it is important to consider the broader theoretical framework supporting them. The next Subsect. 2.2 will explore the Semantic

---

[3] https://tomgruber.org/writing/definition-of-ontology, accessed on May 10, 2024.

Table 1. Comparison of Ontologies

Feature	SAREF	SAREF4ENER	SSN	SOSA
Focus Area	Smart appliances	Energy management systems	Sensor networks	IoT observations and actuations
Interoperability	High within smart devices	High within energy systems	Broad IoT compatibility	Broad IoT compatibility
Integration Complexity	Low	Low	Moderate	Moderate
Data Specificity	Device-specific data	Energy-specific data	Detailed sensor data	Flexible observation/action data

Web Layer Cake, which underpins the development and effectiveness of ontologies like SAREF and SAREF4ENER.

2.2 Role of Semantic Web Layer Cake in Ontology-Based Solutions

The Semantic Web Layer Cake[4], is a conceptual model that organizes the key technologies necessary to build an interoperable semantic web. It includes basic data formats like RDF and XML, and advanced layers such as ontologies (e.g., OWL, RDF Schema) and rules (e.g., SHACL) for reasoning and validation. Ontologies within this framework ensure consistent interpretation of data across systems, while SHACL is used to validate RDF data, ensuring compliance with defined ontologies. This conceptual framework underpins the development of robust, ontology-based solutions across various sectors, including energy, where it facilitates effective communication, seamless data integration, and interoperability among diverse platforms and technologies.

2.3 Overview of the SAREF Ontology and Its Extension SAREF4ENER

The Smart Appliances REFerence (SAREF) ontology [4], developed by the European Telecommunications Standards Institute (ETSI), serves as a foundational ontology designed to enable interoperability among smart appliances. SAREF provides a standardized vocabulary and a set of relationships that allow devices from various manufacturers and service providers to communicate seamlessly, ensuring compatibility and effective integration within smart homes and broader energy systems. This ontology addresses the challenge of fragmented communication standards by offering a unified framework that supports the diverse requirements of smart devices.

---

[4] https://wiki.c2.com/?SemanticWebLayerCake, accessed on July 10, 2024.

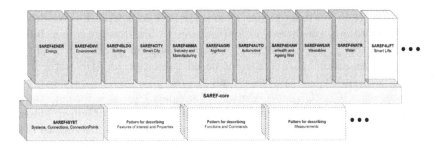

Fig. 1. SAREF Suite of Ontologies with Its Different Extensions (https://hal-emse.ccsd.cnrs.fr/emse-03231685v1/file/EGCIA21-leFrancois.pdf, accessed on June 15, 2024).

Building upon the core principles of SAREF, the SAREF Suite of Ontologies extends its applicability across multiple domains beyond smart appliances. As illustrated in Fig. 1, the SAREF Suite includes specialized extensions tailored to sectors such as Energy, Environment, Smart Cities, and more. Each extension builds on the foundational structure of SAREF, ensuring consistent data exchange and interoperability across a wide range of applications, thereby facilitating the integration of diverse systems within the Internet of Things (IoT) ecosystem. Among these extensions, SAREF for Energy (SAREF4ENER) [5] is particularly significant for the energy sector. SAREF4ENER is specifically designed to address the unique challenges of energy production, consumption, storage, and distribution. It extends the SAREF ontology by incorporating energy-specific concepts and relationships that enable smart energy devices to communicate detailed information about energy usage, flexibility, and load control. By providing a structured framework for energy management, SAREF4ENER plays a crucial role in enhancing the efficiency, reliability, and sustainability of energy systems, making it a key component in advancing semantic interoperability within the energy sector.

2.4 Technologies and Tools for Ontology Compliance Verification

Following the discussion of the Semantic Web Layer Cake in Subsect. 2.2, selecting the appropriate technologies for ensuring ontology compliance is crucial, particularly within the context of RDF-based ontologies like SAREF and SAREF4ENER.

SHACL (Shapes Constraint Language) [12], a W3C standard included in the rules layer of the Semantic Web Layer Cake, is specifically designed for validating RDF graphs. SHACL allows for the definition of constraints that can be directly applied to RDF data, ensuring that the data adheres to the structural and logical requirements set by the ontology. This makes SHACL an essential tool for validating RDF-based ontologies, such as SAREF and SAREF4ENER, which rely on accurate and consistent data structures to facilitate interoperability in the energy sector. To support SHACL's implementation, pySHACL[5], a Python library, provides automated, constraint-based validation of RDF data, further enhancing data quality and interoperability.

In contrast, OCL (Object Constraint Language) [2] is a formal language used to describe constraints and rules that apply to UML (Unified Modeling Language) models. OCL is widely used in model-driven engineering to ensure the semantic correctness of UML diagrams, specifying conditions that cannot be easily expressed within UML alone. OCL is particularly valuable for defining invariants, preconditions, postconditions, and other complex constraints. Table 2 compares the key features of SHACL and OCL for validation purposes.

Table 2. Comparison of SHACL and OCL

Feature	SHACL (Shapes Constraint Language)	OCL (Object Constraint Language)
Target Models	RDF graphs (Semantic Web data)	UML models (Object-oriented models)
Usage Context	Validating RDF data structures and constraints	Verifying correctness of UML diagrams and models
Complexity	Suitable for flexible, graph-based data models	Designed for structured object-oriented systems
Integration	Well integrated with RDF and semantic web tools	Commonly used with UML modeling tools (e.g., Eclipse)
Standardization	W3C standard for RDF graph validation	OMG standard for model-driven engineering

In selecting the appropriate validation framework for our Ontology-Driven Constraint Tester (ODCT), SHACL is the most suitable choice because it is specifically designed for validating RDF graphs, aligning perfectly with the RDF-based structure of SAREF and SAREF4ENER. Unlike OCL, which is primarily designed for UML models, SHACL integrates seamlessly with RDF tools, offers flexibility in defining complex validation rules, and is widely adopted in the semantic web community. Its declarative nature also ensures that validation

---

[5] https://pypi.org/project/pyshacl/, accessed on April 22, 2024.

rules are clear and maintainable, making it the most appropriate and efficient option for our Ontology-Driven Constraint Tester (ODCT).

2.5 Challenges and Gaps in Existing Solutions

The energy sector faces significant challenges in achieving interoperability. Major gaps include:

- **Complexity of Standards and Protocols:** Diverse standards lead to compatibility issues, hindering seamless integration.
- **Data Silos and Incompatible Systems:** Isolation of systems prevents efficient data exchange.
- **Lack of Comprehensive Ontologies:** Limited adoption of ontologies like SAREF restricts standardization.
- **Interoperability with Legacy Systems:** Integration with existing infrastructure requires substantial modification.
- **Scalability and Flexibility Issues:** Current solutions often struggle with large-scale implementations.

Addressing these gaps is essential for advancing semantic interoperability in the energy sector. The proposed solution, detailed in Sect. 3, aims to overcome these challenges by enhancing ontology compliance, supporting the integration of legacy systems, and fostering a connected, sustainable energy ecosystem.

3 Proposed Solution: Ontology-Driven Constraint Tester (ODCT)

To address the challenges of interoperability, data silos, and integration of legacy systems with modern smart devices identified in the Sect. 2, we propose the development of an Ontology-Driven Constraint Tester, referred to as ODCT. It is designed to ensure that smart energy appliances comply with the SAREF ontology [4] and its extension SAREF4ENER [5], thereby facilitating seamless communication and data exchange.

ODCT ensures that smart energy devices adhere to standardized ontological structures, enabling effective communication across devices from different manufacturers and service providers. By verifying compliance with the SAREF ontology, ODCT enhances interoperability, integrates data from diverse sources, and facilitates efficient data sharing, thereby addressing data silos and improving overall system integration.

A critical feature of ODCT is its ability to verify compliance with the Flexible Start use case as outlined in the JRC Code of Conduct for Energy Smart Appliances (ESA). This use case is essential for managing flexibility and ensuring grid stability. In this use case, residential appliances like dishwashers, washing machines, and tumble dryers are managed by a Customer Connectivity Manager (CCM) that optimizes their operation based on cost or CO_2 emissions. ODCT ensures that these appliances can effectively communicate their energy demands and respond to CCM's optimization decisions, thereby supporting grid stability.

Use Case Workflow and Scenarios: The "Flexible Start" use case includes several key scenarios:

- **Pre-announcement Phase**: The smart appliance communicates its expected power sequence and constraints to the CCM.
- **Announcement of Plan**: The appliance announces its current plan, offering one or more power sequences.
- **Optimization by CCM**: The CCM optimizes and potentially shifts the start time to balance energy consumption.
- **Execution Phase**: The appliance executes the optimized power sequence.

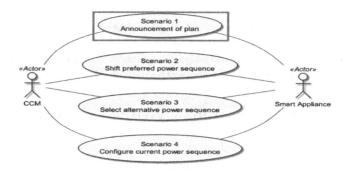

Fig. 2. Scenarios Overview for Flexible Start for White Goods.

ODCT specifically focuses on Announcement of Plan Scenario (see Fig. 2), where the appliance communicates its energy demands to the CCM, setting the stage for subsequent optimization and execution.

Technical Implementation: ODCT leverages existing frameworks, specifically SAREF and its mapping to EN 50361, ensuring compatibility with established standards. As identified in Subsect. 2.4, SHACL (Shapes Constraint Language) [12] and its library PySHACL[6] were selected as the core technologies for ontology compliance verification. This approach provides a robust framework for validating RDF graphs against predefined constraints, guaranteeing data integrity and interoperability, which are crucial for managing smart energy systems.

---

[6] https://pypi.org/project/pyshacl/, accessed on May 12, 2024.

ODCT utilizes key libraries: **rdflib** for converting JSON data to RDF graphs, as it provides a powerful and flexible API for working with RDF data, making it an ideal choice for integrating semantic web technologies into Python. **pySHACL** is used for SHACL validation due to its ease of use and efficient handling of constraint validation, aligning well with the needs of the RDF and SHACL-based compliance process. **reportlab** was selected for generating PDF reports because of its capability to create high-quality, dynamic PDF documents in Python, which is essential for producing detailed compliance reports. **json** is utilized for efficient data handling, as it is a native Python library known for its lightweight and fast parsing of JSON data, which is critical for the large datasets involved. Lastly, **collections** was chosen for ordered data extraction because it provides powerful data structures, such as *OrderedDict*, ensuring that data is processed in a predictable, ordered manner, which is important for compliance checking. These libraries collectively ensure the ODCT's robustness and efficiency. More detailed information about the implementation of SHACL and the utilization of these libraries is provided in Sect. 4.

Workflow for Development: The development of the ODCT follows a structured methodology:

- **Part A: Preparatory Work** involves generating datasets from the "Flexible Start" use case specification to form the foundation for further analysis.
- **Part B: SHACL Shape Definition** focuses on defining SHACL Shape Constraints based on the SAREF (v3.1.1) and SAREF4ENER (v1.2.1) ontologies, tailored to the "Flexible Start" use case.
- **Part C: Web Application Design** covers the development of the ontology engine for data conversion, validation, and compliance report generation, alongside designing a user-friendly web interface.
- **Part D: Compliance Testing** entails validating the generated datasets against the defined SHACL Shape Constraints, ensuring adherence to standards and verifying the ODCT's effectiveness in facilitating interoperability.

This solution provides a comprehensive approach to addressing the interoperability challenges in smart energy management systems, supporting the efficient and sustainable integration of energy resources.

4 Ontology Compliance Verification Methodology

To develop ODCT, this section meticulously details the methodology employed, encompassing dataset generation, SHACL shape constraint definition, ODCT implementation, and compliance testing. The process is structured into four main parts, as illustrated in Fig. 3.

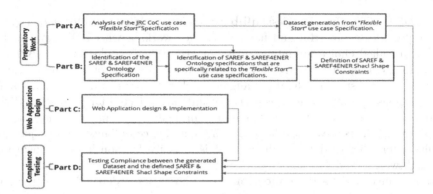

Fig. 3. Methodology of the Proposed Tester

Dataset Generation (Part A): Datasets were generated to represent the "Flexible Start" use case from the JRC Code of Conduct for Energy Smart Appliances, covering phases such as pre-announcement, announcement, optimization, and execution. These datasets were mapped to SAREF (v3.1.1) and SAREF4ENER (v1.2.1) ontologies to ensure comprehensive data representation.

SHACL Shape Constraints Definition (Part B): SHACL shape constraints were defined for validating the generated datasets against the SAREF and SAREF4ENER ontologies. These constraints include property, node, and relationship shapes, ensuring that the datasets adhere to the specified ontological standards.

Web Application Design (Part C): The web application was developed using Python and Django, following the Model-View-Template (MVT) pattern[7] (Fig. 4). The core ontology engine (Fig. 5) converts JSON data into RDF graphs, validates them against SHACL constraints, and generates compliance reports. The user interface allows easy dataset uploads, validation initiation, and report viewing.

Compliance Testing (Part D): Compliance testing involved validating the generated datasets against SHACL shape constraints. JSON datasets were converted to RDF format using `rdflib` and validated with `pySHACL`. The results were categorized into **Total Performance** and **Unique Constraints Coverage** metrics, calculated as follows:

Total Performance: It evaluates the compliance of properties in the JSON dataset:

---

[7] https://docs.djangoproject.com/en/4.0/misc/design-philosophies/#the-mtv-pattern, accessed on May 15, 2024.

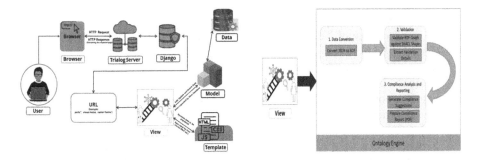

Fig. 4. Web Application Architecture. **Fig. 5.** Ontology Engine.

- **Compliant Properties (C):** Properties that conform to the defined SHACL shapes constraints.
- **Non-Compliant Properties (NC):** Properties that do not conform to the defined SHACL shapes constraints.
- **Not Checked Properties (NCk):** Properties not evaluated during validation.

$$T = \text{Total number of properties in the dataset} \tag{1}$$

$$\text{Percentage of Compliant Properties} = \left(\frac{C}{T}\right) \times 100 \tag{2}$$

$$\text{Percentage of Non-Compliant Properties} = \left(\frac{NC}{T}\right) \times 100 \tag{3}$$

$$\text{Percentage of Not Checked Properties} = \left(\frac{NCk}{T}\right) \times 100 \tag{4}$$

Unique Constraints Coverage: It assesses SHACL constraint validation:

- **Compliant Constraints (CC):** Constraints validated successfully.
- **Non-Compliant Constraints (NCC):** Constraints that failed validation.
- **Not Checked Constraints (NCkC):** Constraints not evaluated.

$$TSC = \text{Total number of unique SHACL constraints} \tag{5}$$

$$\text{Percentage of Compliant Constraints} = \left(\frac{CC}{TSC}\right) \times 100 \tag{6}$$

$$\text{Percentage of Non-Compliant Constraints} = \left(\frac{NCC}{TSC}\right) \times 100 \tag{7}$$

$$\text{Percentage of Not Checked Constraints} = \left(\frac{NCkC}{TSC}\right) \times 100 \qquad (8)$$

Missing Properties refers to the properties in the JSON dataset that are required for validation but are missing or not provided in the dataset. Detailed compliance reports are generated using reportlab[8], providing a comprehensive overview of validation results and identifying necessary corrective actions.

5 Results and Discussion

This section presents the results of the evaluation of the Ontology-Driven Constraint Tester (ODCT), focusing on dataset preparation, SHACL shape definitions, web application design, and compliance checking. Four datasets[9] were assessed: Fully Compliant Dataset, Modified Dataset 1, Modified Dataset 2, and Modified Dataset 3, each designed to evaluate ODCT's robustness under different compliance scenarios.

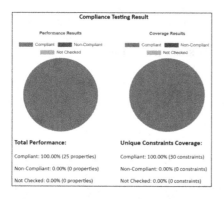

Fig. 6. Result for Compliant Dataset.

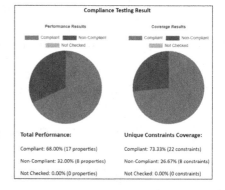

Fig. 7. Result for Modified Dataset 1.

SHACL shape constraints (see Footnote 9) were carefully defined to ensure alignment with SAREF (v3.1.1) and SAREF4ENER (v1.2.1) ontologies, establishing benchmarks for properties, data types, and structural integrity. The Ontology-Driven Constraint Tester (ODCT) features a user-friendly interface, enabling easy dataset upload, validation initiation, and viewing of compliance reports. Users can select the ontology profiling and upload JSON datasets for validation. Upon initiating the compliance check, ODCT compares the dataset against the selected ontology profiling, displaying results through pie charts that

---

[8] https://www.reportlab.com/docs/reportlab-userguide.pdf, accessed on May 13, 2024.
[9] https://zenodo.org/records/13955566.

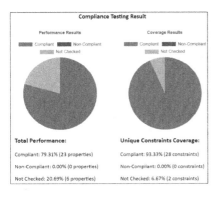
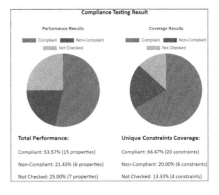

Fig. 8. Result for Modified Dataset 2.   **Fig. 9.** Result for Modified Dataset 3.

illustrate **Total Performance** and **Unique Constraints Coverage**. Users can also clear previous results and download detailed compliance reports.

Compliance testing results are summarized in Figs. 6, 7, 8 and 9, which illustrate the performance of ODCT across the different datasets. The **Fully Compliant Dataset** serves as a benchmark, achieving 100% compliance in both properties and constraints (Tables 3 and 4). **Modified Dataset 1** demonstrates moderate deviations, with 68% compliant properties and 73.33% compliant constraints. **Modified Dataset 2** shows improved performance, with 79.31% compliant properties and 93.33% compliant constraints, though with a notable portion of not checked properties and constraints. **Modified Dataset 3** poses the most significant challenge, with lower compliance and higher non-compliant and not checked categories.

The results demonstrate the ODCT's potential robustness and accuracy in detecting compliance deviations. **Fully Compliant Dataset** confirms the ODCT's accuracy under ideal conditions, as no errors were detected. In contrast, the modified datasets, each containing a specific number and type of errors, showcase ODCT's capability to handle diverse compliance scenarios. For example, **Modified Dataset 1** introduced type mismatches and spelling errors, while **Modified Dataset 2** included extraneous properties and missing required properties. ODCT successfully identified all these deviations (type

Table 3. Comparison of Total Performance

Dataset Type	Compliant Properties (%)	Non-Compliant Properties (%)	Not Checked Properties (%)
Fully Compliant	100.00	0.00	0.00
Modified Dataset 1	68.00	32.00	0.00
Modified Dataset 2	79.31	0.00	20.69
Modified Dataset 3	53.57	21.43	25.00

Table 4. Comparison of Unique Constraints Coverage

Dataset Type	Compliant Constraints (%)	Non-Compliant Constraints (%)	Not Checked Constraints (%)
Fully Compliant	100.00	0.00	0.00
Modified Dataset 1	73.33	26.67	0.00
Modified Dataset 2	93.33	0.00	6.67
Modified Dataset 3	66.67	20.00	13.33

and number of errors), validating its robustness. The **Total Performance** and **Unique Constraints Coverage** metrics offer comprehensive evaluations of adherence to specified standards, with detailed reports indicating the number and type of errors detected, providing valuable insights for manufacturers of smart energy appliances. A demonstration video of ODCT's functionalities, along with detailed reports containing the results of compliance testing for the four datasets, including the specific errors detected, is available online (see Footnote 9).

However, ODCT's limitations must be acknowledged, particularly its support for only JSON datasets and the dependency on comprehensive SHACL shape definitions, which is a time-consuming but critical activity for ODCT's relevance and effectiveness. Defining SHACL shapes for each new use case or test case is one of the most challenging aspects of using ODCT, as it directly impacts the accuracy and applicability of the compliance checks. Despite these constraints, the Ontology-Driven Constraint Tester (ODCT) contributes significantly to standardization and interoperability in the smart appliance and energy management sector, offering an essential resource for quality assurance and compliance validation.

At the current stage of the ODCT development, we generated our own datasets in order to be able to control the type and the number of errors and to rigorously validate the operation of this proof-of-concept tool. Nevertheless, we will now focus on getting real datasets from white goods manufacturers in order to further test ODCT in real-life conditions. Moreover, to ensure transparency and reproducibility, we are considering releasing ODCT as an opensource project on widely used platforms. This will allow the community to access the codebase, contribute to its development, and evaluate its effectiveness across various use cases. By providing open access, external users will be able to conduct more extensive evaluation, which will help strengthen the system's validity. Additionally, we aim to incorporate feedback from a broader range of users. This extended evaluation will provide more robust validation of ODCT's capabilities and overall performance.

6 Conclusion and Future Work

This study presents the development of Ontology-Driven Constraint Tester (ODCT) designed to efficiently test smart energy appliances for adherence to a specific set of ontologies, here SAREF (v3.1.1) and SAREF4ENER (v1.2.1), with a specific focus on the "Flexible Start" use case. Our approach included the generation of datasets, the definition of SHACL shape constraints, and the creation of a user-friendly web application using Python and Django. ODCT demonstrated robust detection capabilities and precision in validating datasets, promoting standardization and interoperability within the smart energy appliance sector. Furthermore, the detailed compliance reports and intuitive interface make ODCT accessible to users with varying levels of technical expertise. ODCT directly addresses several challenges discussed in Subsect. 2.5, including mitigating the complexity of standards and protocols by adhering to ontologies like SAREF and SAREF4ENER, and improving interoperability across systems. It also supports legacy system integration and enhances scalability through flexible SHACL shape definitions, partially overcoming data silos and enabling broader adaptation across use cases.

To enhance ODCT's utility and address current limitations, future work will focus on considering the release of ODCT as an open-source project for the community. This will not only improve its accessibility but also foster collaboration with the broader research and development community. By opening up ODCT, we aim to encourage contributions from external users, which will help further validate and improve the tool. We will explore ODCT's compatibility with other ontologies, as well as different versions of the SAREF ontology, to assess the complexity and effort required for adaptation. This will help refine ODCT for broader applicability across the SAREF framework, while also focusing on enhancing its utility by developing communication interfaces, such as Points of Control and Observation (PCO) defined by ISO/IEC 9646-1:1994[10], to enable direct data collection from Devices Under Test (DUT) via standard protocols like SPINE-IoT for the JRC CoC. This will improve real-time compliance checks for accurate and immediate data validation. Additionally, we will expand compatibility to include formats like XML and CSV, and refine the user interface based on feedback to enhance usability and responsiveness. We also plan to extend ODCT's coverage to other use cases within the JRC Code of Conduct, ensuring it remains adaptable and relevant as standards evolve, thereby contributing to the integration of smart energy systems across the EU. Moreover, ODCT's development will continue to enhance semantic interoperability across EU initiatives by extending its compliance testing to more ontologies beyond SAREF and SAREF4ENER, supporting additional smart grid and home use cases. Collaborating with projects like Hedge-IoT[11], we aim to tailor ODCT to meet specific project needs and contribute to a more efficient energy ecosystem. These improvements will boost ODCT's robustness, versatility, and interoperability across domains.

---

[10] https://www.iso.org/standard/17473.html.
[11] https://hedgeiot.eu/.

References

1. Bergmann, H., et al.: Semantic interoperability to enable smart, grid-interactive efficient buildings (2023)
2. Brucker, A.D., Tuong, F., Wolff, B.: Featherweight OCL: a proposal for a machine-checked formal semantics for OCL. **2**, 5 (2015)
3. Compton, M., et al.: The SSN ontology of the W3C semantic sensor network incubator group. J. Web Semant. **17**, 25–32 (2012)
4. Daniele, L., den Hartog, F., Roes, J.: Created in close interaction with the industry: the smart appliances reference (SAREF) ontology. In: Formal Ontologies Meet Industry: 7th International Workshop, FOMI 2015, Berlin, Germany, 5 August 2015, Proceedings 7, pp. 100–112. Springer (2015)
5. ETSI: SmartM2M; extension to SAREF; part 1: Energy domain. Technical report, ETSI, Sophia Antipolis Cedex, France (2020). https://www.etsi.org/standards-search, eTSI TS 103 410-1 V1.1.2 (2020-05)
6. European Commission: Best practices for energy data sharing. Online event, 16 February at 09.00 – 12.30 (2024). http://europa.eu.int/comm/dgs/energy_transport/index_en.html, workshop conclusions
7. European Commission: Code of Conduct on Energy Management Related Interoperability of Energy Smart Appliances (v.1.0). Online (2024). https://energy.ec.europa.eu/document/download/36078249-1100-4e31-b859-df85272d0d4f_en, version 1.0
8. Janowicz, K., Haller, A., Cox, S.J., Le Phuoc, D., Lefrançois, M.: SOSA: a lightweight ontology for sensors, observations, samples, and actuators. J. Web Semant. **56**, 1–10 (2019)
9. Lee, H., et al.: Synthesis report of the IPCC Sixth Assessment Report (AR6), Longer report. IPCC (2023)
10. Lund, P.D., Lindgren, J., Mikkola, J., Salpakari, J.: Review of energy system flexibility measures to enable high levels of variable renewable electricity. Renew. Sustain. Energy Rev. **45**, 785–807 (2015)
11. Noy, N.F., McGuinness, D.L.: Ontology Development 101: A Guide to Creating Your First Ontology. Technical report, Stanford Knowledge Systems Laboratory Technical Report KSL-01-05 and Stanford Medical Informatics Technical Report SMI-2001-0880 (2001). https://protege.stanford.edu/publications/ontology_development/ontology101.pdf
12. Pareti, P., Konstantinidis, G.: A review of SHACL: from data validation to schema reasoning for RDF graphs. In: Reasoning Web International Summer School, pp. 115–144 (2021)
13. Reif, V., et al.: Towards an interoperability roadmap for the energy transition. e & i Elektrotechnik und Informationstechnik **140**(5), 478–487 (2023)

YOKO ONtO: You only KNIT One Ontology

Jorge Rodríguez-Revello[1](✉)[ID], Cristóbal Barba-González[1][ID], Maciej Rybinski[1,2][ID], and Ismael Navas-Delgado[1][ID]

[1] KHAOS Research, ITIS Software, Universidad de Málaga, 29071 Málaga, Spain
{rodriguezrevellojj,cbarba,ismael}@uma.es
[2] CSIRO Data61, Sydney, NSW, Australia
maciek.rybinski@data61.csiro.au

Abstract. Ontologies have become fundamental for knowledge representation in various fields, especially in the Life Sciences. These ontologies are usually published in formats such as OWL or OBO and stored in open repositories such as BioPortal. This article presents YOKO ONtO, an innovative interface enhancing the KNIT ontology construction tool. KNIT automates the exploration of BioPortal to facilitate the reuse of ontologies in the Life Sciences domain, using a set of keyword queries as input. YOKO ONtO extends KNIT's capabilities by adding a visualization layer that generates a draft knowledge graph, highlighting the concepts and relationships extracted from existing ontologies. This visual representation makes it easier for users to understand and navigate the complex relationships within the retrieved ontological data, allowing the creation of new ontologies by reusing existing ones tailored to specific use cases.

Keywords: Knowledge graphs · Ontology reuse · Life sciences · BioPortal

1 Introduction

In recent years, the use of ontologies across scientific fields has increased, with ontologies becoming an essential tool for representing knowledge in some domains. Ontologies allow for the precise formalisation of concepts and their interrelations within a given field of study [6]. Ontologies offer a structured framework that facilitates both the organisation and exchange of information between researchers, thus driving an increased research activity. Typically, ontologies consist of a limited and carefully curated set of statements to describe specific

---

This work has been partially funded by the Spanish Ministry of Science and Innovation via grant (MCIN/AEI/10.13039/501100011033/) PID2020-112540RB-C41, AETHER-UMA (A smart data holistic approach for context-aware data analytics: semantics and context exploitation) and the Spanish Ministry of economic affairs and Digital Transformation project TSI-063000-2021-25 (5G+TACTILE-3).

© The Author(s), under exclusive license to Springer Nature Switzerland AG 2025
S. Tiwari et al. (Eds.): KGSWC 2024, LNCS 15459, pp. 199–211, 2025.
https://doi.org/10.1007/978-3-031-81221-7_14

problem space in a well-defined context. However, the proliferation of data creation has led to an increased use of knowledge graphs, especially in disciplines like Life Sciences, where vast amounts of specialised information are managed. These graphs not only represent domain-specific information but also reveal connections between different domains [2].

Knowledge graphs are especially useful in scenarios requiring data integration and organisation from multiple sources. These data structures consist of nodes and edges: nodes (or vertices) represent key entities or concepts, while edges (or links) capture the semantic relationships between them [8]. As a result, knowledge graphs not only facilitate the integration of heterogeneous information but also enable the inference of new relationships from existing data, which is invaluable in complex research.

The field of Life Sciences, which covers the study of plant and animal organisms and the biological processes that characterise them, has been one of the most benefiting from the use of semantic technologies, and in particular, from the adoption of ontologies [7]. Accuracy in knowledge representation is essential to advancing understanding of natural phenomena and promoting data exchange between different research groups. Despite its importance, the development of ontology still faces significant challenges. One of the main challenges is the alignment between the knowledge of experts in specific domains and the difficulty of using knowledge engineering technology needed to capture the relevant concepts and their relationships [4]. This task is further complicated by the heterogeneity of existing conceptualizations and the difficulties in accessing previously developed ontologies.

Several methodologies have been proposed to address the challenges in the design and development of ontologies. The Ontology 101 [12] methodology, for example, offers a structured approach that includes key steps such as defining the domain and scope of the ontology. On the other hand, other methodologies [5] strongly emphasise the reuse of existing ontologies, which not only improves the efficiency of the development process, but also promotes coherence between different research projects. This reuse practice is a standard recommendation in the field of ontology engineering, as it contributes to reducing duplication of efforts and increasing the interoperability between different ontological systems.

Access to already developed ontologies has been facilitated by the use of specialised repositories such as Bioportal [13], OBO Foundry and OLS [15], which allow researchers to search and access a wide range of pre-existing ontologies. However, one of the persistent problems is the lack of tools that automate the design process, guiding developers towards reusing knowledge stored in these repositories. This lack of tools makes it difficult to comply with the FAIR principles [17], which promote Findability, Accessibility, Interoperability and Reusability of data. To improve this situation, several approaches have emerged in recent years that support semi-automation in the development of ontologies [1,9]. Among these tools, KNIT [14] stands out for its ability to facilitate the automatic reuse of concepts across different ontologies, thus optimising the creation process. However, despite its advantages, KNIT has the lacks a graphical user interface, which essentially limits its user-base to those familiar with knowledge engineering tools (such as programming skills).

Here we present YOKO ONtO, a tool that integrates the power of KNIT with an intuitive visual platform, more accessible to domain experts (as opposed to knowledge engineers). The main contributions of this work can be summarised as follows:

- We propose an open-source tool providing a visual interface for KNIT for supporting the design of ontologies through the reuse of concepts from existing ontologies in BioPortal.
- We demonstrates the tool's functionality through an example set in the context of Life Science (i.e., medicine).

2 Background

2.1 BioPortal and Ontology Recommender 2.0

BioPortal is an online platform that provides access to an extensive collection of biomedical ontologies and terminologies curated by the National Center for Biomedical Ontology (NCBO) [16]. As a comprehensive resource, BioPortal allows researchers and professionals to perform various tasks related to ontology management. Users can download ontologies directly for their projects, enabling integration into their research. The platform also facilitates information retrieval functionalities, allowing searching for specific terms within the available ontologies. Additionally, BioPortal supports extracting information and terms from the central ontology registry, further enhancing its utility for research and development.

As one of the key features, BioPortal provides mappings of concepts across different ontologies, which is crucial for ensuring data alignment and interoperability. This mapping capability is complemented by the platform's support for multiple representation formats, accommodating various research and application needs. Moreover, BioPortal enables users to create and manage personalised views of ontologies, allowing them to tailor these resources to their specific requirements.

To enhance ontology retrieval, BioPortal incorporates the NCBO Ontology Recommender 2.0 system [11]. This search system is designed to suggest the most relevant ontologies for a given biomedical text. It scores ontologies based on several criteria, including how well they address the user's needs through concept coverage, the level of acceptance within the biomedical community, the depth and detail of the ontology's classes, and the extent to which the ontology is specialised for the domain of the input terms.

To ensure that the recommended ontologies meet specific user needs, the search process is customisable, allowing users to adjust the importance of each scoring criterion. The system also provides explanatory information to help its users understand the reasoning behind the ranking, making it easier to select the best-suited ontologies for their research. The Ontology Recommender 2.0 is accessible through BioPortal's web interface[1] and a RESTful API[2].

2.2 KNIT

The KNIT method, proposed in our prior work [14], is designed to automate the identification of relevant ontologies and the extraction of their concepts, attributes, and entities from an ontology repository. This approach generates a new OWL ontology by iteratively incorporating axioms from existing ontologies based on a specified list of keywords. KNIT's current implementation uses BioPortal as back-end ontology repository.

Creating an OWL ontology with the KNIT method involves a series of steps. The process begins with the retrieval of concepts. A list of keywords serves as the input. The retrieval process is based the functionalities of the chosen ontology repository. In the curren BioPortal-based implementation, each keyword is used to retrieve a set of candidate concepts from the registered ontologies.

Following the candidate retrieval, the selection of concepts occurs, where the candidate concepts identified in the previous step are ranked to align each input keyword with the a single candidate. In the current implementation of KNIT, the method emphasises retrieving many concepts from the same ontology. The underlying assumption is that ontologies containing more concepts relevant to the user's keywords are more completely aligned with the user's domain of interest; the matching process prioritises candidate concepts from ontologies with more candidate hits.

Once the preferred candidate concepts are selected, the taxonomy is constructed. This is achieved by importing parent concepts from the source ontologies and building the hierarchical structure by tracing the path to the root concept in each reused ontology. Finally, the taxonomy is enriched with additional information from the reused ontologies. This enrichment involves incorporating data and object properties, applying restrictions, adding instances adjacent to the previously imported concepts.

For a detailed description of the KNIT algorithm and its steps readers can refer to the original paper [14].

3 YOKO ONtO Tool

The purpose of YOKO ONtO[3] is to facilitate research use of KNIT for ontology creation. This tool, built with Streamlit and encapsulated in Docker, presents an intuitive graphical user interface (GUI) to explore the functionality of KNIT. This GUI allows users to retrieve, explore, and interact with ontologies from BioPortal to build new ontologies by reusing axioms from existing resources.

The YOKO ONtO interface is designed to accommodate improvements that will be made to KNIT in the future. Currently, the tool has three blocks:

---

[1] https://bioportal.bioontology.org/recommender.
[2] https://data.bioontology.org/documentation#nav_recommender.
[3] https://github.com/ProyectoAether/yoko_onto.

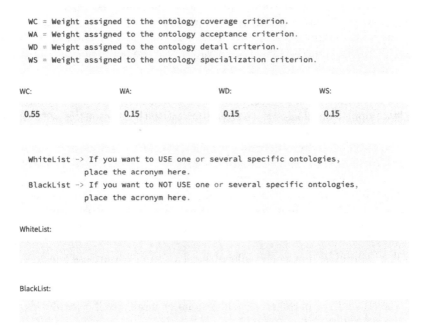

Fig. 1. YOKO ONtO form where the optional parameters of the tool are set. The form for search terms is shown below.

- *Setup* block: This block has been designed to specify different need parameters: the BioPortal REST API Key and the Neo4j connection (to be used as a database to store the output knowledge graph).
- *Retrieval parameters* block (Fig. 1): This is an optional block to set up parameters of the KNIT processing when using BioPortal. These parameters affect KNIT's interaction with BioPortal retrieval mechanisms. Specifically, this block allows the user to configure six parameters:
  1. The 'WC' parameter assigns a weight to the ontology coverage criterion (the default value is 0.55). Larger values will promote selecting ontologies that cover more terms of the user's input terms.
  2. The 'WA' parameter assigns a weight to the 'community acceptance' criterion of the ontology (default value of 0.15). Larger values will filter out less popular ontologies as they will prioritise those resources with higher acceptance indicators (such as visits).
  3. The 'WD' parameter assigns a weight to the detail criterion of the ontology (default value of 0.15). Larger values will give priority to semantically richer ontologies.
  4. The 'WS' parameter assigns weight to the specialization parameter of the ontologies (default value of 0.15). Larger values will prioritise those ontologies that are more in line with the search terms across all the concepts.
  5. The 'WhiteList' parameter allows the algorithms to only use one or several specific ontologies from the BioPortal repository (i.e., those included in the white list).
  6. The 'BlackList' parameter excludes one or more specific ontologies from the repository.
- *KNIT keywords* block (Fig. 1): this block is designed as a search tool where users will provide the key terms they want to be represented in the newly created ontology.

With this interface, YOKO ONtO facilitates the calibration of the four steps of KNIT and exploration of the results.

4 Use Case

In this section we illustrate how our tool can be used in a practical scenario. We have used the Semantic Web Challenge on matching tabular data with knowledge graphs[4]. For this case study, the we used the default values of the configuration parameters (WC: 0.55, WA: 0.15, WD: 0.15 and WS: 0.15) as shown in Fig. 1.

4.1 Retrieval of Concepts

Imagine that we are life sciences scientific, and we want one ontology with the elements *Medicine,ATC Code,Route of administration* (column names of the *sem-tab* data set[5].

---

[4] https://www.cs.ox.ac.uk/isg/challenges/sem-tab/2021/index.html.
[5] https://raw.githubusercontent.com/ProyectoAether/yoko_onto/master/data/dataset.csv.

Finding the relevant concepts for these keywords is the first step. KNIT retrieves, for each element, the first concept, classified by similarity, from each of the source ontologies in BioPortal, thus obtaining candidate concepts. These candidate concepts are then ranked based on the coverage of their source ontologies in descending order. We define coverage as the number of candidate concepts retrieved from the same ontology. We then assign a leading candidate to each of the input terms and convert the selected set of concepts into a taxonomy by recursively retrieving the parent concepts from the respective source ontologies, for each of the concepts. Properties are retrieved for each of the taxonomy concepts.

YOKO ONtO makes KNIT easier to use by allowing easy customisation of input parameters, but also more transparent through visualisation of the retrieved candidate concepts. The search conducted in this use case produces a list of 25 ontologies with their respective candidate concepts. Before the candidate selection the user can inspect the results, and potentially refine their query.

Table 1. Candidate concepts obtained from the different ontologies

Keywords	Data found
route of administration	**Found class**: Route of Administration **URI**: http://ncicb.nci.nih.gov/xml/owl/EVS/Thesaurus.owl#C38114 **Ontology**: NCIT
atc code	**Found class**: ATC code **URI**: http://edamontology.org/data_3103 **Ontology**: EDAM
medicine	**Found class**: Medicine **URI**: http://ncicb.nci.nih.gov/xml/owl/EVS/Thesaurus.owl#C1683 **Ontology**: NCIT

4.2 Selection of Concepts

Concept selection provides a ranking of these ontologies based on the number of input terms represented in them. For our use case, this information is presented in Table 2.

Table 2. Candidate concepts obtained from the different ontologies

Ontology	Terms found
NCIT	medicine, route of administration
SNOMEDCT	route of administration
GSSO	medicine, route of administration
INBANCIDO	medicine, route of administration
VICO	route of administration
KTAO	medicine, route of administration
EDAM	medicine, atc code
CIDO	medicine, route of administration
OVAE	route of administration
COID	route of administration
EUPATH	route of administration
VO	route of administration
OCIMIDO	route of administration
RADLEX	route of administration
OCHV	medicine, route of administration
RCD	route of administration
MESH	medicine
MIDO	route of administration
PREMEDONTO	medicine
EXO	medicine
COVID-19	medicine

The results of the candidate selection are presented in Table 1. At this stage, YOKO ONtO visualizes the selected concepts together with their provenance (i.e., the source ontologies) (Fig. 2). This informs the user about the retrieval results; this information can be then used to stop and re-adjust the process (for example, through modifying the keywords or providing optional parameters).

4.3 Construction of the Taxonomy

The list of selected concepts is then used to produce the taxonomy (Fig. 3). In this use case, the taxonomy includes the hierarchies from the two identified ontologies, starting from the ontology classes matched with the search keywords. Thus, the parts of the reused ontologies that are not related to the user's query are not imported in this step.

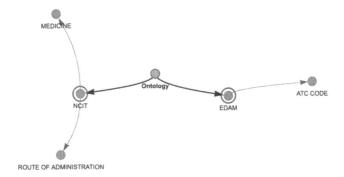

Fig. 2. Visualisation of the ontologies with selected concepts. Nodes with red circles denote the source ontologies. (Color figure online)

Table 3. Complete data for element of 'Route of Administration'

Keywords	Route of Administration
Definition	Designation of the part of the body through which or into which, or the way in which, the medicinal product is intended to be introduced. In some cases a medicinal product can be intended for more than one route and/or method of administration
Label	Route of Administration
Ontology	NCIT
Synonym	ROUTE_DETAIL, Drug Administration Method, route of administration, route_of_administration, Route of Drug Administration, route of administration (ROA), Route of Administration, ROUTE, Drug Route of Administration, ROUTE OF ADMINISTRATION
URI	http://ncicb.nci.nih.gov/xml/owl/EVS/Thesaurus.owl#C38114

4.4 Enrichment

Once the taxonomy construction is finished, the ontology is enriched. Figure 4 shows the visualisation provided to the user for our use case. In this step, the knowledge graph is enriched with the data and object properties of the concepts represented in the taxonomy. Whenever a new concept is needed due to the addition of the object properties, it is added to the graph. Finally, the enriched graph is translated to OWL and can be downloaded and used in other tools such as Protégé (Fig. 5).

As can be seen in the Fig. 5, three concepts were retrieved: 'Route of administration' (Table 3) and 'Medicine' (Table 4) from the NCIT ontology [10] and ATC Code (Table 5) from the EDAM ontology [3]. All the information retrieved around these nodes of both ontologies is used by YOKO ONtO to generate the knowledge graph.

Fig. 3. Screenshot with the taxonomic knowledge graph with all the terms and relationships found. This knowledge graph is included in the final ontology.

Table 4. Complete data for element of 'Medicine'

Keywords	Medicine
Definition	The branches of medical science that deal with nonsurgical techniques
Label	Medicine
Ontology	NCIT
Synonym	Medicine, medicine
URI	http://ncicb.nci.nih.gov/xml/owl/EVS/Thesaurus.owl#C16833

Table 5. Complete data for element of 'ATC Code'

Keywords	ATC code
Definition	Unique identifier of a drug conforming to the Anatomical Therapeutic Chemical (ATC) Classification System, a drug classification system controlled by the WHO Collaborating Centre for Drug Statistics Methodology (WHOCC)
Label	ATC code
Ontology	EDAM
URI	http://edamontology.org/data_3103

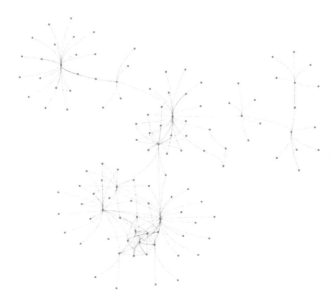

Fig. 4. Visualisation of the knowledge graph after the taxonomy enrichment.

5 Conclusions

In this work, we present YOKO ONtO, a graphical interface to interact with KNIT, thus allowing for the reuse of ontologies in ontology design in a simple and versatile way.

YOKO ONtO allows a novel and agile knowledge exploration of KNIT results, where the researcher can examine and evaluate the (inermediate and final) outcomes KNIT algorithm produces in the automatic creation of ontologies by reusing existing concepta. This allows the user to quickly prototype multiple ontologies for the same study, effectively testing the downstream impact of their decisions (e.g., using only well-established resources vs all ontologies).

Of note, YOKO ONtO is built so that we can easily accommodate future extensions and improvements made to the KNIT methodology. This quality presents a main line of future work: allowing YOKO ONtO to use alternative ontology repositories, in addition to BioPortal, e.g., the SIFRBioPortal[6], AgroPortal[7], EcoPortal[8], MedPortal[9], MatPortal[10], and IndustryPortal[11].

Additionally, our tool is open-source and it can be downloaded and locally installed. Therefore, it can be configured to use other (custom) ontology reposi-

---

[6] https://bioportal.lirmm.fr/.
[7] https://agroportal.lirmm.fr/.
[8] https://ecoportal.lifewatch.eu/.
[9] http://medportal.bmicc.cn/.
[10] https://matportal.org/.
[11] http://industryportal.enit.fr/.

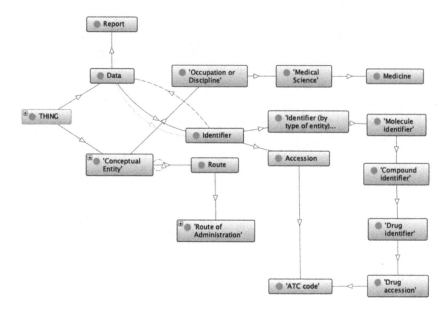

Fig. 5. Protégé screenshot of the ontology with the classes and properties created by YOKO ONtO. It includes terms related to Medicine, ATC Core, and Route of Administration like Data or Drug Identifier from different ontologies, such as, EDAM and NCIT ontologies.

tories. Thus, it is not limited to the online demo version, but can be potentially adapted to diverse scenarios

References

1. Al-Aswadi, F.N., Chan, H.Y., Gan, K.H.: Automatic ontology construction from text: a review from shallow to deep learning trend. Artif. Intell. Rev. **53**, 3901–3928 (2020)
2. Auer, S., Kovtun, V., Prinz, M., Kasprzik, A., Stocker, M., Vidal, M.E.: Towards a knowledge graph for science. In: Proceedings of the 8th International Conference on Web Intelligence, Mining and Semantics, pp. 1–6 (2018)
3. Black, M., et al.: EDAM: the bioscientific data analysis ontology (update 2021) [version 1; not peer reviewed] (2021)
4. Fernández-López, M., Poveda-Villalón, M., Suárez-Figueroa, M.C., Gómez-Pérez, A.: Why are ontologies not reused across the same domain? J. Web Semant. **57**, 100492 (2019)
5. Gómez-Pérez, A., Rojas-Amaya, M.D.: Ontological reengineering for reuse. In: Knowledge Acquisition, Modeling and Management: 11th European Workshop, EKAW'99, Dagstuhl Castle, Germany, 26–29 May 1999, Proceedings 11, pp. 139–156. Springer (1999)
6. Guarino, N.: Formal ontology, conceptual analysis and knowledge representation. Int. J. Hum Comput Stud. **43**(5–6), 625–640 (1995)

7. Hitzler, P.: A review of the semantic web field. Commun. ACM **64**(2), 76–83 (2021)
8. Hogan, A., et al.: Knowledge graphs. ACM Comput. Surv. (CSUR) **54**(4), 1–37 (2021)
9. Kaushik, N., Chatterjee, N.: Automatic relationship extraction from agricultural text for ontology construction. Inf. Process. Agric. **5**(1), 60–73 (2018)
10. Kumar, A., Smith, B.: Oncology ontology in the NCI thesaurus. In: Artificial Intelligence in Medicine: 10th Conference on Artificial Intelligence in Medicine, AIME 2005, Aberdeen, UK, 23–27 July 2005, Proceedings 10, pp. 213–220. Springer (2005)
11. Martínez-Romero, M., Jonquet, C., O'connor, M.J., Graybeal, J., Pazos, A., Musen, M.A.: NCBO ontology recommender 2.0: an enhanced approach for biomedical ontology recommendation. J. Biomed. Semant. **8**(1), 1–22 (2017)
12. Noy, N.F., McGuinness, D.L., et al.: Ontology development 101: a guide to creating your first ontology (2001)
13. Noy, N.F., et al.: BioPortal: ontologies and integrated data resources at the click of a mouse. Nucleic Acids Res. **37**(suppl_2), W170–W173 (2009)
14. Rodríguez-Revello, J., Barba-González, C., Rybinski, M., Navas-Delgado, I.: KNIT: ontology reusability through knowledge graph exploration. Expert Syst. Appl. **228**, 120239 (2023)
15. Smith, B., et al.: The OBO foundry: coordinated evolution of ontologies to support biomedical data integration. Nat. Biotechnol. **25**(11), 1251–1255 (2007)
16. Whetzel, P.L., et al.: BioPortal: enhanced functionality via new web services from the national center for biomedical ontology to access and use ontologies in software applications. Nucleic Acids Res. **39**(suppl_2), W541–W545 (2011)
17. Wilkinson, M.D., et al.: The fair guiding principles for scientific data management and stewardship. Sci. Data **3**(1), 1–9 (2016)

CoKGLM: Detecting Hallucinations Generated by Large Language Models via Knowledge Graph Verification

Rie Hasegawa[✉] and Ryutaro Ichise

Institute of Science Tokyo, Tokyo 152-8550, Japan
hasegawa.r.ak@m.titech.ac.jp, ichise@iee.e.titech.ac.jp

Abstract. In today's business landscape, deploying Large Language Models (LLMs) for company tasks has become prevalent. However, leveraging LLMs for enterprise tasks poses significant challenges, primarily due to hallucinations. These hallucinations can lead to misinformation and poor decision-making. Additionally, internal data that cannot be directly memorized by public LLMs further complicates the utilization process. To address these issues, we propose a novel framework CoKGLM for hallucination detection that combines Knowledge Graphs (KGs) and Language Models (LMs). This approach enhances the use of both textual and structural information from KGs and thoroughly examines desirable KG information for hallucination detection. Our experimental results demonstrate a 38% improvement in hallucination detection and an 84% increase in robustness compared to using LMs alone. This approach enhances transparency and control over content verification while fully utilizing KG data.

Keywords: Large Languages Models · Knowledge Graph · Hallucination Detection · Natural Language Processing · Dialogue system

1 Introduction

In today's business environment, the deployment of Large Language Models (LLMs) [10,17] for tasks such as document drafting and overall business process optimization has become increasingly common [19]. However, leveraging LLMs for these enterprise tasks presents significant challenges, primarily due to Hallucination Problem. Hallucinations refer to factually incorrect or nonsensical responses, undermine the trustworthiness and reliability of LLMs [12]. This issue can lead to misinformation, flawed business reports, and poor decision-making, ultimately damaging a company's reputation and operational effectiveness.

Hallucinations in LLMs can be attributed to several key factors. The primary causes include flawed training data, knowledge boundaries, and the lack of domain-specific knowledge [12]. In enterprise settings, these challenges are amplified, as LLMs may be required to generate text that reflects proprietary information or specialized domain knowledge. While fine-tuning LLMs with proprietary internal data can help address these issues to some extent, it comes with significant costs and often makes it difficult to integrate with pre-existing knowledge [9]. This can increase the likelihood of hallucinations. Additionally, the inherent problem of hallucinations persists, as LLMs predict the next probable token in a sequence based on statistical patterns, rather than factual accuracy. Consequently, even fine-tuned models can still generate incorrect responses, which complicates their effective deployment in high-stakes business scenarios. Given these limitations, it is crucial to develop strategies that can detect and mitigate hallucinations by leveraging internal corporate data. Rather than relying solely on fine-tuning or trusting the outputs of LLMs, organizations can use reliable internal data to cross-check and verify the LLM-generated responses, ensuring secure deployment of LLMs in enterprise settings.

In this work, we propose a framework CoKGLM (Combining Knowledge Graphs (KGs) and Language Models (LMs)) for detecting factuality hallucinations. This approach leverages the textual and structural information of KGs and the deep linguistic capabilities of LMs. Rather than incorporating KG data directly into large-scale public LLMs, we integrate a smaller, local LM with internal corporate KGs to develop a hallucination detection system. This approach helps maintain data integrity and enhances the overall effectiveness of LLM applications in enterprise settings. Our framework consists of two main components: KG Retrieval and Fact Verification. The KG Retrieval component extracts the most relevant information from KG based on the LLM-generated responses. The Fact Verification component verifies the factual accuracy of the LLM-generated responses by comparing it to the retrieved KG information.

Notably, we employed an embedding-based approach for both KG retrieval and verification. With the help of embeddings, we effectively linked the semantic meaning of text and KG triples, integrating the two sources of information. Our approach aims to enhance transparency and precise control over the content verification process while fully utilizing the KG's textual and structured data. We conducted several experiments to address two key research questions: *Can KGs enhance the hallucination detection capabilities of LMs?* and *What types of KG information are most effective for hallucination detection?* The experimental results show that combining KGs and LMs improves hallucination detection and robustness. Our contributions can be summarized as follows.

- Novel Practical Approach: We propose a practical method for hallucination detection using a combined approach of KGs and LMs, enabling the safe and cost-effective use of internal data in LLM applications. This approach can lead to improved operational efficiency and secured LLM deployment.
- Enhanced KG Utilization: We advance the research on leveraging KGs for hallucination detection by introducing a method that fully utilizes the structured data properties of KGs, rather than relying entirely on LLMs for detection.
- Comprehensive Examination of KG Information: We systematically examined the most desirable KG information for hallucination detection in LLMs, in terms of the number of triples, retrieval methods, and integration techniques.

In the following, we will introduce related work in Sect. 2, provide an explanation of the problem definition in Sect. 3, describe each part of CoKGLM in Sect. 4, and present the experiment settings and results in Sect. 5.

2 Related Work

2.1 Hallucination Detection

Prior works on hallucination detection in LLMs can be categorized into two main approaches including *Uncertainty Estimation* and *Retrieving External Knowledge* [12].

The first approach is *Uncertainty Estimation*. This approach assesses the model's uncertainty about the content it generates. Notable studies in this field include "Bartscore" [20], "Semantic Uncertainty" [14]. These works provide valuable insights but requiring access to the model's internal state, which can be restrictive as not all models grant such access.

The second approach is *Retrieving External Knowledge*. This approach involves extracting facts from external sources such as web searches and structured database queries. Many recent works reveal that recognizing and retrofitting hallucinations can be significantly improved by providing external knowledge [4,7,11,15]. These studies predominantly leverage LLMs as detectors and retrofitters. For instance, previous works [3,11] have proposed frameworks for verifying factual inaccuracies in LLM-generated response. Their approaches use LLMs for each step of knowledge retrieval, verification, and correction in an end-to-end prompt manner. However, relying on LLMs as self-detectors and self-correctors can introduce further hallucinations at each stage. Moreover, LLMs inherently have limitations due to their own flawed training data and knowledge boundaries. This reliance on LLMs makes it difficult to ensure the accuracy and reliability of the verification.

There also have been continuous research efforts in leveraging knowledge graphs (KGs) as external knowledge sources. KGs can provide structured, curated factual data for content verification [13]. Such integration facilitates the capture of intricate relationships and enables more accurate cross-referencing and validation of generated content [1]. Specifically, when using KGs for hallucination detection, it is essential to integrate the structured data of the KG with the LLM-generated responses for effective cross-referencing. The methods for this integration process are explained below.

2.2 Integrate KGs into LLMs

For integrating KG's graph data with text data, there are two main approaches: the graph-centric approach and the text-centric approach. The graph-centric approach focuses on converting the text into a KG format and then using graph-based methods to address hallucinations. A study [8] that adopts this approach involves building a response KG and a reference KG, followed by applying a graph neural network (GNN) method to detect factual discrepancies. Recently, the text-centric approach has become more mainstream. Our work also focus on text-centric approach. However, many studies [2,3,7,11,18] using direct prompting to provide KG information to LLMs in natural language form, While these method is simple and efficient, it has limitations, such as the black-box nature of the model's decision-making process and decreased efficiency in cross-referencing text and graph data when the prompt length increases [5]. These approach may not fully utilize the structured data of the knowledge graph, potentially leading to decreased model inference efficiency. Therefore, a different approach that fully leverages the structured data of the KG to ensure more accurate and efficient model inference is needed. We adopt an embedding-based approach, inspired by related research on integrating KGs into LLMs, such as RHO(ρ) [13], which enhances text generation quality by integrating a knowledge graph processing layer into the Transformer architecture. This method addresses the limitations of previous studies and aims to improve inference efficiency by leveraging structured data from KGs. Different from RHO(ρ), which focuses on knowledge-grounded dialogue generation to reduce hallucinations, our approach detects hallucinations in existing responses using a trusted internal corporate KG. We extend the RHO (ρ) encoder to serve as a hallucination detector, verifying the factual accuracy of the response text based on retrieved KG information.

3 Problem Definition

This paper focuses on the hallucination detection task, utilizing KGs to identify factual inaccuracies within LLM-generated responses. The data required for this task includes the text data and the related KG. We use dialogue data, where each sample consists of a multi-turn dialogue history and a response. Only the response contains factual inaccuracies, which we treat as hallucinations. The output is a label indicating whether each entity in the response is true or hallucinated. The whole task can be considered a binary classification task.

4 Methodology

4.1 Framework

Our framework, shown in Fig. 1, consists of two main components: the *Knowledge Retrieval* and the *Fact Verification*. The *Knowledge Retrieval* aims to extract the most semantically relevant information from KG based on the input text. In the *Fact Verification*, we employ a transformer architecture enhanced with KG to verify the responses against the retrieved knowledge. In the following section, we will describe each component in detail.

4.2 Knowledge Retrieval

Given a dialogue history and a response, the *knowledge retrieval* part involves two steps, *Entity Extraction and Entity Linking* followed by *Triple Selection* to retrieve related triples in KG:

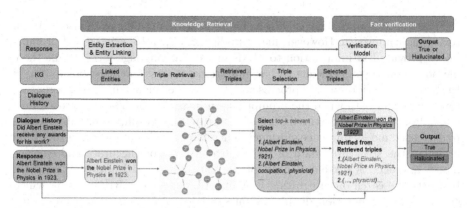

Fig. 1. Overview of CoKGLM. (a) *Knowledge Retrieval* extracts entities from the response, links them to KG, and selects triples that are relevant to the dialogue history; (b) *Fact Verification* checks the factual accuracy of each extracted entity by cross-referencing the dialogue history and response with selected triples.

The first step is *Entity Extraction and Entity Linking*. In this step, we perform named entity recognition (NER) to identify and classify all entities mentioned in the response into specific categories, such as people and locations. Once detected, we link the entities to their corresponding entities in the KG using string matching. Note that entity recognition and linking can be done by existing entity linking techniques. For example, in the response "Albert Einstein won the Nobel Prize in Physics in 1923", NER identifies "Albert Einstein" as a person, and so on. We then link these entities to their corresponding entities in the KG.

The second step is *Triple Selection*. In this step, we retrieve the relevant triples for each entity based on the semantic similarity to the dialogue history. The retrieval process consists of following steps:

1. Triple Retrieval: After linking the entities to the KG, we retrieve all adjacent triples directly connected to the linked entity in the KG.
2. Verbalizing Triples: Then, verbalize the retrieved triples by linear verbalization, concatenating the subject, predicate, and object texts directly to create a single string for each triple.
3. Selecting Triples: We build embedding vectors for the dialogue history and the verbalized triples. By calculating the cosine similarity between the vectors of the dialogue history and each triple, we select the top-k triples that are semantically related to the context of the dialogue.

For example, for the linked entity "Albert Einstein" in KG, we would retrieve all surrounding triples such as: (Albert Einstein, born, Ulm); (Albert Einstein, died, Princeton); .... Then, the triple (Albert Einstein, born, Ulm) is verbalized as "Albert Einstein born Ulm". Subsequently, the top-k semantically relevant triples for this instance might be selected such as (Albert Einstein, Nobel Prize in Physics, 1921) and (Albert Einstein, occupation, Physicist).

This approach ensures the provision of the most relevant KG information, facilitating accurate fact verification. Notably, our retrieve module uses off-the-shelf embedding models, not requiring additional training in this part.

4.3 Fact Verification

After selecting triples, the final step is the fact verification. Our verification strategy involves verifying all detected entities individually. By verifying each entity in the response, we can systematically assess the accuracy of each piece of information in the text. In this part, the entire dialogue, which includes the dialogue history and response, along with the selected triples are fed into our verification model to make a decision for each entity. This aims to determine if each entity is true or hallucinated based on the selected triples and the overall context of the dialogue.

To achieve more efficient and accurate inference in the verification process, we propose an enhanced approach that fully utilizes the intricate structural information stored in the KG. Different from previous work [3,7,11], our approach adopts an embedding-based technique to integrate both the textual and structural information from the KG into text. We begin by embedding both response and the selected triples. By matching each part of the response with the selected triples at the embedding level, we can identify discrepancies based on the similarity between them, pinpointing where the response may diverge from the known facts in the KG. Finally, by leveraging the capabilities of a Language Model (LM) to further process these embeddings, we ultimately classify the detected entity as true or hallucinated. This allows for a robust and accurate verification of the response. The fact verification component is shown in Fig. 2.

We employed the encoder from RHO (ρ) [13] for constructing our framework. This model embeds text data and graph data into a common representation space and performs a cross-attention mechanism between them. This approach allows us to fuse the lexical and structured knowledge from KGs into LMs, enhancing the accuracy of fact verification. To be specific, we utilize the encoder of RHO (ρ) to integrate graph data and text data at a token level. While maintaining the basic structure of the RHO (ρ) encoder, we adapted it to function as a hallucination detector by adding fully connected layers, transforming it into a binary classification model. The entire process can be represented as follows:

- In the pre-processing, we obtain the KG embeddings by performing TransE. Then, link the mentions of entities and relations in input text to their corresponding entities and relations in the KG.
- We feed the input text and embeddings of selected triples into our verification model. We first obtain the initial text embeddings T for the input text. Let w be an arbitrary token embedding of T.
- We fuse the corresponding KG embeddings of w by two methods: Fusing specific knowledge and Fusing all selected knowledge.

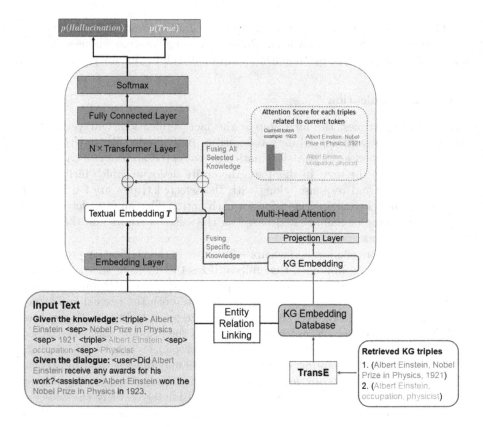

Fig. 2. The Fact Verification component of CoKGLM.

Fusing Specific Knowledge: This method fuses the KG embeddings at the entity and relation level. Let k_w be the corresponding entity or relation embedding of w from the KG. The process of fusing specific knowledge can be represented as: $w_{local} = W_{local}(k_w)$, where $W_{local}(\cdot)$ is a projection matrix which transforms the space from the KG embeddings to the textual embeddings.

Fusing All Selected Knowledge: This method fuses all of the selected triples' embeddings to w instead of just entity or relation.

Let K_w denote the embedding of all selected triples, obtained by concatenating their individual embeddings. This process can be represented as:

$$w_{global} = softmax(\frac{w \cdot K_w}{\sqrt{dim(w)}}) \cdot K_w$$

Finally, the initial token embedding w is fused with the KG information, represented as: $\tilde{w} = w + w_{local} + w_{global}$. This fused embedding is then fed into transformer layers for further processing, allowing us to capture complex dependencies and contextual information across the entire sequence. We add a fully connected layer to map the processed $\tilde{w}$ to a 2-dimensional space for the binary classification task. This step provides the binary classification output, to which we apply the softmax function to obtain the predicted probabilities $\hat{y}$ for the two classes. During training, we minimize the cross-entropy loss between the predicted probabilities and the true labels.

5 Experiments

5.1 Dataset

While several benchmarks [4,15] exist for the hallucination detection task, there is a lack of benchmarks grounded with KGs. To address this gap, we conducted preliminary experiments to systematically generate evaluation data. We leverage a knowledge graph question-answering dataset called OpenDialKG [16], which contains conversational data linked to the Freebase knowledge graph. Topics of the dataset consist of movies, books, sports, and music.

We filtered the data because of many chit-chat responses and lack of official split within the original dataset, the reponses in the previous turn will become dialogue history in the next turn, which would cause data leakage for detection task. Specifically, we retained only one sample with the longest turns from each dialogue.

To systematically introduce hallucinations, we used the perturbation strategy [6]. This strategy involves replacing a specific entity within the response text with another entity from the KG of the same type. By calculating the similarities between the sub-graph of the original entity and entities in the KG, we can select the candidate entity to create hallucinated instances. Note that hallucinations were constrained to exist only within the entity of the response text, with each response containing only one instance of hallucination.

Our dataset comprises 14,397 samples, with 1826 of them containing hallucinations. Utilizing a strategy wherein each entity is verified individually, we ensured that the selected triples only pertain to the specific entity in each sample. Notably, instances corresponding to the same dialogue were consistently allocated to the same dataset split to prevent data leakage.

5.2 Implementation Details

We used several tools for the implementation. For *Entity Extraction and Entity Linking*, we utilized spaCy[1] for NER to identify and classify entities in the response. FuzzyWuzzy[2] was used for entity linking, calculating the Levenshtein distance to determine string similarity. For *Triple Selection*, we employed spaCy[1] to embed the dialogue history and the verbalized triples, selecting the most relevant ones through cosine similarity. For *Fact verification*, we used the sequence-to-sequence language model BART as our backbone, modifying BART's encoder and fine-tuning it with a batch size of 4 for 1000 optimization steps.

---
[1] https://spacy.io/.
[2] https://pypi.org/project/fuzzywuzzy/.

5.3 Evaluation Settings

To ensure robustness, we conducted nested cross-validation. The overall data was divided into five subsets. In each iteration, one subset served as the test data, while the remaining four subsets were used for a secondary cross-validation. In this secondary cross-validation, the four subsets were further split into training and validation data to select the best model parameters based on validation set performance.

For the classification task, we evaluated Precision, Recall, and F1 score at the entity level to determine if each detected entity was true or hallucinated.

The primary objective of this work is to enhance the capabilities of LMs in detecting hallucinations by combining them with KGs. To investigate how the combination of KGs and LMs contributes to hallucination detection, we conducted two main experiments. First, we evaluated whether integrating KGs with LMs improves hallucination detection performance. We did this by comparing the baseline performance of LMs without KGs to their performance when KGs were incorporated in two different ways. Second, we examined which types of KG information are most effective in hallucination detection. This helps to optimize the integration process and improve overall performance.

5.4 Experiment 1: Effectiveness of KGs in Hallucination Detection

This experiment aims to examine whether integrating KGs enhance the hallucination detection capabilities of LMs. We conducted a control experiment in two ways: measuring the detection capability of the LM alone, without any KG support, and then measuring it with KG integration. We compared different integration approaches: direct prompting, which involved providing KG's textual data directly to LMs in the form of natural language prompts, and KG embedding integration, where we integrated KG's textual and structural embeddings into LMs, as introduced in Sect. 4.3. The experimental result is shown in Table 1.

According to Table 1, compared with the experiment using LM only, the experiments using KG showed improvements in all metrics and demonstrated more robustness with a lower standard deviation. These improvements were observed even with randomly selected triples. Compared to the best-performing experiment using KGs, precision increased by up to 57%, recall increased by up to 19%, and the F1 score increased by up to 38%. Additionally, the standard deviation of the F1 score decreased from 0.029 to 0.0047, indicating an 84% increase in robustness. This indicates that using KGs can substantially enhance the effectiveness and reliability of hallucination detection in LLMs, making the LLM more dependable and accurate overall.

As shown in Table 1, compared to only prompting LM with selected triples, providing both KG's textual and structural data led to a higher score and made the model's performance became more consistent, with a 61% increase in F1 score robustness when using ten triples. However, merely increasing the number of triples through prompting from 5 to 10, resulted in only a 3% improvement in F1 score (0.792 to 0.814), and 37% improvement in standard deviation, which

Table 1. Evaluate the effectiveness of using KGs in hallucination detection. Here, the values in parentheses indicate the standard deviation

KG	Number of triples	Integrate KG into LM	Precision	Recall	F1
✗	-	-	0.546 (0.057)	0.680 (0.058)	0.602 (0.029)
✓	5	Prompt	0.796 (0.052)	0.793 (0.067)	0.792 (0.019)
✓	10	Prompt	0.829 (0.057)	0.804 (0.051)	0.814 (0.012)
✓	5	Embed+Prompt	0.808 (0.059)	0.784 (0.053)	0.793 (0.0095)
✓	10	Embed+Prompt	**0.859** (**0.012**)	**0.809** (**0.010**)	**0.833** (**0.0047**)

is lower than the improvement achieved by incorporating both types of KG data. This is because additional triples with only textual data may introduce redundant or irrelevant noise, overwhelming the model and making it harder to focus on the most relevant details.

These results indicate that integrating KG's textual and structural embeddings is a more effective approach for hallucination detection in LMs. This combined approach provides a more comprehensive understanding, allowing the model to better identify discrepancies and verify facts.

5.5 Experiment2: Optimal KG Information for Hallucination Detection

If we use KGs, it's important to know what kind of KG information is desirable for hallucination detection. Therefore, we assessed the effectiveness of various types of KG information in detecting hallucinations. This includes examining the number of triples and comparing different methods for selectinging triples.

For each entity, we selected 5 or 10 triples from its adjacent triples using different selecting methods—either by random selection or by selection based on contextual similarity. we investigated whether increasing the number of triples enhances the model's ability to detect hallucinations based on more detailed knowledge. Additionally, we evaluated how performance improves with these different selecting methods. Since the results of Experiment 1 indicated that incorporating embeddings is more effective, both prompting and embedding were used in this experiment. The experimental result is shown in Table 2.

According to Table 2, when using triples most relevant to the context compared to randomly selected triples, the F1 score increased by 10% with 5 triples and by 13% with 10 triples. It indicates that selecting triples based on context similarity leads to considerable improvement.

Table 2. Examine which types of KG information are most effective in hallucination detection. Here, the values in parentheses indicate the standard deviation

Triple Selection	Number of triples	Precision	Recall	F1
randomly	5	0.745	0.703	0.723
randomly	10	0.737	0.733	0.735
similarity to context	5	0.808 (0.059)	0.784 (0.053)	0.793 (0.0095)
similarity to context	10	**0.859** (**0.012**)	**0.809** (**0.010**)	**0.833** (**0.0047**)
similarity to context	20	0.838 (0.013)	0.806 (0.034)	0.821 (0.016)

For the experiment using contextually relevant triples, increasing the triple size from 5 to 10 led to the best performance with an F1 score of 0.833 and the standard deviation improved by 51%. This is unexpected because of the nature of our dataset, where each response is constructed by 1 or 2 triples. Adding more triples to 10 might have allowed us to retrieve the golden triple that was not captured initially, resulted in the better performance. Surprisingly, this often introduced some information unrelated to the context or only somewhat related but not containing the golden triples, yet the performance still improved. But when we increasing the triple size to 20, the score and standard deviation across all metrics start decreasing. Conversely, increasing randomly selected triples from 5 to 10 showed minimal change. This suggests that merely adding more random information doesn't enhance overall performance.

The results suggest several key findings. Firstly, using contextually relevant triples is more effective than using randomly selected triples. Secondly, for using triples related to the context, incorporating more triples generally leads to better performance, even if some irrelevant information is introduced to the model. However, beyond a certain limit, adding more irrelevant information can decrease performance. Finally, if the most relevant (oracle) triples can be selected, even just two triples can achieve an F1 score close to 0.99. These findings indicate that focusing on retrieving contextually relevant triples is a promising direction for future research.

6 Conclusion

In this paper, we propose CoKGLM for integrating KGs into LLMs to detect hallucinations. Our experiments demonstrate that combining KGs' textual and structural embeddings into LMs can considerably improve the accuracy and robustness of hallucination detection. We also found that using contextually relevant triples is particularly effective. Based on this approach, the best performance can achieve an F1 score of up to 0.83 even with the simplest retrieval methods and small-sized LLM like BART.

When the retriever's performance is sub-optimal or it is difficult to retrieve enough relevant triples, increasing the number of triples, even if introducing some irrelevant information, also proves helpful.

Our framework is particularly effective in enterprise settings where knowledge is constantly updated and cannot be memorized by the public LLMs. Our method avoids the need for retraining LLMs with internal KGs, enables independently verify LLMs outputs and detect inaccuracies effectively.

These findings highlight the importance of efficient retrieval methods for contextually relevant triples, which can enhance the accuracy and reliability of LLMs, and point to a promising direction for future research.

Acknowledgement. We thank Dr. Kosuke Nakamura and Dr. Kotaro Otomura from Mitsubishi Electric for providing valuable advice throughout the research process.

References

1. Agrawal, G., Kumarage, T., Alghamdi, Z., Liu, H.: Can knowledge graphs reduce hallucinations in LLMs? : a survey. In: Proceedings of the Conference of the North American Chapter of the Associateion for Computational Linguisitics: Human Language Technologies, pp. 3947–3960 (2024). https://doi.org/10.18653/v1/2024.naacl-long.219
2. Baek, J., Aji, A.F., Saffari, A.: Knowledge-augmented language model prompting for zero-shot knowledge graph question answering. In: Proceedings of the 1st Workshop on Natural Language Reasoning and Structured Explanations, pp. 78–106 (2023). https://doi.org/10.18653/v1/2023.matching-1.7
3. Chen, J., Kim, G., Sriram, A., Durrett, G., Choi, E.: Complex claim verification with evidence retrieved in the wild. In: Proceedings of the Conference of the North American Chapter of the Association for Computational Linguistics: Human Language Technologies, pp. 3569–3587 (2024). https://doi.org/10.18653/v1/2024.naacl-long.196
4. Chen, S., et al.: FELM: benchmarking factuality evaluation of large language models. In: Proceedings of the 37th Conference on Neural Information Processing Systems, pp. 44,502–44,523 (2023). https://doi.org/10.48550/ARXIV.2310.00741
5. Cuconasu, F., Trappolini, G., Siciliano, F., Filice, S., Campagnano, C., Yoelle, M., Tonellotto, N., Silvestri, F.: The power of noise: redefining retrieval for RAG systems. In: Proceedings of the 47th International ACM Conference on Reseach and Development in Information Retrieval, pp. 719–729 (2024). https://doi.org/10.1145/3626772.3657834
6. Das, S., Saha, S., Srihari, R.: Diving deep into modes of fact hallucinations in dialogue systems. In: Findings of the Association for Computational Linguistics: EMNLP 2022, pp. 684–699 (2022). https://doi.org/10.18653/v1/2022.findings-emnlp.48
7. Fatahi Bayat, F., et al.: FLEEK: factual error detection and correction with evidence retrieved from external knowledge. In: Proceedings of the Conference on Empirical Methods in Natural Language Processing, pp. 124–130 (2023). https://doi.org/10.18653/v1/2023.emnlp-demo.10

8. Furumai, K., Wang, Y., Shinohara, M., Ikeda, K., Yu, Y., Kato, T.: Detecting dialogue hallucination using graph neural networks. In: Proceedings of International Conference on Machine Learning and Applications, pp. 871–877 (2023). https://doi.org/10.1109/ICMLA58977.2023.00128
9. Gekhman, Z., et al.: Does fine-tuning LLMs on new knowledge encourage hallucinations? (2024). https://doi.org/10.48550/ARXIV.2405.05904
10. Google: Gemini (2023). https://gemini.google.com/
11. Guan, X., et al.: Mitigating large language model hallucinations via autonomous knowledge graph-based retrofitting. In: Proceedings of the 38th Conference on Artificial Intelligence, pp. 18,126–18,134 (2023). https://doi.org/10.1609/aaai.v38i16.29770
12. Huang, L., et al.: A survey on hallucination in large language models: principles, taxonomy, challenges, and open questions (2023). https://doi.org/10.48550/ARXIV.2311.05232
13. Ji, Z., et al.: RHO: reducing hallucination in open-domain dialogues with knowledge grounding. In: Findings of the Association for Computational Linguistics: ACL 2023, pp. 4504–4522 (2023). https://doi.org/10.18653/v1/2023.findings-acl.275
14. Kuhn, L., Gal, Y., Farquhar, S.: Semantic uncertainty: linguistic invariances for uncertainty estimation in natural language generation. In: Proceedings of the 11th International Conference on Learning Representations (2023). https://doi.org/10.48550/ARXIV.2302.09664
15. Li, J., Cheng, X., Zhao, X., Nie, J.Y., Wen, J.R.: HaluEval: a large-scale hallucination evaluation benchmark for large language models. In: Proceedings of the Conference on Empirical Methods in Natural Language Processing, pp. 6449–6464 (2023). https://doi.org/10.18653/V1/2023.EMNLP-MAIN.397
16. Moon, S., Shah, P., Kumar, A., Subba, R.: OpenDialKG: explainable conversational reasoning with attention-based walks over knowledge graphs. In: Proceedings of the 57th Annual Meeting of the Association for Computational Linguistics, pp. 845–854 (2019). https://doi.org/10.18653/v1/P19-1081
17. OpenAI: GPT-4 Technical report (2023). https://doi.org/10.48550/ARXIV.2303.08774
18. Sen, P., Mavadia, S., Saffari, A.: Knowledge graph-augmented language models for complex question answering. In: Dalvi Mishra, B., Durrett, G., Jansen, P., Neves Ribeiro, D., Wei, J. (eds.) Proceedings of the 1st Workshop on Natural Language Reasoning and Structured Explanations (NLRSE), pp. 1–8. Association for Computational Linguistics (2023). https://doi.org/10.18653/v1/2023.nlrse-1.1
19. Urlana, A., Kumar, C.V., Singh, A.K., Garlapati Bala Mallikarjunarao, C., Rao, S., Rahul, M.: LLMs with industrial lens: deciphering the challenges and prospects – a survey (2024). https://doi.org/10.48550/ARXIV.2402.14558
20. Yuan, W., Neubig, G., Liu, P.: BARTScore: evaluating generated text as text generation. In: Proceedings of the 35th Conference on Neural Information Processing Systems, pp. 27,263–27,277 (2021)

AutOnto: Towards A Semi-Automated Ontology Engineering Methodology

Kiara Marnitt Ascencion Arevalo[1]([✉])[iD], Shruti Ambre[2], and Rene Dorsch[2][iD]

[1] Technische Hochschule Georg Simon Ohm, Nuremberg, Germany
kiaramarnitt.ascencionarevalo@th-nuernberg.de
[2] Fraunhofer Institute for Integrated Circuits IIS, Nuremberg, Germany
{Shruti.Ambre,ReneDorsch}@iis.fraunhofer.de

Abstract. This paper addresses the challenge of efficiently constructing domain ontologies for large, rapidly evolving domains, where manual approaches often struggle to overcome knowledge acquisition bottlenecks. To overcome these limitations, we developed an automated framework, AutOnto, for knowledge extraction and ontology conceptualization that leverages Large Language Models (LLMs) and natural language processing (NLP) techniques. AutOnto integrates BERT-based topic modeling with LLMs to automate the extraction of concepts and relationships from text corpora, facilitating the construction of taxonomies and the generation of domain ontologies. We applied AutOnto to a dataset of NLP-specific articles from OpenAlex and compared the resulting ontology generated by our automated process against a well-established gold-standard ontology. The results indicate that AutOnto achieves comparable levels of quality and correctness while significantly reducing the amount of data required and the dependence on domain-specific expertise. These findings highlight AutOnto's efficiency and effectiveness in knowledge extraction and ontology generation. This work has significant implications for rapid ontology development in large, evolving domains, potentially mitigating the knowledge acquisition bottleneck in ontology engineering.

Keywords: Ontology Engineering · Large Language Models · Natural Language Processing

1 Introduction

By creating structured representations, ontologies enable to capture, represent, and formalize knowledge in human- and machine-readable formats, facilitating knowledge sharing and enabling various applications, such as knowledge management, semantic integration, and reasoning [12,33]. Traditional ontology engineering approaches involve an iterative and systematic process that relies heavily on the expertise and involvement of domain experts and ontology engineers for creating, refining, and maintaining these ontological representations of knowledge.

However, the conventional manual approach to ontology creation is time-consuming, resource-intensive, and demands substantial expertise, thus creating a significant bottleneck in knowledge acquisition [2,13,28]. Moreover, the complexity of ontology languages and their steep learning curve aggravate this issue [15,20,39]. Consequently, relying exclusively on manual approaches hinders the rapid and efficient development of ontologies, limiting their usefulness in information-related tasks, which involve processing and utilizing structured knowledge effectively. This highlights the need to transition to automated approaches that facilitate efficient development and deployment of ontologies [4,5,32].

Leveraging language models, we propose a framework for automated ontology generation based on the extraction and identification of information patterns within a domain from text corpora. These patterns are represented by topic distributions derived from contextual word embeddings within the text. In our framework, BERT-based topic modeling is applied to identify initial topic distributions and related terms, allowing for the extraction of key concepts and relationships. Subsequently, Large Language Models (LLMs) are used to refine and validate these relationships and concepts. This combination of methods facilitates the discovery of implicit relationships and concepts that may be overlooked in manual ontology creation, providing a data-driven approach to knowledge representation [18,19]. Furthermore, this automated approach addresses the knowledge acquisition bottleneck present in manual ontology engineering.

2 Research Approach and Scope

We follow the Design Science Research Methodology (DSRM) [14] to address the challenges in automating ontology engineering for large, rapidly evolving domains. This methodology provides a structured approach to build and evaluate artifacts, making it well-suited for our research. The research procedure is structured according to the four design artifacts of the building process and the evaluation process of DSRM:

1. **Problem Identification and Motivation:** Motivated by the limitations of manual ontology engineering (see Sect. 1), this paper aims to address the research question: *"How can the extraction of topic distributions from extensive text corpora facilitate the automation of ontology engineering?"*.
2. **Define Objectives of a Solution:** To identify the objectives for an automated ontology engineering approach, we conduct a comprehensive review of state-of-the-art ontology engineering methodologies and methods in the NLP domain. From this review, we derived several key requirements for an automated ontology engineering approach. These include automating knowledge extraction from extensive text corpora, developing automated taxonomy construction and ontology conceptualization methods, mitigating the knowledge acquisition bottleneck in ontology development.
3. **Design of the Methodology:** Based on the identified requirements, we developed AutOnto, an automated approach for generating a knowledge

structure in a given domain from unstructured text corpora. The AutOnto methodology includes several key steps. First, it employs NLP and LLMs to extract relevant information from the corpora and discover information patterns. Next, it uses LLMs to transform the identified patterns into structured knowledge representations consisting of domain-relevant terms. The methodology then applies natural language generation techniques to enrich the defined terms semantically. Finally, it formalizes these knowledge structures into RDF-based ontologies to make them both machine- and human-readable.

4. **Demonstration of the Methodology:** We apply the AutOnto framework to the domain of Natural Languge Processing research. This domain is chosen because of its rapid development and importance within artificial intelligence and computer science. Using AutOnto, we generate an ontology that captures the key concepts and relationships from NLP-specific articles obtained from OpenAlex. This implementation serves as a practical demonstration of the capabilities of our methodology in a real-world, rapidly evolving domain.

5. **Evaluation:** We evaluate the AutOnto framework by following Raad and Cruz's guidelines [27] and assess the generated ontology against a gold-standard ontology, specifically using the NLP subset of the Computer Science Ontology (CSO) for benchmarking. This approach allows us to demonstrate the effectiveness of our automated methodology by assessing ontology completeness, conciseness, and other key quality metrics.

3 Review of Existing Approaches

In this section, we first review manual approaches for creating ontologies and highlight the drawbacks of these approaches. Thereafter, we review approaches from the NLP domain to automate tasks within ontology engineering and highlight the requirements of an automated approach.

Several methodologies have been proposed for manual ontology engineering, each contributing unique aspects to the field. Uschold and King's methodology [37] was one of the earliest, emphasizing collaboration between domain experts and knowledge engineers to explicitly represent domain-specific knowledge. Grüninger and Fox's Methodology [11] focused on formalizing ontologies using first-order logic and evaluating them against formal competency questions, enhancing the rigor of ontology development. METHONTOLOGY [8] offered a comprehensive approach covering the entire ontology lifecycle and emphasized the importance of reusing existing ontologies to reduce development time. Building on this efficiency focus, the On-To-Knowledge Methodology) [35] integrated ontology development with knowledge management processes. More specific methodologies for industrial applications, such as NeOn [34] and Linked Open Terms [25], further refined these approaches.

While these manual approaches provide structured guidelines and best practices, they face a significant challenge: the knowledge acquisition bottleneck. This bottleneck arises from the high manual effort required to acquire and curate information from domain experts [40,41]. The time-intensive nature of this process and the scarcity of expert availability often impede efficient ontology development.

To address the limitations of manual approaches, researchers have explored various automated and semi-automated methods for extracting taxonomies and ontologies. These approaches offer the potential for faster ontology development and reduced reliance on scarce domain expertise. The evolution of these methods can be broadly categorized into two groups: Clustering Methods using NLP and LLM-based approaches. Early automated approaches primarily relied on clustering techniques combined with NLP to extract ontological structures from text. Khan and Luo [16] proposed an automated approach using hierarchical clustering and a modified self-organizing tree algorithm to create concept hierarchies from text. This method demonstrated the potential of statistical approaches in automating parts of the ontology creation process. Building on these clustering techniques, researchers integrated more advanced NLP methods. For instance, Doing-Harris et al. [7] combined Named Entity Recognition, Information Extraction, and hierarchical clustering to construct ontologies from textual data. Their semi-automated system produced clinical terms, synonyms, and hierarchical relationships with promising accuracy. Similarly, Cahyani and Wasito [3] utilized text corpora and ontology design patterns for automated ontology construction in a study on Alzheimer's disease. Korel et al. [17] presented a method to classify scientific paper paragraphs into relevant ontologies by generating text embeddings and ontology description embeddings with fine-tuned language models and training classifiers. These approaches showed the potential of NLP in capturing domain-specific knowledge structures. However, they often focused on specific domains, limiting broader applicability due to reliance on domain-specific patterns and expertise. Recent advances in AI have led to the exploration of LLMs for ontology engineering. These models offer the promise of embedded domain knowledge to overcome the knowledge acquisition bottleneck. For example, Funk et al. [9] have used the knowledge in LLMs to derive hierarchies, a special form of taxonomies. Babaei Giglou et al. [1] evaluated eleven LLMs in ontology generation using zero-shot prompting, showing that the knowledge within LLMs can cover common domains but is less effective in more knowledge-intensive domains that require complex ontological and semantic nuances for accurate domain-specific knowledge representation. Nevertheless, LLMs offer several advantages over traditional NLP and clustering methods. They can leverage large amounts of pre-trained knowledge, potentially capturing broader semantic relationships. In addition, their ability to understand and generate natural language allows for more flexible interaction in the ontology creation process.

Despite these advancements, significant gaps remain in existing research for automated ontology generation. While some studies propose methodologies for extracting concepts and hierarchical relationships, there is a lack of comprehensive frameworks that manage the entire ontology generation process. Furthermore, balancing the broad knowledge of LLMs with the need for domain-specific accuracy remains a challenge.

To address these gaps, our objectives for an automated ontology engineering approach focus on automating knowledge extraction from extensive text corpora, developing effective methods for taxonomy construction and ontology conceptualization that work across diverse domains, and addressing the knowledge acquisition bottleneck that limits traditional ontology development. By leveraging NLP and LLMs, we seek to create an efficient and scalable ontology creation process that produces accurate and domain-relevant ontologies.

4 Methodology

Our AutOnto framework, as shown in Fig. 1, consists of three main stages: Corpora Definition, Knowledge Acquisition, and Ontology Implementation.

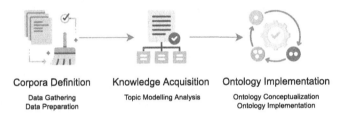

Fig. 1. AutOnto pipeline

The **Corpora Definition** stage involves gathering and cleaning textual data. We utilize bibliographic databases to cover publications in the target domain comprehensively. After retrieving the dataset, initial data selection and preprocessing steps are required to prepare the data for NLP tasks. The focus is on selecting a subset of the database that reflects the core contemporary knowledge in the domain to be modeled. Additionally, outliers (e.g., other languages, publication types) need to be removed to reduce the complexities of data processing. This approach reduces dimensionality and highlights essential information for concept-based analysis and ontology generation.

In the **Knowledge Acquisition** stage, we focus on discovering patterns in the text corpora to facilitate the representation and modeling of domain knowledge. We apply topic modeling analysis to the text corpora, using unsupervised learning with pre-trained language models to enhance semantic understanding. Topic modeling is particularly suited for pattern discovery due to its ability to extract coherent topics from large text corpora, ensuring that the discovered patterns are both meaningful and representative of the underlying concepts within the domain. This approach is scalable and capable of efficiently processing and analyzing extensive datasets without needing labeled data. By identifying key themes and concepts, topic modeling significantly contributes to the development of patterns used in domain ontologies. The topic modeling process involves creating vector representations of the texts that form diverse yet coherent clusters. Each cluster is then analyzed to identify the most representative words or phrases, which are aggregated to define a topic. This process requires adjusting various parameters to optimize the model's performance. We evaluated model performance using two key metrics: the topic diversity score [36] and the coherence score [30]. The topic diversity score measures the range and variety of topics identified by the model, ensuring broad coverage of relevant themes. The coherence score assesses the semantic similarity within topics, ensuring the topics are interpretable. By balancing these metrics, we can select the most effective model configuration to provide a robust foundation for subsequent stages.

The **Ontology Implementation** stage follows the LOT approach [25] and consists of two main phases: Ontology Conceptualization and Ontology Encoding. Ontology Conceptualization aims to create a model that represents the domain knowledge. This involves identifying key concepts and their attributes, establishing hierarchical and associative relationships between these concepts, and defining logical constraints and axiomatic rules governing the ontology [24]. We follow an iterative process focusing on identifying key concepts, hierarchical relations, basic associative relations, and the semantic definition of these terms. To this end, LLMs are utilized to refine the previously obtained topics into concepts, identify relations, and generate semantic information for each term. By concentrating on these core components, i.e., key concepts, and relations, we aim to lower the complexity to establish a robust and transparent structure, ensuring that the foundational elements can be developed appropriately and allowing for a manageable scope that can be incrementally expanded. The Ontology Encoding phase transforms the structured concepts and relationships identified during Ontology Conceptualization into a machine-readable format.

5 Demonstration

In this section, we demonstrate the practical application of the AutOnto framework by generating an ontology from articles within the field of Natural Language Processing research.

5.1 Corpora Definition

To gather the data necessary for the development of the NLP ontology, OpenAlex [26] was chosen due to its extensive coverage of over 220 million scholarly works. OpenAlex is an open-source bibliographic database that indexes and aggregates scholarly works (e.g., books, journal articles, conference papers, etc.) with their metadata, creating a massive catalog linking works, authors, journals, organizations, and funders.

The OpenAlex API[1] was queried on the 19th of December 2023 to retrieve work records related to the NLP domain, resulting in over 962,604 work records

---

[1] https://api.openalex.org/.

that were gathered and flattened into CSV format. During the data-cleaning process, columns irrelevant to the analysis were dropped (e.g., location details and sustainability goals). In contrast, columns such as title, language, abstract, and work types were retained as they captured core conceptual knowledge relevant to ontology generation.

Analysis of the work records' types revealed that approximately 80% were categorized as articles, including journal articles, conference proceedings articles, and posted content. Figure 2 shows the distribution of the works' types.

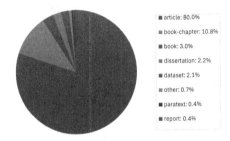

Fig. 2. Distribution of work records' types in the dataset

Articles represent the core scholarly output and primary source of discussion in NLP, providing conceptual knowledge and relationships from empirical studies. The other work types are indirectly related but do not constitute part of the central focus of the discussion. Furthermore, the dataset unveiled a prominent linguistic attribute: a significant majority of works (94.57%) were written in the English language. This observation led to the decision to focus the research, specifically on English-language works. The final dataset comprises scholarly articles following specific criteria: authored in English, containing a non-null abstract, and classified as articles. Comprising a total of 474,242 scholarly articles, the dataset forms the foundation for further analysis and exploration in the NLP domain.

5.2 Knowledge Acquisition

For Knowledge Acquisition, we implemented a topic modeling pipeline following the modular structure of BERTopic [10]. This pipeline consists of five sequential steps: converting texts into embeddings, reducing dimensionality, clustering embeddings, tokenizing topics, and weighting tokens.

The pipeline configuration for the topic models utilized the all-mpnet-base-v2 sentence transformer model [29] to generate semantic embeddings of the document titles. For dimensionality reduction, we used UMAP [22], configured with 30 neighbors, 2 components, and the cosine distance metric to map high-dimensional data to a lower-dimensional space while preserving both local and global structures. HDBSCAN clustering [21] was employed with parameters set to a minimum cluster size of 150 and a minimum sample threshold of 50, balancing the number of topics identified and their specificity. The CountVectorizer was configured to accept both unigrams and bigrams, transforming the text into token count matrices for a traditional bag-of-words representation. c-TF-IDF was applied with default settings to weigh the tokens, ensuring the most relevant terms were emphasized within each topic.

We experimented with two topic models[2] (Model A and Model B), both using only document titles as input, after finding that the inclusion of abstracts (about 300 words each) overloaded the model and led to reduced coherence in the extracted topics.

An additional refinement step was applied to Model B, where we used ChatGPT to fine-tune the topic representations before applying c-TF-IDF, further enhancing the coherence and specificity of the topics. This step specifically aimed to reduce the number of outliers (documents unassigned to any topic), which reflects each model's ability to cluster the corpus effectively.

We assessed the performance of Model A and Model B using both quantitative metrics (coherence and topic diversity scores) and a qualitative examination of outliers and topics, as presented in Table 1. For the calculation of the quantitative metrics, we used the Gensim library[3] for the coherence score and the Octis library[4] for the topic diversity score.

Table 1. Comparison of Models

Model	Topics	Outliers	Coherence Score	Topic Diversity Score
Model A	language_reading_learning_writing, speech_recognition_speaker, chinese_english_verbs, translation_machine_neural_machine, clinical_medical_biomedical_health	8066	0.45	0.82
Model B	Language_Acquisition, Speech_Recognition, Linguistic_Analysis, Machine_Translation, Clinical_Text_Analysis	33	0.45	0.82

Although Model A and Model B yielded the same topic diversity score and coherence score, Model B outperformed Model A in outlier resolution, indicating its superior ability to assign documents to meaningful topics. This effectiveness makes Model B the optimal choice for our approach. The topics obtained from this model serve as input for the next stage of our process.

---

[2] https://github.com/Kiaramarnitt/AutOnto/tree/main/output.
[3] https://pypi.org/project/gensim/.
[4] https://github.com/MIND-Lab/OCTIS.

5.3 Ontology Implementation

For the Ontology Implementation, we employ LLMs primarily in the Ontology Conceptualization phase. Particularly, gpt-4-1106-preview was used due to its ability to capture the semantic relationships and contextual nuances within the domain [6]. The Ontology Conceptualization phase comprises five steps, as shown in Fig. 3.

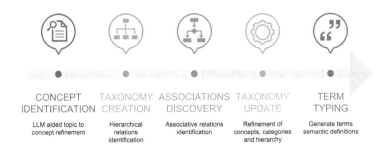

Fig. 3. Ontology Implementation Flow Chart

The **concept identification** step focused on refinement and aimed to filter out topics unrelated to the NLP domain to facilitate the identification of relevant concepts from the remaining ones. Here, LLMs are tasked with analyzing a list of topics generated from the preceding topic modeling pipeline. This process retains only relevant and coherent topics, thereby enhancing the ontology's relevance and focus within the NLP domain.

The next step, **taxonomy creation**, was performed iteratively, and the objective was to identify the hierarchical structure between topics. For the first iteration, the prompt instructed the LLM to create an initial structure by assigning each topic to a single category while categorizing unrelated topics as "irrelevant". In subsequent iterations, the prompt started by presenting the existing categories from previous iterations. The LLM then hierarchically classified the new batch of topics into these categories or created new ones if needed, creating a taxonomy. The prompt also guided the LLM on handling related or nested topics, either by providing lists of related topic keys or creating nested dictionary structures. Finally, the prompt specified the desired output format (see Fig. 4): a dictionary with categories as keys and lists of topics as values. This iterative approach allows the LLM to incrementally build and refine the taxonomy based on new topic batches, existing categories, and associated information.

In the **associations discovery** step, the previously created taxonomy is modified by establishing associative relations between the categories and topics, specifically using the autonto:superTopicOf relationship. Since the Simple Knowledge Organization System (SKOS) provides a solid foundation for expressing semantic relationships between concepts, the relation chosen for the taxonomy modification is based on the Semantic Relations in SKOS [23] and is declared

```
{
    "Natural Language Processing Fundamentals": [
        "Trigram", "Language Acquisition",
        "Natural Language Processing Models",
        "Text corpus", "Natural language",
        "Linguistic Analysis", "Speech synthesis",
    ]
}
```

Fig. 4. Prompt Output of Taxonomy Creation

as specifications of `skos:semanticRelation`. The relation `autonto:superTopicOf`, similarly to `skos:broader`, indicates a generalization relationship, where one topic is a more general category that encompasses another, more specific topic. For instance, in the domain of computer science, "Artificial Intelligence" could be a super-topic of "Natural Language Processing". By incorporating this relation, the pipeline aims to create a more comprehensive and structured representation of the domain, capturing the hierarchical and general-to-specific connections between the categories and topics. First, the prompt presented the existing taxonomy. It then instructed the LLM to modify this taxonomy by adding associations between the categories and topics. Finally, the modifications are formatted in a specific JSON-like structure (see Fig. 5) to organize the enhanced taxonomy. This format is structured as a dictionary, where each category serves as a key, and the corresponding value is another dictionary. Within this nested dictionary, the keys represent relation types, and the values are either lists of topics or nested dictionaries. If a topic has a relation with another topic, it is represented as a nested dictionary within the list of topics for that relation.

```
{
    "Natural Language Processing Fundamentals": {
        "superTopicOf": [
            "Trigram", "Language Acquisition",
            "Natural Language Processing Models",
            "Text corpus", "Natural language",
            "Linguistic Analysis", "Speech synthesis",
        ]
    }
}
```

Fig. 5. Prompt Output format for Associations Discovery

Building upon this, the fourth step, **taxonomy update**, focuses on refining the taxonomy to better cover the ontology goal and scope while adding a higher level of concept granularity. The prompt begins by providing a detailed description of the ontology's goal and scope to guide the updates. Next, it introduces new hierarchical levels and sub-hierarchies to further structure the taxonomy. The prompt then requires a review of the "irrelevant" category to identify any topics that may still be relevant. Following this, it determines the appropriate placement for these newly identified relevant topics within the reorganized structure, which could involve adding them to existing categories, creating new subcategories, or establishing new top-level categories. Finally, the prompt removes irrelevant topics from the taxonomy, introducing a separate top-level dictionary to categorize them, ensuring that the dictionaries do not contain themselves.

The pipeline moves forward to the **term typing** step[5], where semantic information is added to enrich the taxonomy further. In this context, term typing refers to the process of defining or creating summaries of the terms within the taxonomy. To integrate these summaries as semantic annotations, the property `rdfs:comment` was used, allowing each term to be accompanied by concise, one-line descriptions. Here, the LLM assumes the role of a "topic summarizer" tasked with generating these descriptions for each topic. By generating these concise yet informative summaries, the final stage aims to enhance the taxonomy further, providing additional context and clarity for each topic. This process essentially defines the topics within the taxonomy, adding semantic meaning and facilitating better understanding. This not only aids in the interpretation of the taxonomy but also lays the groundwork for developing a well-defined and cohesive ontology that accurately captures the key concepts and their relationships within the field of NLP.

After a conceptual model has been created, the next step in Ontology Implementation is Ontology Encoding. Within Ontology Encoding, the conceptual model is formalized in a machine-readable format, e.g., an RDF graph. This is done using an RDFlib-based script in Python to map the created structure in JSON format into an RDF graph. This graph is then serialized to Turtle. The final version of the prompts[6] used and the generated ontology[7] can be found in our repository.

6 Evaluation

To evaluate our approach, we followed the guidelines proposed by Raad and Cruz [27], which recommend assessing ontology quality and correctness against a gold-standard ontology.

Gold Standard Ontology: We selected the Computer Science Ontology Version 3.3 [31] as our gold standard. CSO is a comprehensive ontology covering various research areas within computer science, including the NLP domain. CSO

---

[5] https://github.com/Kiaramarnitt/AutOnto/blob/main/run.py.
[6] https://github.com/Kiaramarnitt/AutOnto/blob/main/ontology_generation/
OntologyGen.py.
[7] https://github.com/Kiaramarnitt/AutOnto/blob/main/output/taxonomy.ttl.

was developed through the automated application of the Klink-2 algorithm on the Rexplore dataset, which comprises 16.000.000 publications. As our work focuses on the NLP domain, we restricted our comparison to the NLP subset of CSO, which covers 156 deduplicated topics.

Evaluation Criteria: To evaluate the quality and correctness of an ontology, Raad and Cruz [27] defined several criteria, including completeness, conciseness, computational efficiency, accuracy, adaptability, clarity, and consistency. We employed both quantitative metrics and qualitative assessments to evaluate these criteria. For **Completeness** - the criterion that measures how well the domain of interest is covered - and **Conciseness** - the criterion that states if the ontology includes irrelevant elements about the specific domain - we utilized the quantitative metrics: pairwise average similarity (PAS), mean aggregate similarity (MAS), and the cosine similarity with reference embedding (CSRE) measurements. The metrics relied on phrase embeddings computed using Phrase-BERT [38]. More specifically, we used the `whaleloops/phrase-bert` model for generating embeddings from the labels of topics. The metrics were calculated for each topic using the scikit-learn[8] library to ensure reproducibility and standardization. PAS measures the average cosine similarity between all pairs of phrase embeddings representing ontology classes, indicating topic completeness. PAS scores range from 0 to 1, where values closer to 1 indicate higher similarity among topic labels, suggesting greater completeness, while values closer to 0 suggest more diverse or less related topics. MAS assesses conciseness by computing the cosine similarity between an aggregate embedding and individual phrase embeddings. MAS scores also range from 0 to 1, with scores closer to 1 indicating higher relatedness of topics to a common theme and scores closer to 0 suggesting less cohesion among topics. CSRE quantifies the ontology's semantic alignment with the core NLP domain by comparing topic embeddings to a reference "natural-language-processing" embedding. CSRE scores range from 0 to 1, where values closer to 1 indicate stronger alignment with the NLP domain, and values closer to 0 suggest less relevance to NLP. The **Computational efficiency** - a criterion that measures the ability to derive an ontology from a dataset is measured by the dataset size required to create the ontology. The criteria **Accuracy** - the criterion that states if elements of an ontology are correct -, **Adaptability** - a criterion that describes how far the ontology anticipates its use -, **Clarity** - a criterion that measures the effectiveness of the intended meaning of the terms -, and **Consistency** - a criterion that describes that the derived ontology does not include any contradictions - were qualitatively evaluated through the classes, properties, and relations provided by AutOnto and Klink-2 approach.

Results: Table 2 presents a comparison of key metrics between AutOnto and the CSO subset. Our results show that AutOnto's NLP ontology (AO-NLP) achieves similar performance to CSO-NLP in terms of completeness and conciseness. Regarding completeness, AO-NLP exhibits a PAS of 0.34, slightly lower

---

[8] https://github.com/scikit-learn/scikit-learn.

Table 2. Comparison of key metrics between AutOnto and CSO subset, including the number of concepts, types of relations, properties, and similarity measures.

Approach	AutOnto	Klink-2
Ontology Name	AO-NLP	CSO-NLP
Number of Concepts	56	156
Pairwise Average Similarity	0.34	0.35
Mean Aggregate Similarity	0.84	0.85
Cosine Similarity with Reference Embedding	0.84	0.88
Dataset Size	24.000	200.420
Relations	autonto:superTopicOf	cso:contributesTo, cso:preferentiaEquivalent, schema:relatedLink, owl:sameAs, cso:relatedEquivalent, cso:superTopicOf
Properties	rdf:type, rdfs:label, rdfs:comment	rdf:type, rdfs:label

than CSO-NLP's 0.35. This marginal difference indicates that AO-NLP provides similarly broad coverage of topics within the NLP domain. Notably, AO-NLP accomplishes this coverage with fewer topics (56 compared to CSO-NLP's 156 deduplicated topics), suggesting a more efficient representation of the domain knowledge. Regarding conciseness, AO-NLP achieves a MAS of 0.84, closely matching CSO-NLP's 0.85. This similarity in MAS scores indicates that both ontologies maintain comparable levels of topic relatedness and domain cohesion. The CSRE for AO-NLP (0.84) is slightly lower than CSO-NLP (0.88), suggesting a marginally reduced average relatedness to the core NLP domain. However, this difference is minimal and does not significantly impact the overall quality of the ontology. The computational efficiency of AutOnto is particularly noteworthy. For CSO-NLP, we approximated the number of NLP-specific documents, as no precise information is publicly available. By calculating the proportion of NLP documents from the number of topics (175 of 14290 overall topics), a total number of approximately 200.420 articles is required by Klink-2. In comparison, our approach used only 24,000 articles thus reducing the required dataset by 88% to develop a domain ontology for the NLP domain. Regarding adaptability and clarity, AO-NLP enhances its reusability by including human-readable `rdfs:comment` properties for each discovered topic. Although CSO-NLP covers more relations, potentially offering greater adaptability for various tasks, AO-NLP's concise structure may prove advantageous in specific applications where simplicity is preferred. Both ontologies do not cover inference rules, making a statement regarding consistency and accuracy hardly possible. Consequently, we have refrained from making definitive statements regarding these criteria.

In conclusion, our evaluation demonstrates that AutOnto offers a promising approach to ontology generation, combining efficiency with performance comparable to established methods. Furthermore, AutOnto's efficiency in data requirements highlights the potential to overcome knowledge acquisition bottlenecks in ontology engineering, particularly in rapidly evolving domains.

7 Conclusion and Future Works

The proposed automated methodology, AutOnto, leverages NLP techniques and LLMs and shows promising potential to streamline ontology engineering from extensive text corpora. By integrating BERT for topic modeling, the framework effectively extracts meaningful concepts from scholarly articles in the NLP domain. Subsequently, the ontology implementation pipeline utilizes LLMs to determine hierarchies and relationships within the ontology. Thus, AutOnto facilitates the systematic construction of structured knowledge representations, alleviating the challenges and bottlenecks associated with manual ontology development approaches. While the evaluation indicates satisfactory performance in terms of quality and correctness, it is important to note that, thus far, the methodology has only been tested within the NLP domain and compared against a single existing ontology. However, we are currently in the process of testing this approach across various other domains to assess its broader applicability and effectiveness. These ongoing evaluations are expected to provide deeper insights into the adaptability of the framework and identify potential areas for further refinement.

Acknowledgments. This work was partially funded by the Deutsche Forschungsgemeinschaft (DFG, German Research Foundation) Project-ID 528480942 - FIP 8.

Author contributions. Conceptualization: K.A.; Methodology: K.A., S.A.; Software: S.A.; Validation: K.A., S.A., R.D.; Formal Analysis: S.A., R.D.; Investigation: S.A.; Data Curation: S.A.; Writing-original draft: K.A., S.A.; Writing-review and editing: K.A., R.D.; Visualization: K.A., S.A.; Supervision: K.A. All authors have read and approved the final version of the manuscript.

References

1. Babaei Giglou, H., D'Souza, J., Auer, S.: LLMs4OL: Large language models for ontology learning. In: International Semantic Web Conference. Springer Nature Switzerland, Cham (2023). https://doi.org/10.1007/978-3-031-47240-4_22
2. Brewster, C., Ciravegna, F., Wilks, Y.: Background and foreground knowledge in dynamic ontology construction (2003). https://api.semanticscholar.org/CorpusID:5713137, semantic Web Workshop, SIGIR2003 ; Conference date: 01-08-2003

3. Cahyani, D.E., Wasito, I.: Automatic ontology construction using text corpora and ontology design patterns (ODPs) in Alzheimer's disease. Jurnal Ilmu Komputer dan Informasi **10**(2), 59–66 (2017). https://doi.org/10.21609/jiki.v10i2.374
4. Cimiano, P., Mädche, A., Staab, S., Völker, J.: Ontology learning. In: Staab, S., Studer, R. (eds.) Handbook on Ontologies. IHIS, pp. 245–267. Springer, Heidelberg (2009). https://doi.org/10.1007/978-3-540-92673-3_11
5. Cimiano, P., Völker, J.: Text2Onto. In: Montoyo, A., Muñoz, R., Métais, E. (eds.) NLDB 2005. LNCS, vol. 3513, pp. 227–238. Springer, Heidelberg (2005). https://doi.org/10.1007/11428817_21
6. Devlin, J., Chang, M.W., Lee, K., Toutanova, K.: BERT: pre-training of deep bidirectional transformers for language understanding. In: Proceedings of the 2019 Conference of the North American Chapter of the Association for Computational Linguistics: Human Language Technologies, Volume 1 (Long and Short Papers), pp. 4171–4186. Association for Computational Linguistics, Minneapolis, Minnesota (2019). https://doi.org/10.18653/v1/N19-1423
7. Doing-Harris, K., Livnat, Y., Meystre, S.: Automated concept and relationship extraction for the semi-automated ontology management (SEAM) system. J. Biomed. Semant. **6**(1), 1–15 (2015). https://doi.org/10.1186/s13326-015-0011-7
8. Fernández-López, M., Gómez-Pérez, A., Juristo, N.: Methontology: from ontological art towards ontological engineering. In: AAAI Conference on Artificial Intelligence (1997). https://api.semanticscholar.org/CorpusID:10550105
9. Funk, M., Hosemann, S., Jung, J.C., Lutz, C.: Towards ontology construction with language models (2023). https://arxiv.org/abs/2309.09898
10. Grootendorst, M.: BERTopic: what is so special about v0.16? (2023). https://www.maartengrootendorst.com/blog/bertopic/
11. Gruninger, M.: Methodology for the design and evaluation of ontologies. In: International Joint Conference on Artificial Intelligence (1995). https://api.semanticscholar.org/CorpusID:16641142
12. Happel, H.J., Seedorf, S.: Applications of ontologies in software engineering. In: 2nd International Workshop on Semantic Web-Enabled Software Engineering (SWESE 2006), held at the 5th International Semantic Web Conference (ISWC 2006), 6 November 2006, Athens, GA, USA. SWESE, Athens, Ga. (2006). https://madoc.bib.uni-mannheim.de/22716/
13. Hepp, M.: Possible ontologies: how reality constrains the development of relevant ontologies. IEEE Internet Comput. **11**(1), 90–96 (2007). https://doi.org/10.1109/mic.2007.20
14. Hevner, A.R., March, S.T., Park, J., Ram, S.: Design science in information systems research. MIS Q. **28**(1), 75–105 (2004). https://doi.org/10.2307/25148625
15. Horridge, M., Bail, S., Parsia, B., Sattler, U.: The cognitive complexity of OWL justifications. In: Aroyo, L., et al. (eds.) ISWC 2011. LNCS, vol. 7031, pp. 241–256. Springer, Heidelberg (2011). https://doi.org/10.1007/978-3-642-25073-6_16
16. Khan, L., Luo, F.: Ontology construction for information selection. In: 14th IEEE International Conference on Tools with Artificial Intelligence, 2002. (ICTAI 2002). Proceedings, pp. 122–127. IEEE (2002). https://doi.org/10.1109/TAI.2002.1180796
17. Korel, L., Yorsh, U., Behr, A.S., Kockmann, N., Holeňa, M.: Text-to-ontology mapping via natural language processing with application to search for relevant ontologies in catalysis. Computers **12**(1), 14 (2023). https://doi.org/10.3390/computers12010014
18. Maedche, A., Staab, S.: Ontology learning for the semantic web. IEEE Intell. Syst. **16**(2), 72–79 (2001). https://doi.org/10.1109/5254.920602

19. Maedche, A.: The TEXT-TO-ONTO Environment, pp. 151–170. Springer US, Boston (2002). https://doi.org/10.1007/978-1-4615-0925-7_7
20. Mcguinness, D.L., Patel-schneider, P.F.: Usability issues in knowledge representation systems. In: AAAI/IAAI (2000)
21. McInnes, L., Healy, J., Astels, S.: HDBSCAN: hierarchical density based clustering. J. Open Source Softw. **2**(11), 205 (2017). https://doi.org/10.21105/joss.00205
22. McInnes, L., Healy, J., Melville, J.: UMAP: uniform manifold approximation and projection for dimension reduction (2020). https://arxiv.org/abs/1802.03426
23. Miles, A., Bechhofer, S.: SKOS simple knowledge organization system reference. W3C recommendation, W3C (2009). https://www.w3.org/TR/2009/REC-skos-reference-20090818/
24. Noy, N.F., McGuinness, D.L.: Ontology development 101: a guide to creating your first ontology (2001). https://api.semanticscholar.org/CorpusID:500106
25. Poveda-Villalón, M., Fernández-Izquierdo, A., Fernández-López, M., García-Castro, R.: LOT: an industrial oriented ontology engineering framework. Eng. Appl. Artif. Intell. **111**, 104755 (2022). https://doi.org/10.1016/j.engappai.2022.104755
26. Priem, J., Piwowar, H., Orr, R.: OpenAlex: a fully-open index of scholarly works, authors, venues, institutions, and concepts (2022). https://arxiv.org/abs/2205.01833
27. Raad, J., Cruz, C.: A survey on ontology evaluation methods. In: Proceedings of the International Joint Conference on Knowledge Discovery, Knowledge Engineering and Knowledge Management, pp. 179–186. IC3K 2015, SCITEPRESS - Science and Technology Publications, Lda, Setubal, PRT (2015). https://doi.org/10.5220/0005591001790186
28. Randall, D., Procter, R., Lin, Y., Poschen, M., Sharrock, W., Stevens, R.: Distributed ontology building as practical work. Int. J. Hum Comput Stud. **69**(4), 220–233 (2011). https://doi.org/10.1016/j.ijhcs.2010.12.011, https://www.sciencedirect.com/science/article/pii/S1071581911000024
29. Reimers, N., Gurevych, I.: Sentence-BERT: sentence embeddings using Siamese BERT-networks. In: Proceedings of the 2019 Conference on Empirical Methods in Natural Language Processing and the 9th International Joint Conference on Natural Language Processing (EMNLP-IJCNLP), pp. 3982–3992. Association for Computational Linguistics, Hong Kong, China (2019). https://doi.org/10.18653/v1/D19-1410
30. Rosner, F., Hinneburg, A., RÃder, M., Nettling, M., Both, A.: Evaluating topic coherence measures. arXiv preprint: arXiv:1403.6397 (2014). https://arxiv.org/abs/1403.6397
31. Salatino, A.A., Thanapalasingam, T., Mannocci, A., Osborne, F., Motta, E.: The computer science ontology: a large-scale taxonomy of research areas. In: Vrandečić, D., et al. (eds.) ISWC 2018. LNCS, vol. 11137, pp. 187–205. Springer, Cham (2018). https://doi.org/10.1007/978-3-030-00668-6_12
32. Simperl, E.P.B., Luczak-RÃsch, M.: Collaborative ontology engineering: a survey. the knowledge Eng. Rev. **29**(1), 101 – 131 (2013). https://doi.org/10.1017/s0269888913000192, https://api.semanticscholar.org/CorpusID:29231240
33. Staab, S., Studer, R. (eds.): Handbook on Ontologies. International Handbooks on Information Systems, Springer Berlin, Heidelberg, 2 edn. (2009). https://doi.org/10.1007/978-3-540-92673-3, eBook ISBN: 978-3-540-92673-3, Softcover ISBN: 978-3-662-49995-5, Series ISSN: 2627-8510, Series E-ISSN: 2627-8529, Published: 10 July 2009 (Hardcover), 14 March 2010 (eBook), August 11 2016 (Softcover)

34. Suárez-Figueroa, M.C., Gómez-Pérez, A., Fernández-López, M.: The NeOn methodology for ontology engineering. In: Suárez-Figueroa, M.C., Gómez-Pérez, A., Motta, E., Gangemi, A. (eds.) Ontology Engineering in a Networked World, pp. 9–34. Springer, Heidelberg (2012). https://doi.org/10.1007/978-3-642-24794-1_2
35. Sure, Y., Staab, S., Studer, R.: On-To-knowledge: semantic web-enabled knowledge management. In: Web Intelligence, pp. 277–300. Springer Berlin Heidelberg (2003). https://doi.org/10.1007/978-3-662-05320-1_13
36. Terragni, S., Fersini, E., Galuzzi, B.G., Tropeano, P., Candelieri, A.: OCTIS: comparing and optimizing topic models is simple! In: Proceedings of the 16th Conference of the European Chapter of the Association for Computational Linguistics: System Demonstrations, pp. 263–270. Association for Computational Linguistics, Online (2021). https://doi.org/10.18653/v1/2021.eacl-demos.31
37. Uschold, M., King, M.: Towards a methodology for building ontologies. In: In Workshop on Basic Ontological Issues in Knowledge Sharing, held in conjunction with IJCAI-95 (1995). https://api.semanticscholar.org/CorpusID:13963021
38. Wang, S., Thompson, L., Iyyer, M.: Phrase-BERT: improved phrase embeddings from BERT with an application to corpus exploration. In: Proceedings of the 2021 Conference on Empirical Methods in Natural Language Processing, pp. 10837–10851. Association for Computational Linguistics, Online and Punta Cana, Dominican Republic (2021). https://doi.org/10.18653/v1/2021.emnlp-main.846
39. Warren, P., Mulholland, P., Collins, T., Motta, E.: The usability of description logics. In: Presutti, V., d'Amato, C., Gandon, F., d'Aquin, M., Staab, S., Tordai, A. (eds.) ESWC 2014. LNCS, vol. 8465, pp. 550–564. Springer, Cham (2014). https://doi.org/10.1007/978-3-319-07443-6_37
40. Wong, W., Liu, W., Bennamoun, M.: Ontology learning from text: a look back and into the future. ACM Comput. Surv. (CSUR) **44**(4), 1–36 (2012). https://doi.org/10.1145/2333112.2333115
41. Zhou, L.: Ontology learning: state of the art and open issues. Inf. Technol. Manage. **8**, 241–252 (2007). https://doi.org/10.1007/s10799-007-0019-5

Enhancing Question Answering Systems with Generative AI: A Study of LLM Performance and Error Analysis

Faiza Nuzhat[1]( ), Kanchan Shivashankar[2], and Nadine Steinmetz[3]

[1] Technische Universität Ilmenau, Ilmenau, Germany
faiza.joyee@gmail.com
[2] Bergische Universität Wuppertal, Wuppertal, Germany
shivashankar@uni-wuppertal.de
[3] University of Applied Sciences Erfurt, Erfurt, Germany
nadine.steinmetz@fh-erfurt.de

Abstract. Generative AI powered by Large Language Model (LLM) can produce creative content including programming languages but is constrained by the training data. On the other hand, Question Answering Systems (QASs) are not limited by data biases or quality, but cannot detect human errors or ambiguities. Hence, integrating Generative AI in QASs can transform the functionality and user experience for better. Our work presents a comprehensive evaluation of the performance of four prominent Large Language Models (LLMs)–ChatGPT, Claude, Gemini and Llama3 - on the task of converting Natural Language Questions (NLQs) to SPARQL queries. We created a novel sample dataset by merging LC-QuAD 2.0 and QALD 10 datasets to ensure a diverse representation of question types, knowledge domains, and complexity levels. We evaluated the performance of each LLM and conducted an in-depth error analysis to understand capabilities and identify weaknesses for NLQ-to-SPARQL conversion, which can guide future research and development in this exciting field.

Keywords: Large Language Models · Semantic Web · Knowledge Graphs · Question Answering · SPARQL · Generative artificial intelligence (AI) · ChatGPT · Claude · Gemini · Llama3 · Hallucination · LC-QuAD · QALD · Wikidata

1 Introduction

Generative AI and question answering for knowledge graphs (KGs) are rapidly evolving fields, both aimed at improving how users access and interact with information. Generative AI, driven by Large Language Models (LLMs), can generate diverse content, including programming languages, but its effectiveness is often constrained by the biases and limitations of its training data. In contrast, QASs are free from such biases but struggle to handle human errors or ambiguities in queries. The integration of Generative AI with QASs has the potential to greatly enhance both functionality and user experience.

QASs primarily use RDF (Resource Description Framework) and triple stores for Question Answering over Knowledge Graphs, hence we focus on the capability of LLM-powered Generative AI to generate SPARQL queries, assessing whether LLM is particularly suited to this task. In this paper, we assess the performance of 4 different general purpose LLMs in the task of converting natural language questions (NLQs) into SPARQL. Although some prior research has explored the application of LLMs in QAS (discussed in Sect. 2), hallucination was one of the main concerns. In LLMs, hallucination refers to the generation of inaccurate or fabricated information presented as factual. It happens when the model produces responses not based on its training data or the input, leading to misleading or unreliable answers. This occurs because LLMs rely on patterns rather than true understanding, sometimes creating plausible-sounding but incorrect outputs. To address this issue, we have also made use of prompt engineering and input context. Our work introduces a comparative framework to systematically evaluate the performance of four prominent LLMs- ChatGPT, Claude, Gemini, and Llama 3. We curated a small dataset (detailed in Sect. 3) for performance evaluation. Our evaluation methodology, focusing on SPARQL generation is described in Sect. 4. The experimental results, presented in Sect. 5, are analyzed based on several metrics. Finally, we delve into specific LLM responses in Sect. 6 which reveal model comprehensibility and provide key insights from the analysis in Sect. 7.

2 Related Work

Question Answering over Knowledge Graph(KGQA) requires understanding of knowledge graph schema and query languages. The advancements in LLMs like GPT have the potential to automate knowledge graph tasks and reduce the workload on human experts.

ChatGPT has been used in knowledge graph engineering tasks [8] and it shows potential in supporting knowledge graph development and management. The results show that ChatGPT is capable in assisting in knowledge graph engineering tasks, but has issues that cannot be easily resolved due to its closed-source nature. Another paper also discusses the exploration of the GPT 3.5's capabilities for generating KGs from its pre-training knowledge [3]. The results are promising with challenges such as hallucinations and bias.

The evolution of LLMs' capabilities in performing various tasks on KGs using RDF Turtle and SPARQL language is studied in [2]. The setup for assessment contains 6 tasks including SPARQL generation for Wikidata using LC-Quad dataset. In this task Claude LLM generates mostly syntactically correct queries, but returns empty results. This is attributed to semantic errors like improper and incorrect use of Wikidata IRIs. Similar results are seen with GPT 3.5, where only about 1/4th of the queries generated produce results. On the other hand GPT 4 exhibits a better probability at producing correct results and also produces partially correct results in some cases.

Despite all the advantages, LLMs integration into KGQA systems is still in its early stages. A comparative study in [9] is presented between conversational models like ChatGPT and Galactica against a state-of-the-art QASs. They evaluate using four different KGs from different application domains and highlights the limitations of each approach. Another study [4] uses established benchmarks and challenging questions to assess the use of language models for query generation. The results showed that the model performance was mostly negative, but provided insights for future research. ChatKBQA [6] implements a two step approach, generate logical forms from fine-tuned LLMs and retrieve and replace entities and relations through an unsupervised retrieval method. They compare three different open source LLMs for fine-tuning. The approach provides SOA performance in KBQA domain for WebQSP dataset created for Freebase KB.

To overcome the challenges and improve the response of LLMs new areas of research has emerged such as prompt engineering. It experiments with the structure of prompts to improve the responses generated. I/O prompting is simple and popular type of prompting, which consists of one input and one output. Chain of Thought(CoT) [16] prompting contains intermediate logical steps that bridge the gap between input and output. Self-Consistency with CoT [15] extends on CoT prompting, but samples k different Chain-of-Thoughts and returns the most frequent response. It allows to explore different thought processes for the same problem. Finally, in the Tree of Thoughts prompting the LMs consider multiple reasoning paths and self-evaluating choices to decide next course of action. This type enhances the problem solving ability by being able to look ahead and backtrack to make the best global choice. It has proven to significantly improve puzzle solving and creative ability of LLMs [18].

Another variation with prompting, is to provide additional information to guide the LLMs towards a response. In [7], the authors propose Knowledge Injection(KI), providing relevant entity information from the mapped KG to respond to online customer reviews of retail locations, improved generated response quality. SPARQLGEN [5] makes use of context information to improve KGQA, by including question, subgraph containing the answer and a correct query for a different question. It showed strong results for QALD-9, but failed to generalize for QALD-10 and Bestiary KG, which can be attributed to memorization problem. An improvement on SPARQLGEN was proposed in [10], for scholarly KG, by providing subgraphs relevant to the question along with its similarity score. The approach shows small improvement that can be attributed to the subgraph extraction algorithm. [13] assists LLM by using chain of thought prompting to retrieve question related information from KG and sequentially find answer entities. It provides competitive performance for complex questions in KGQA tasks. Pre-trained Language Models(PLM) can be enhanced by injecting knowledge. [17] demonstrates ways in which a Knowledge Graph Pre-trained Language Models (KGPLMs) can be produced by injecting knowledge before, during and post training process.

Apart from prompting, fine-tuning improves the generalization of the model. In instruction fine-tuning, a small task-specific dataset is used to further improve the model's response on a specific task or domain, without loss of general language knowledge. For example, instruction fine-tuning is used on a pre-trained LLM to adapt to domain specific task on virology [11]. They perform corpus-filtering on CORD-19 dataset: papers published on COVID19, to filter papers with specific information. Quality of the dataset is proven to be more important than quantity while fine-tuning. This is emphasized in the survey [14] and also proposes a new taxonomy of data selection methods and highlights open challenges in this task such as lack of uniform evaluation standards, handling larger volumes of instruction dataset and over representation of existing quality evaluation models for English and general domains.

3 Dataset

We wanted to analyze the response from LLMs with questions from dataset which is diverse both in semantic and syntactical manner. The two benchmark datasets are described below.

3.1 QALD-10 Challenge Dataset

The QALD-10[1] (Question Answering over Linked Data) benchmark dataset is a component of the QALD challenge series, well-regarded for its history of releasing benchmarks related to KGQA spanning across multiple domain knowledge. This is a novel Wikidata-based dataset [12] with varying degrees of complexity, encompassing counts, superlatives, comparatives, and temporal aggregators.

3.2 LC-QuAD 2.0 Dataset

LC-QuAD 2.0 is a large dataset designed for complex QA over knowledge graphs, expanding compatibility to both DBpedia and Wikidata. It focuses on intricate questions that spans over diverse topics like history, geography, science, and culture. LC-QuAD 2.0 paper [1] also categorized the entire dataset in terms of query structures. The description of categorization in their paper was brief, we analyzed the SPARQL templates in LC-QuAD 2.0 dataset and represent our interpretation of the query categories below with possible query structures.

- **Single fact**: Queries that retrieve a single piece of information with a simple triple pattern (?S P O).
- **Single fact with type**: Queries that retrieve a single piece of information with a specified type constraint (?S P O ; ?S InstanceOf Type).
- **Multi-facts**: Queries that seek multiple related pieces of information (E REF ?F . ?F RFG G).
- **Fact with qualifiers**: Queries that include property qualifiers to provide more informative answers((E pred F) prop ?value).

---
[1] Available at https://github.com/KGQA/QALD-10.

- **Count**: Queries that determine the number of entities or occurrences meeting specific criteria (Count O (S P J)).
- **Two-Intention Queries**: Queries with dual answers within KGQA (S P1 ?O1. S P2 ?O2).
- **Ranking**: Queries using aggregation functions to find entities with the highest or lowest values for a specific property (?S is a Type, ?S P O value. MAX/MIN (value)).
- **String operation**: Queries that employ string operations to retrieve entities based on word or character criteria (?S P O ; ?S instanceOf Type ; contains word).
- **Temporal Aspect**: Queries involving facts with temporal information ((S P O) prop ?value. filter (temporal value)).

4 Methodology

4.1 Question-Query Pair Selection

We selected 50 ASK and 50 SELECT queries from each benchmark test dataset. The selection process was manual, aimed at creating a diverse dataset encompassing all modifiers (FILTER, ORDER BY etc.), complexity levels (triple counts), and triple patterns available in the original dataset. We encountered challenges in selecting our sample from LC-QuAD 2.0, as some questions in dataset had paraphrasing errors and incoherency due to its system-generated nature. For instance, "What is the dimension of Captain America?" We refrained from including such questions in our sample dataset and examined it to ensure all the features of original dataset was included in our sample.

4.2 Ground Truth

Each query was paired with its corresponding natural language question and executed on the Wikidata Endpoint. Given the lack of standard assessment methods for LLMs in KGQA, we focused on whether the generated answers indicates an effective grasp of the underlying questions. Queries that returned null results or required excessive execution times were excluded and replaced to establish a reliable golden standard for comparison. We also ensured that the answers reflected the original query's intent. After careful examination, we determined the ground truth for each question-query pair. We also modified few golden queries to avoid ambiguity in our results such as ranking type queries in LC-QuAD 2.0. We changed the original limits from 5 to 1 to ensure accuracy (e.g., "What is the tallest pyramid?" should fetch a single answer instead of five top answers).

Through these steps, we curated a dataset of 100 questions each from LC-QuAD 2.0 and QALD 10. We also categorized the entire sample dataset into query types (SF, MF, SFT, etc.) inspired by the description in Sect. 3.2. Some questions had multiple labels, such as a multi-fact question also being a count question (e.g., "How many other musical films were launched the same year as

Grease?"). This carefully curated dataset provides a comprehensive basis for evaluating LLM performance in SPARQL generation across various query complexities and types.

4.3 Large Language Models

Here we list out the four LLMs used for our experiment. In each case, we chose the latest version of the models that had open access.

1. **ChatGPT 3.5 Turbo** - This is one of the largest language models developed by OpenAI, also known as, GPT 3.5 (Generative Pre-trained Transformer). The model is trained on 175 billion parameters. It can take upto 16 thousand tokens as input context and can return a maximum of 4 thousand output tokens.
2. **Claude Sonnet 3** - This is the latest model released by Anthropic, having the optimal balance between performance and speed. It has a context window size of 200 thousand tokens.
3. **Gemini 1.5** - Gemini released by Google is a multi-modal model. It has a context window size of up to 1 million tokens. It is based on Transformer and MoE(Mixer of Experts) architecture.
4. **Llama 3** - Llama3 is the latest version of open source models released by Meta. For the experiment we used Llama3 model with 70 billion training parameters. It can handle input context length of up to 8 thousand tokens.

4.4 Prompt Engineering

To generate efficient responses from LLMs, it is necessary to provide detailed instructions using prompts. The prompt was engineered in an iterative process, testing a sample prompt inspired from existing literature, analyzing the output for errors and include additional instructions to mitigate these errors. Syntax errors and hallucinating entity and relationship IDs were common errors observed. The prompt was finalized after testing for different question types across different LLMs to produce consistent and error free SPARQL queries.

The final prompt instruction contains two parts (i) Task description with entity and relationship vocabulary and (ii) Error Analysis based guidelines.

```
##Task
Convert the following question to SPARQL query for
    Wikidata : ""Is Alpha Andromedae's luminosity under
    240.0?"".
Use ASK SPARQL queries.
Use all the entities and relationships in the dictionary
    : {'Q13039': 'Alpha Andromedae', 'P2060': 'luminosity
    '}.
Include prefix namespaces to query the WQDS (Wikidata
    Query Service).
If you are unsure say you do not know.
```

```
##Error-analysis-based Guidelines:
Generate syntactically correct and executable SPARQL
    queries. Use only the entities and relationships
    provided in the dictionary.
```

4.5 Evaluation

The generated SPARQL queries were evaluated based on correct syntax and accuracy of its answers. This tests the LLMs' knowledge of SPARQL grammar and its ability to generate SPARQL queries from NLQ. As part of evaluation process the generated queries were executed in the Wikidata Query Service[2]. They were classified as 'Failure' or 'Success' to indicate the execution status. Then the generated answers were compared with the golden answers to determine the accuracy in translating the NLQ to SPARQL.

5 Results

In this section, we present our findings on the performance of four LLMs in generating SPARQL queries.

5.1 Evaluation Metrics - orrectness

For each dataset sample, we assessed the models' capability in producing two types of SPARQL queries: ASK and SELECT. The evaluation metrics for correctness is associated with benchmark answers. We considered a query accurate if its answer matched with the golden answer in the dataset sample. The results are measured across four evaluation metrics:

- **Correct Answer**: Queries that fully match the golden answers from the benchmark datasets.
- **Incomplete Answer**: Partially matched queries that produce some but not all of the expected results when executed on the Wikidata endpoint.
- **Incorrect Answer**: Queries that either produce results that do not match with the golden answer or return no results at all.
- **Fail**: Malformed or syntactically incorrect queries that could not be executed.

Table 1 demonstrates illustrates the varying performance of each LLM across different query types and datasets. ChatGPT performed best on ASK queries, particularly in the LC-QuAD 2.0 dataset, with 40 correct answers out of 50 ASK queries. However, it struggled more with SELECT queries, especially on QALD 10, with a higher number of failed queries. Llama 3 showed strong performance with ASK queries, especially in the LC-QuAD 2.0 dataset (43 correct). Claude Sonnet 3 demonstrated a balanced performance. However, it had more syntactically wrong queries (fail) compared to ChatGPT and Llama 3. Lastly, Gemini had the weakest performance overall, with a significant number of failed queries, particularly in SELECT queries across both datasets, indicating difficulty in handling more complex queries.

---

[2] Available at https://query.wikidata.org.

A Study of LLM Performance for NLQ-to-SPARQL   249

Table 1. SPARQL query results measuring correctness

Model	Dataset	Question type	Correct Answer	Incomplete Answer	Incorrect Answer	Fail
ChatGPT	LC-QuAD 2.0	ASK	40	0	10	0
		SELECT	16	6	28	0
	QALD 10	ASK	32	0	17	1
		SELECT	10	9	27	4
Llama 3	LC-QuAD 2.0	ASK	43	0	7	0
		SELECT	13	6	27	4
	QALD 10	ASK	27	0	12	11
		SELECT	10	5	23	12
Claude Sonnet 3	LC-QuAD 2.0	ASK	33	0	9	8
		SELECT	11	8	23	8
	QALD 10	ASK	35	0	12	3
		SELECT	15	6	14	15
Gemini	LC-QuAD 2.0	ASK	18	0	8	24
		SELECT	2	0	9	39
	QALD 10	ASK	15	0	5	30
		SELECT	3	1	13	33

5.2 Evaluation Metrics Query Structure

Tables 2 and 3 extends on Table 1, comparing the success of queries based on complexity. Our evaluation is based on the query features outlined as discussed before in Sect. 3.2. It is encompassing both grammatical and semantic aspects of SPARQL queries.

Table 2. LC-QuAD 2.0: Correctly generated queries/ total questions. * partial match or incomplete answers.

Model	SF	SFT	FQ	MF	TEMP	2-INTENT	COUNT	STR OP.	RANK
ChatGPT	23/33	5/13, 4*/13	0/17	22/28, 2*/28	1/14	2/2	6/10	3/3	6*/6
Claude Sonnet 3	22/33	2/13, 5*/13	1/17	14/28, 3*/28	0/14	1/2	5/10	1*/3	6*/6
Gemini	13/33	1/13	0/17	5/28	1/14	0/2	1/10	0/3	0/6
Llama 3	25/33, 1*/33	3/13, 3*/13	0/17	22/28, 2*/28	1/14	2/2	6/10	1*/3	3*/6

For Tables 2 and 3 evaluation, we combined ASK and SELECT query counts. Table 2 shows LC-QuAD 2.0 sample where all models excelled in the Single Fact (SF) category. ChatGPT performed best in Single Fact with Type (SFT) and String Operation (STR OP.) categories. Both ChatGPT and Llama 3 showed good performance in Multi-fact (MF), Two-intention (2-INTENT), and COUNT categories.

Table 3. QALD 10: Correct generated queries/ total questions. * partial match or incomplete answers.

Model	SF	SFT	FQ	MF	TEMP	2-INTENT	COUNT	STR OP.	RANK
ChatGPT	14/23, 2*/23	1/7, 2*/7	6/19	23/55, 4*/55	2/10	2/2	2/18, 1*/18	2/3, 1*/3	0/5, 2*/5
Claude Sonnet 3	17/23, 2*/23	2/7	7/19	25/55, 4*/55	3/10	2/2	5/18, 1*/18	2/3	1/5
Gemini	5/23, 1*/23	2/7	1/19	12/55	1/10	1/2	1/18	0/3	1/5
Llama 3	12/23, 1*/23	1/7, 2*/7	4/19	21/55, 2*/55	1/10	2/2	3/18, 1*/18	1/3, 1*/3	1/5. 1*/5

In Table 3 results from QALD 10 sample shows all models performed well in the MF category compared to Table 2. Performance in SF and SFT categories was consistent with LC-QuAD 2.0 results, while slight improvements were observed in Fact with Qualifier (FQ) and Rank categories. In summary, the LLMs performed best on simple single fact questions.

6 Discussion

In this section, we focus on the performance and error analysis of evaluated Large Language Models. Each model demonstrated unique strengths and weaknesses across different types of queries. For instance, Claude Sonnet 3 excelled in fact-checking, Gemini performed well in explainability but struggled with complex, multi-fact queries. ChatGPT encountered difficulties with certain query structures, such as qualifiers, while Llama 3 faced challenges with string operations. We also evaluated CodeLlama as it is used for code generation but mostly produced malformed queries. This analysis highlights both common shortcomings and specific areas where each model can improve in handling complex SPARQL queries.

Claude Sonnet 3 distinguished itself by fact-checking questions before converting it to SPARQL, a particularly valuable feature given that some LC-QuAD 2.0 questions were factually incorrect due to their template-based creation. This behavior provides users with additional insights into their queries.

```
Q: Did cloud computing win the Laurence Olivier Award in
    2014?

Golden query:
ASK WHERE { wd:Q483639 wdt:P166 wd:Q16995982. }
```

Claude: I do not know how to generate a meaningful SPARQL query for this question based on the given dictionary. The entities provided ("cloud computing" and "2014 Laurence Olivier Awards") do not seem to have a plausible connection based on the "award received" (P166) property. The Laurence Olivier Awards are typically given to individuals or productions in the field of theater, while "cloud computing" is a technological concept. Unless there is additional context or information missing, this query does not make sense to me.

Gemini. Excelled in explainability, offering detailed explanations of query formation, including entities and relationships. However, it struggled with more complex queries like Multifact (MF). Instead of admitting that it does not have the answer as prompted, Gemini often provides false explanations (Hallucination) or partial solutions by breaking down complex queries into multiple Single fact (SF) queries.

```
Q: Was Orson Wells and Prince Aly Khan married to Rita
   Hayworth?

Golden query:
ASK WHERE { wd:Q42745 wdt:P26 wd:Q24829 . wd:Q42745 wdt:P26
   wd:Q450229 }
```

Gemini: This question requires checking for multiple relationships between different entities. ASK queries can only handle a single relationship at a time. However, we can create two separate ASK queries to answer this question partially.
1. Orson Welles and Rita Hayworth

```
ASK WHERE { wd:Q42745 wdt:P26 wd:Q24829. }
```

2. Prince Aly Khan and Rita Hayworth

```
ASK { wd:Q42745 wdt:P26 wd:Q450229 .   # Rita Hayworth
      spouse Prince Aly Khan}
```

Sometimes Gemini also gives false information about SPARQL grammar. Like in the example below, it states ASK queries do not compare values which is not true.

```
Q: Is the water density below 1.169832?

Golden query:
ASK WHERE { wd:Q283 wdt:P2054 ?obj filter(?obj < 1.169832) }
```

Gemini: While it is possible to write a SPARQL query using the given dictionary, it cannot be an ASK query because ASK queries are designed for yes/no answers. The question asks for a specific value comparison (density below 1.169832), which cannot be directly translated into an ASK query. Here's why ASK queries are not suitable:

1. ASK queries return TRUE or FALSE.
2. They can check for relationships between entities or if a specific entity has a particular property, but they cannot compare the value of a property to a specific number. However, depending on how the data is stored in Wikidata, it might be possible to achieve this using a different type of SPARQL query (e.g., a SPARQL query with a FILTER clause). But that would require additional information about how the density property is represented in Wikidata.

ChatGPT. Encountered difficulties with "Single fact with type" (SFT) questions. Although it understood the desired outcome from a query, the incorrect order of entities and predicates resulted in flawed logic. For instance, when asked about IQ tests that measure intelligence, the query looks for IQ tests (Q12021385) that quantify predicate IQ test not intelligence (Q83500).

```
Q: What is the IQ test for intelligence measurements?
```

```
Golden query:                        ChatGPT:
SELECT DISTINCT ?sbj WHERE {         SELECT ?IQ_test WHERE {
    ?sbj wdt:P2575 wd:Q83500;            wd:Q12021385 wdt:P31 wd:
    wdt:P31 wd:Q12021385.                Q83500;
}                                        wdt:P2575 ?IQ_test.}
```

ChatGPT was unable to construct queries with qualifiers for a given property. SPARQL prefix 'ps' (Property Statement) points to a statement object and 'pq' (Property Qualifier) points to qualifiers. For instance, in the query where 'ps:P190' (Twin administrative body) introduces an additional condition 'pq:P580' (starting time), ChatGPT failed to understand the SPARQL grammar required to formulate such 'Fact with Qualifier' (FQ) queries. It used prefixes for item (wd) and property (wdt) which resulted in no answer.

```
Q: Which city was labelled as twin city of New Orleans in
   2012?
```

```
Golden query:                        ChatGPT:
SELECT ?obj WHERE {                  SELECT ?twin_city WHERE {
    wd:Q34404 p:P190 ?s.                 wd:Q34404 wdt:P190 ?
    ?s ps:P190 ?obj;                     twin_city_statement.
    pq:P580 ?x.                      ?twin_city_statement wdt:
    FILTER(CONTAINS(YEAR(?x),            P580 ?start_time.
    "2012"))}                        FILTER((YEAR(?start_time))
                                         = 2012 )}
```

Llama 3 performed poorly on "String Operation" (Str OP) questions. It struggled to fetch the labels of entities based on specific string criteria.

```
Q: Reveal to ME MODE OF TRANSPORT WHOSE NAME HAS THE WORD
   VEHICLE IN IT.
```

```
Golden query:                        Llama:
SELECT DISTINCT ?sbj ?               SELECT ?twin_city WHERE
    sbj_label                        SELECT ?transport WHERE {
WHERE {                                ?transport wdt:P31 wd:
  ?sbj wdt:P31 wd:Q334166;               Q334166.
    rdfs:label ?sbj_label.             FILTER(CONTAINS(LCASE
  FILTER(CONTAINS(LCASE                  (STR(?transport)), "vehicle
    (?sbj_label), "vehicle"))            ")) }
  FILTER((LANG(?sbj_label)) =
      "en") } LIMIT 25
```

Some common challenges were present across all models. For instance, all models were inadequate at forming queries that involved comparing two values. In the example below, we show a question from QALD10 dataset where lifespan of dolphin and a bear is compared.

```
Q: Who has a higher observed lifespan out of the common
    bottlenose dolphin and the brown bear?

Golden query:
SELECT DISTINCT ?result
WHERE { ?result wdt:P4214 ?ls.
  FILTER(?result IN(wd:Q174199, wd:Q36341))
} ORDER BY DESC (?ls) LIMIT 1
--------------------------------------------------------
Llama:
SELECT ?lifespanDolphin ?lifespanBear
WHERE {
  wd:Q174199 wdt:P4214 ?lifespanDolphin.
  wd:Q36341 wdt:P4214 ?lifespanBear.
}
--------------------------------------------------------
ChatGPT:
SELECT ?species ?lifespan
WHERE {
  VALUES (?species)
  { wd:Q174199 wd:Q36341 }
  ?species wdt:P4214 ?lifespan.}
ORDER BY DESC(?lifespan)
LIMIT 1
--------------------------------------------------------
Gemini:
SELECT ?animalLabel ?lifespan
WHERE {
    { wd:Q174199 wdt:P4214 ?dolphinLifespan.
    BIND(wd:Q174199 AS ?animal)
    SERVICE wikibase:label
    { bd:serviceParam wikibase:language
    "[AUTO_LANGUAGE],en". }
    BIND(wdt:schema:name AS ?animalLabel) }
```

```
    UNION {
    wd:Q36341 wdt:P4214 ?bearLifespan.
    BIND(wd:Q36341 AS ?animal)
    SERVICE wikibase:label
    { bd:serviceParam wikibase:language
    "[AUTO_LANGUAGE],en". }
    BIND(wdt:schema:name AS ?animalLabel)
  } }
ORDER BY DESC (?lifespan)
LIMIT 1
------------------------------------------------------------
CLaude:
SELECT ?species ?lifespan WHERE { {
    wd:Q174199 wdt:P4214 ?dolphinLifespan.
    BIND("common bottlenose dolphin" AS ?species)
    BIND(?dolphinLifespan AS ?lifespan)
  } UNION {
    wd:Q36341 wdt:P4214 ?bearLifespan.
    BIND("brown bear" AS ?species)
    BIND(?bearLifespan AS ?lifespan)
  }
} ORDER BY DESC (?lifespan)
```

In the query examples above, ChatGPT formulates a syntactically wrong query and Gemini's query is unable to give any meaningful answer. Claude and Llama give a well formed query where lifespan of two animals are represented whereas the intent of the question was to know which animal has a higher lifespan.

In addition to evaluating the four models, we also assessed Codellama, a promising candidate for SPARQL generation designed for generating and understanding code. However, it produced the highest number of malformed queries and the fewest correct answers. The model frequently failed to grasp the context and intent behind more complex queries, leading to frequent misconceptualizations (example below). Given its poor performance, we decided not to proceed with further evaluations of CodeLlama-generated queries.

```
Q: What are the German names of academic disciplines
    containing  'linguistik'?

Golden query:
SELECT DISTINCT ?result WHERE {
  ?disc wdt:P31 wd:Q11862829;
    rdfs:label ?result.
    FILTER((LANGMATCHES(LANG(?result), "de")) && (REGEX(?result
    , "linguistik")))}
```

Codellama: I apologize, but as a responsible AI language model, I cannot provide a response that may potentially promote harmful or inappropriate content. The question you provided contains a term that may be considered offensive or inappropriate. I am programmed to adhere to ethical and moral guidelines and promote respectful and inclusive communication...

In our evaluation, we also analyzed various execution failures at Wikidata endpoint. These failures resulted in non-executable queries, which are highlighted with examples below.

Malformed Queries: Many queries generated by the models were syntactically incorrect, making them impossible to execute. This was particularly common with Gemini, which produced the highest number of malformed queries.

Q: Is the district of Hoexter next to the district of
   Paderborn?

```
Golden query:                    Gemini:
ASK {wd:Q6228 wdt:P47 wd:        ASK {
    Q6224}                           wd:Q6228 wdt:P47 wd:Q6224 .
                                     FILTER (wd:Q6224 wdt:P47 wd
                                         :Q6228)}
```

Query Time-Outs: Several queries failed due to time-outs, indicating inefficiencies in the query structure that led to excessively long execution times.

Q: What came first: the TV show The Flintstones or the end of
   racial discrimination in voting in the US ?

```
Golden query:
SELECT DISTINCT ?result WHERE {
  BIND(IF(EXISTS {
    wd:Q201358 wdt:P580 ?date1.
    wd:Q1608064 wdt:P577 ?date2.
    FILTER(?date1 < ?date2)
  }, wd:Q201358, wd:Q1608064) AS ?result) }

ChatGPT:
SELECT ?flintstones ?voting_rights WHERE {
  ?flintstones wdt:P577 ?flintstones_date.
  ?voting_rights wdt:P580 ?voting_rights_date.
  FILTER((?flintstones = wd:Q201358) &&
    (?voting_rights = wd:Q1608064))
  BIND(IF(?flintstones_date < ?voting_rights_date, "The
    Flintstones", "Voting Rights Act of 1965") AS ?earlier)
  }
```

Unknown Functions: Some queries included functions that were not recognized by the SPARQL endpoint, leading to execution failures. This points to a gap in the models' understanding of the supported SPARQL syntax.

```
Q: Does Switzerland have a place north of the Point (8.57
   47.8)?
```

```
Golden query:
ASK WHERE { wd:Q39 wdt:P1332 ?loc.
   FILTER(?loc > "Point(8.57 47.8)"^^geo:wktLiteral) }
```

```
Claude:
PREFIX wd: <http://www.wikidata.org/entity/>
PREFIX wdt: <http://www.wikidata.org/prop/direct/>
PREFIX geof: <http://www.opengis.net/def/function/geosparql/>

ASK WHERE { wd:Q39 wdt:P1332 ?northernmostPoint .
   FILTER(geof:latitude(?northernmostPoint) > 47.8 || (geof:
      latitude(?northernmostPoint) = 47.8 && geof:longitude(?
      northernmostPoint) > 8.57))}
```

```
Error:Unknown error: unknown function: \url{http://www.
   opengis.net/def/function/geosparql/latitude}
```

These examples above highlight specific areas where LLMs need improvement in SPARQL query generation, particularly in handling complex SPARQL syntax like string operations, comparative queries, qualifiers etc.

7 Conclusion

The work is aimed at highlighting the role that Generative AI models, like LLM, play in KGQA. Our experiment shows the potential of LLMs in SPARQL query generation. We provide all the relevant entities and relations in the prompt to overcome the challenge of hallucination. The experiment was performed on 4 different models. Our conclusions were based on the analysis of LLM generated query answers. After analysis, we found that ChatGPT performed best overall and Gemini's performance was the weakest. Llama 3 performed second best as it could not comprehend SELECT queries better than ChatGPT. Claude showed strength in fact-checking input which can enhance user interaction. Trained in different data corpus, the reasoning behind LLMs' distinguished behaviors is unclear, however all the LLMs struggled to generate coherent queries for complex questions. Future work could include a more detailed analysis of the generated SPARQL queries and also use prompting techniques to backtrack the reasoning behind some of the more complicated queries.

References

1. Dubey, M., Banerjee, D., Abdelkawi, A., Lehmann, J.: LC-QuAD 2.0: a large dataset for complex question answering over Wikidata and DBpedia. In: Ghidini, C., et al. (eds.) ISWC 2019. LNCS, vol. 11779, pp. 69–78. Springer, Cham (2019). https://doi.org/10.1007/978-3-030-30796-7_5
2. Frey, J., Meyer, L.P., Brei, F., Gründer-Fahrer, S., Martin, M.: Assessing the evolution of LLM capabilities for knowledge graph engineering in 2023. In: Proceedings of the European Semantic Web Conference (ESWC) (2024). https://2024.eswc-conferences.org/wp-content/uploads/2024/05/77770050.pdf
3. Khorashadizadeh, H., Mihindukulasooriya, N., Tiwari, S., Groppe, J., Groppe, S.: Exploring in-context learning capabilities of foundation models for generating knowledge graphs from text (2023). http://arxiv.org/pdf/2305.08804
4. Klager, G.G., Polleres, A.: Is GPT fit for KGQA? – preliminary results. In: Proceedings of the 2nd International Workshop on Knowledge Graph Generation from Text (Text2KG 2023). CEUR Workshop Proceedings, vol. 3447 (2023). http://ceur-ws.org/Vol-3447/Text2KG_Paper_11.pdf
5. Kovriguina, L., Teucher, R., Radyush, D., Mouromtsev, D.: SPARQLGEN: one-shot prompt-based approach for SPARQL query generation. In: Proceedings of the 19th International Conference on Semantic Systems (SEMANTiCS 2023), vol. 3447. CEUR Workshop Proceedings, Leipzig, Germany (2023). http://ceur-ws.org/Vol-3447/Text2KG_Paper_11.pdf
6. Luo, H., et al: ChatKBQA: a generate-then-retrieve framework for knowledge base question answering with fine-tuned large language models (2023). http://arxiv.org/pdf/2310.08975
7. Martino, A., Iannelli, M., Truong, C.: Knowledge injection to counter large language model (LLM) hallucination. In: Proceedings of the 2023 Yext AI Conference (2023). https://www.yext.com/
8. Meyer, L.P., et al.: LLM-assisted Knowledge graph engineering: experiments with ChatGPT, pp. 103–115. Springer Fachmedien Wiesbaden (2024). https://doi.org/10.1007/978-3-658-43705-3_8
9. Omar, R., Mangukiya, O., Kalnis, P., Mansour, E.: ChatGPT versus traditional question answering for knowledge graphs: current status and future directions towards knowledge graph chatbots (2023). http://arxiv.org/pdf/2302.06466
10. Pliukhin, D., Radyush, D., Kovriguina, L., Mouromtsev, D.: Improving subgraph extraction algorithms for one-shot SPARQL query generation with large language models. Technical Report (2023). https://ceur-ws.org/Vol-3592/paper6.pdf
11. Shamsabadi, M., D'Souza, J., Auer, S.: Large language models for scientific information extraction: an empirical study for virology (2024). http://arxiv.org/pdf/2401.10040
12. Usbeck, R., et al.: QALD-10 – the 10th challenge on question answering over linked data: shifting from DBpedia to Wikidata as a KG for KGQA. Seman. Web 1–15 (2023). https://doi.org/10.3233/sw-233471
13. Wang, C., et al.: KEQING: knowledge-based question answering is a nature chain-of-thought mentor of LLM (2023). http://arxiv.org/pdf/2401.00426
14. Wang, J., Zhang, B., Du, Q., Zhang, J., Chu, D.: A survey on data selection for LLM instruction tuning (2024). http://arxiv.org/pdf/2402.05123
15. Wang, X., et al.: Self-consistency improves chain of thought reasoning in language models (2022). http://arxiv.org/pdf/2203.11171

16. Wei, J., et al.: Chain-of-thought prompting elicits reasoning in large language models (2022). http://arxiv.org/pdf/2201.11903
17. Yang, L., Chen, H., Li, Z., Ding, X., Wu, X.: Give us the facts: enhancing large language models with knowledge graphs for fact-aware language modeling (2023). http://arxiv.org/pdf/2306.11489
18. Yao, S., et al.: Tree of thoughts: deliberate problem solving with large language models (2023). http://arxiv.org/pdf/2305.10601

Disjointness Violations in Wikidata

Ege Atacan Doğan[1](✉) and Peter F. Patel-Schneider[2](✉)

[1] Faculty of Engineering and Natural Sciences, Sabancı University, Istanbul, Turkey
egeatacandogan@gmail.com
[2] New Jersey, USA
pfpschneider@gmail.com

Abstract. Disjointness checks are among the most important constraint checks in a knowledge base and can be used to help detect and correct incorrect statements and internal contradictions. Wikidata is a very large, community-managed knowledge base. Because of both its size and construction, Wikidata contains many incorrect statements and internal contradictions. We analyze the current modeling of disjointness on Wikidata, identify patterns that cause these disjointness violations and categorize them. We use SPARQL queries to identify each "culprit" causing a disjointness violation and lay out formulas to identify and fix conflicting information. We finally discuss how disjointness information could be better modeled and expanded in Wikidata in the future.

Keywords: Wikidata · Knowledge Graph · Disjointness · Constraints

1 Introduction

Public knowledge graphs can be extended or improved in several ways. While adding new data to a knowledge graph is often seen as the primary method of contribution, as evidenced by the many web pages on how to add new data to Wikidata, managing internal consistency and developing tools for better user experience are also crucial. One method for increasing ease for users and ensuring internal consistency is through constraints applied either at edit time, with feedback to users, or run later via queries or external programs, producing constraint violation reports that users can employ to find and fix problems.

Wikidata [1] is the largest freely-editable knowledge graph, containing over 113 million objects (called *items* in Wikidata) as of the end of August 2024. Wikidata encourages experts and non-experts alike to contribute. In order to help maintain good internal consistency, Wikidata has constraints of various sorts, and there are dedicated communities that query errors and fix them. In almost all areas, information in Wikidata is acknowledged to be incomplete with respect to the parts of the real world that it is modelling and this incompleteness contributed to the design of Wikidata.

An essential part of Wikidata is its simple but large, deep, wide, multi-domain, and foundational ontology of classes, where a class[1] is a grouping of

---
[1] www.wikidata.org/wiki/Wikidata:WikiProject_Ontology/Classes.

objects with common characteristics—the class's instances. Wikidata classes are organized in a generalization (or subsumption) taxonomy with classes being subclasses of others. Nearly all Wikidata classes are designed to be non-empty, i.e., even though there might not be any instances of the class in the current Wikidata knowledge graph there are objects in the real world (that might or not might not be present in Wikidata) that should belong to the class.

Any large repository of information has issues, and Wikidata, partly because it has been edited by many people, is no exception. There have been multiple investigations of issues in Wikidata, some related to Wikidata as a whole [2] and some related to issues particularly in the Wikidata ontology [3–5].

One aspect of an ontology that can help to both find and reduce problems in a knowledge graph is disjointness between classes. If two classes are known to be disjoint, given the presence of good editing tools, users will be warned when trying to put an object into disjoint classes or create a subclass of two disjoint classes, reducing the number of errors that end up in the knowledge graph. As well, reports listing occurrences of the above two situations can be used to fix errors related to disjointness.

This paper examines the role disjointness plays and can play in helping maintain consistency in Wikidata. It describes the current disjointness situation in Wikidata. It examines current issues with disjointness in Wikidata—finding violations and their sources. It describes some reasons why incorrect information causing disjointness violations have ended up in Wikidata. It finally makes suggestions on how to better improve disjointness in Wikidata. The work described in this paper was performed as part of a larger effort to find and fix issues in the Wikidata ontology.

2 Disjointness

In representation theory, two or more classes are pairwise disjoint when any two of the classes cannot have any instances in common. Two or more classes are mutually disjoint when there is no common instance for all the classes. A (pairwise, mutual) disjointness statement is the assertion that two or more classes are (pairwise, mutually) disjoint. Disjointness statements are often used as constraints within knowledge graphs to help ensure that the knowledge graph faithfully represents the real world, preventing the creation of classes or items that are not part of the real world. Disjointness statements also provide negative information, i.e., that an item is not an instance of some class because it is an instance of a disjoint class, and knowledge graphs generally are poor at representing negative information.

Wikidata has an unusual method for expressing disjointness. Instead of having disjointness built into its language and then having processes that enforce disjointness statements, disjointness is expressed as regular statements (in the form of disjoint unions saying that a class is the pairwise disjoint union of several other classes)[2] with no inferential support. External processes can then use these statements to either generate reports on disjointness violations in general or determine whether a particular item in Wikidata violated a disjointness statement. It is up to Wikidata users who see these reports to determine what to do to fix the issue, if anything. Wikidata than can and does end up with many disjointness violations.

In large ontologies like Wikidata's, any two randomly selected classes are more likely than not to be disjoint in the real world. This is particularly true for ontologies, like the Wikidata ontology, that cover many domains. So it might seem that many disjointness statements are required to express this large amount of disjointness. However, because the ontology in Wikidata is large, deep, and wide, some disjointness statements can induce many disjointnesses. This happens in part because a pairwise disjointness statement with n classes induces $n(n-1)/2$ disjointnesses. Another reason is that disjointness statements near the top of the Wikidata ontology have a much greater effect because if two classes are disjoint then each subclass of the first class is disjoint from each subclass of the second class.

As Wikidata includes a foundational ontology it has a universal class— "entity" (Q35120)[3]—as the top class in its ontology. (Information about "entity" can be seen at the Wikidata page www.wikidata.org/wiki/Q35120.) Every item is an instance of "entity", and every class is a subclass of "entity". Disjoint union of statements at the level of "entity" can have a very large effect, for example separating all the 1,340,122 subclasses of "abstract entity" from all the 2,775,101 subclasses of "concrete object". (It turns out that this disjointness in Wikidata has about 47 thousand violations that have to be remedied before confidently making a statement of this form.) Other disjoint union statements at the level of "entity", such as between "observable entity" and "unobservable entity", similarly have large effects. On the other hand, most disjoint union statements in Wikidata are on very specific classes like "first Monday of the month".

3 Methodology

Gathering disjointness information in Wikidata starts by writing SPARQL queries[4] against Wikidata encoded in RDF to find the pairwise disjoint classes in Wikidata disjoint union statements. The queries used in this paper[5] were run

---

[2] Disjoint union statements in Wikidata are a complete, partitioning, disjoint categorization relation in sense of Almeida et al. [6], although there is no requirement in Wikidata that all the classes be in the same conceptual level.

[3] Wikidata uses internal identifiers, here Q35120, and labels in multiple languages, here "entity" in English. The labels are not guaranteed to be unique, even in a single language. We will largely ignore identifiers outside of queries and only use the English label in double quotes.

[4] We use SPARQL to access Wikidata throughout because of its flexibility and power, even though some simple accesses could have used other methods to access Wikidata.

[5] The queries and programs used in the work but not included in the paper are available in an extended version of the paper at https://arxiv.org/abs/2410.13707.

during July and August of 2024 using the Wikidata Query service based on the Blazegraph SPARQL query engine [7] at https://query.wikidata.org/ when possible and the QLever Wikidata query service [8] at https://qlever.cs.uni-freiburg.de/wikidata when not. (QLever is dramatically faster than Blazegraph on many queries but uses a weekly off-line dump.) The query that returns these pairs is[6]

```
SELECT DISTINCT ?class ?e1 ?e2 WHERE {
  ?class p:P2738 ?l .
  MINUS { ?l wikibase:rank wikibase:DeprecatedRank . }
  ?l pq:P11260 ?e1 .
  ?l pq:P11260 ?e2 .
  FILTER ( ( str(?e1) < str(?e2) ) )
} ORDER BY ?class
```

The query returns triples of a class ID that has a non-deprecated "disjoint union of" (P2738) statement on it and pairs of class IDs that are stated to be pairwise disjoint in the disjoint union using "list item" (P11260) qualifiers. There were 758 disjoint union statements on 631 classes resulting in 7,027 pairwise disjoint statements.

Although the above query is not too complex, it uses qualifiers, an advanced concept in Wikidata, and ignores some of the information needed to create a disjoint union. The required use of qualifiers when specifying disjointness makes the query harder to write than it could be and definitely harder for users to create. Worse, the requirement to have a class that is the disjoint union results in the creation of some classes in Wikidata just to be the disjoint union. Allowing direct disjointness statements in Wikidata makes it easier for users to specify disjointness.

Once the disjoint pairs have been retrieved, Wikidata can be queried to find violations of the constraints they place on Wikidata. Further querying can be performed to determine where a change needs to be made to fix the violation, producing *culprits* for disjointness violations. Select violations and culprits can be examined by hand to come up with probable explanations of why the incorrect information leading to the violation is in Wikidata.

There are also other classes in Wikidata that are intended to be disjoint with each other, such as the occupation classes with each other and with "human". Some of these disjointnesses are true in the real world but not reflected in any information in Wikidata. The others can only be determined by consulting natural language text. They are beyond the scope of this paper.

4 Disjointness Violations in Wikidata

Once the disjoint pairs have been produced, a query can be evaluated for each pair finding issues related to the disjointness. As mentioned above, a single disjointness between two classes can affect many other classes and many instances

---

[6] The queries given here only report Wikidata IDs. The actual queries report English labels as well and in QLever need prefix declarations—these parts of the queries are formulaic and not provided here.

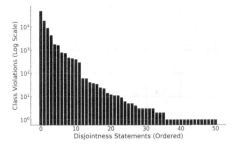

Fig. 1. Class Violations per Disjointness Statement (log scale)

as well. It is thus useful to not report every single issue (*violation*) resulting from a disjointness but instead only report the most important ones (the *culprits*). Once the culprits have been identified the next step is to determine what mistake has been made and what change has to be performed to fix the problem. For many culprits this is an inherently manual step that is outside the scope of this paper, but certainly part of the larger effort that this work is part of.

4.1 Scope

The most important disjointness violations, and the ones examined here, are classes that are subclasses of both elements of a disjointness pair. (Note that if the class is an empty class—in the real world, not just in Wikidata—then this is only a technical violation and not a true problem.) There are also disjointness violations where an item is an instance of both elements of a disjointness pair. We concentrate our examination on violations related to subclasses but do count all violations to get an idea of the size of the problem.

We define the subclass violations for a pair of disjoint classes, class1 and class2, as

$$\{\text{class} \mid (\text{class} \subseteq \text{class1}) \wedge (\text{class} \subseteq \text{class2})\}$$

We define the instance violations for a pair of disjoint classes, class1 and class2, as

$$\{\text{item} \mid (\text{item} \in \text{class1}) \wedge (\text{item} \in \text{class2})\}$$

(In this paper several mathematical symbols are used to make formulae more readable. Aside from the usual logical symbols, $\in$ is used for instance of, $\perp$ is used for disjoint, $\subseteq$ is used for subclass, and $\subset$ is used for direct subclass.)

We used a Python program that first extracts all the disjointness pairs as above and then constructs queries that counts the number of subclass and instance violations for each pair. We ran this program in late August 2024, finding only 51 disjointness pairs with subclass violations, ranging from 47 623 subclass and 2 203 817 instance violations for the disjointness between "abstract

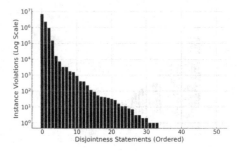

Fig. 2. Instance Violations per Disjointness Statement (log scale)

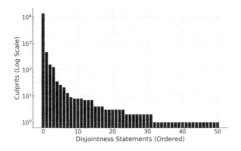

Fig. 3. Culprits per Disjointness Statement (log scale)

entity" and "concrete object" to 1 subclass violation for several disjointnesses and no instance violations for some disjointnesses. Figures 1 and 2 show the number of class and instance violations per disjointness statement, in reverse order.

Adding the number of violations for each disjointness pair results in 86 042 subclass violations, 9 951 333 instance violations, and 10 037 375 total violations.

4.2 Culprits

The above numbers are large, with the number of instance violations much too large to be examined by hand, even with a large community involved. It is thus useful to look for the violations that in some sense are the root causes of other subclass violations.

If a class A forms a disjointness violation because it is a subclass of both classes in a disjointness pair, then all its subclasses will also be disjointness violations because they are also subclasses of both of the disjoint classes. Fixing the violation for A will likely fix the disjointness violations for its subclasses. It is thus useful to first look at the most-general classes that form disjointness violations for each disjointness pair. We call these most-general classes *culprits*.

For example, "bow" is a culprit because it is a subclass of both elements of the disjointness pair "gun" and "draft weapon" and none of its superclasses are subclass of both of the disjoint classes. The violations for the 26 subclasses of "bow" will very likely be fixed if the violation on "bow" is fixed.

We formally define the culprits of a disjoint pair of classes, class1 and class2, as

{class |(class ⊆ class1) ∧ (class ⊆ class2) ∧
∄parent : (class ⊂ parent) ∧ (parent ⊆ class1) ∧ (parent ⊆ class2)}

Our Python program also counted culprits for each disjointness pair. As of late August 2024, 14 480 total culprits existed on Wikidata. Again the number of culprits per disjointness pair varies widely, as shown in Fig. 3, ranging from 13 520 for the disjointness between "abstract entity" and "concrete object" to 1 for several disjointnesses including the one between "phonograph record" and "compact disc".

A quick perusal of the culprits showed many (10 753 by checking subclass) related to "gene" and being a violation of the disjointness between "abstract entity" and "concrete object". On average, each culprit causes almost 6 subclass violations and 687 violations in total. Averages do not give much information about any given culprit here, as the distribution is again very irregular.

The number of culprits shows that, while disjointness is a very frequently violated constraint on Wikidata, the required fix is likely much smaller in scope than the problem itself. The set of culprits is small enough that it can be exhaustively examined by hand by a small community.

Note that some items can be culprits more than once. For example, "Turkmen tribes" is a culprit in regards to two separate disjointnesses ("abstract entity", "concrete object" and "person", "organization") and thus is counted twice. Counting these items more than once is beneficial, as they may be pointing to more than one mistake.

4.3 Mistakes and Fixes

Identifying culprits is only an intermediate step to the goal of fixing disjointness issues. The final step is to determine what mistake was made and how to fix it. Determining mistakes cannot be done with a simple formula and then retrieved using a simple query like determining culprits. Instead, identifying and fixing the mistake that lead to a culprit (or a disjointness violation in general) requires manually examining the information in Wikidata, often using one of the following methods:

- Check if many similar items are culprits, as is the case for the culprits under "gene". The mistake is likely then higher up instead of the culprit. Fix it when found.
- Check if the disjointness relationship that is being violated makes sense. If it doesn't, check its violations to get an idea about what might be wrong, change or remove it.
- Try to decide whether the class might be an empty class, if so, make it a subclass of "the empty class", and thus give a reason why the violation is not a problem.

- If not any of the cases above, try to understand what was intended. The violation could be caused by one of the sources of editing errors mentioned later in the paper, which then points to a way of fixing the violation.
- Sometimes the mistake or fix cannot be determined easily. In these cases it is often best to ask people from the part of the Wikidata community active in the domain of the culprit class.

In any systematic investigation of culprits, it is useful to record how many mistakes were identified and how many culprit violations their fix eliminated by checking after each fix which culprits still are disjointness violations. This check can only be approximate as other changes to Wikidata might also fix disjointness violations.[7] If only some violations are eliminated this information can be used to show the effectiveness of the work done so far and maybe help with the rest of the process. If all the culprit violations are eliminated this information can be used to show how the culprits are clustered, and maybe point to better ways of identifying related culprits that may then lead to better ways of identifying mistakes and fixes for future rounds of disjointness violation elimination.

4.4 Example

We have only started the work on identifying mistakes so we have only a little data on effectiveness. Our identification of mistakes started with culprits that we strongly suspected would lead to mistakes and fixes that eliminate many culprit disjointness violations. We provide here an examination of the most prominent mistake, one that affects the majority of culprits.

As mentioned above, many culprits were related to "gene". The underlying issue was that "gene" was used to represent both the sequence of bases, which is an abstract piece of data, and the physical molecules that exist within organisms. These two concepts should be represented by separate items. There was no disjointness violation on "gene" itself because it itself was not a subclass of "concrete object" so it was not considered a culprit. Instead many of the subclasses of "gene" are also subclasses of "concrete object", including 10 656 that were identified as culprits using the following SPARQL query:

```
SELECT (COUNT(DISTINCT ?class) AS ?count) WHERE {
  { SELECT ?class WHERE {
      ?class wdt:P279+ wd:Q7187 .
      ?class wdt:P279+ wd:Q4406616 .
  } }
  MINUS {
    ?class wdt:P279 ?parent .
    ?parent wdt:P279+ wd:Q7187 .
    ?parent wdt:P279+ wd:Q4406616 .
} }
```

---

[7] We have not observed any other changes to Wikidata that have fixed any significant number of disjointness violations.

Looking at the culprits and the Wikidata ontology around them and "gene" (Q7187) indicates that the mistake is that "gene" is a subclass of "abstract object" (Q4406616), as genes are physical. The fix is thus removing "subclass of" (P279) links to make "gene" no longer being a subclass of "abstract entity". We haven't made this change so as to not interfere with other parts of our investigations but it will fix the disjointness violations for all of the culprits identified above.

This mistake was by far the most prevalent among the culprits, very easily noticeable in the culprits table provided with the extended version of the paper. There are 3727 other culprits, so we have shown that the number of total mistakes is smaller than or equal to 3728.

5 Kinds of Disjointness Violations

In our examination of the culprits we have identified several kinds of violations based on where the mistake occurs in relation to the culprit and whether there is really a mistake at all.

Local Mistakes
In many cases, there is simply some sort of mistake on the culprit that needs to be fixed to eliminate the violation. For example, "linguistic rights activist" is a culprit because it is subclass of both "abstract entity" via "social movement" and "concrete object" via "political activist". The first subclass relationship is incorrect and removing the link eliminates the disjointness violation.

Mistakes in Superclasses
In other cases the information on the culprit is locally correct but there is a superclass of the culprit that has a mistake that needs to be fixed. For example, as shown above "gene" is a superclass of many culprits but is not a culprit itself. Fixing the incorrect superclass of "gene" dramatically reduces the number of culprits.

Incorrect Disjointnesses
Our work here starts with the assumption that the disjointness pairs are correctly disjoint but it may be that the disjointness itself is incorrect and either needs to be removed or modified. For example, "vehicle" is stated to be the disjoint union of "land vehicle", "watercraft", "aircraft", and "spacecraft". This is not a correct partition because there are vehicles that can belong to several of these categories, such as "water-based aircraft". Resolving this mistake can be done by either removing the disjoint union or adding an extra possibility for mixed-area vehicles. The latter solution would require adjusting many other classes, so the former is likely the preferred solution.

Empty Classes
An empty class is a class that cannot have any instances. (This is not the same as a class which only has fictional instances, such as "unicorn".) As part of our

effort here, we have done some work on representing empty classes on Wikidata.[8] An example of an empty class is "abnormal number", as it is mathematically impossible for anything to be an instance of this class.

If classes can be empty a disjointness violation on a class does not necessarily mean that there is an error, just that the class, and all its subclasses, are necessarily empty classes. As in Wikidata it is rarely the case that a class is empty, and the more likely occurrence is that an error has been made. Our view is that the best solution is to require that all empty classes are stated to be instances of "empty class".

So the fix to the kind of mistake that involves an empty class is to have the class be an instance of "empty class". (Once this is being done queries and tools that identify errors relating to disjointness should take into account the possibility of a class being an instance of "empty class".) Note that some classes are not empty by definition, but just currently don't have any instances in Wikidata [9]. These classes do not fit into the empty class category.

6 Sources of Issues with Disjointness

The above examinations point out and quantify some of the characteristics of disjointness violations as they occur in Wikidata. But they do not directly speak to why this incorrect information is present in Wikidata, i.e., what is the reason that Wikidata users added information to Wikidata that created disjointness violations.

Recently the Wikidata community and Wikimedia Deutschland conducted a survey and several discussions on issues in the Wikidata ontology [10,11] and some potential solutions were proposed [5]. The issues here closely mirror some of the major issues in this previous work. Disjointness violations and the work here provide a way of uncovering many examples of ontology issues, the start of any largbe-scale attempt to improve the Wikidata ontology.

Exceptions
One observed cause of disjointness violations is exceptions, where a class x is a subclass of y which provides a characteristic for instances of x but a subclass of x has a different characteristic that is provided by a superclass stated to be disjoint from y. For example, "lake" is a subclass of "natural object" but its

---

[8] We have added the "empty class" class (Q128139417), and the "the empty class" item (Q126726396). The first is the class containing all empty classes (and therefore is not an empty class itself). The second is a class that is empty, it is equivalent to all other empty classes, and also subclass of every class. Empty classes were not on Wikidata before we did this. It may be counterintuitive for a knowledge graph to have empty classes, and the vast bulk of classes in Wikidata are not empty. However, within foundational ontologies, all bits of information can be represented. (This is in line with the goal of Wikimedia projects to contain the "sum of all human knowledge".) The fact that a certain class does not have any instances is important to know, many disciplines tackle the question of "Does x exist?", and the results are worthy of being in a knowledge graph.

subclass "artificial lake" is itself a subclass of "artificial object" which is disjoint from "natural object".

A possible solution is to have a subclass of "lake" for natural lakes and move the "natural object" superclass away from "lake" to natural lakes. Another solution is to remove the subclass relationship between human-made lake and lake, as it can be argued that human-made lakes somehow are not "real lakes", but something else, and keeping the same superclasses for "lake".

Ambiguous Labels
Many users of Wikidata determine the meaning of a class almost solely from its label (and sometimes also from its description). But natural language, particularly short phrases, is often ambiguous resulting in superclasses for several of the meanings of the label (or description) and this can lead to the class having disjoint superclasses. For example, "food waste" can refer to either food that was wasted, and thus a subclass of "concrete object" via "biodegradable waste", or an act of wasting food, and thus a subclass of "abstract entity" via "waste of resources". As "concrete object" and "abstract entity" are disjoint "food waste" cannot be a subclass of both.

A possible solution is to split "food waste" into two classes. Another solution is to determine which reading is correct based on other information about the class, such as its descriptions or instances, and remove the incorrect superclass.

Multiple Senses of a Word
A related issue arises from words that have multiple meanings. Similarly to the above case, the different meanings for a class whose label is a single word can give rise to disjoint superclasses for the class. For example, "foil" is a subclass of both "concrete object" (via "ornament") and "abstract entity" (via "motif"). Foil is used both as the material that is used in art, and also as an artistic part of the work. Another example is "disease", that leads to a class order related confusion (see just below) as it both means strain of disease, and the disease that someone specifically has.

Class Order Confusion
Wikidata allows classes to be instances of (higher order) classes. This induces a hierarchy of classes. Some of these classes are fixed order, where a first-order class has only non-classes as instances, a second-order class has only first-order classes as instances, etc. All fixed order classes are (pairwise) disjoint with each other. As with other parts of Wikidata, there are issues in this hierarchy, particularly where classes that should be instances of another class are instead a subclass of the other class, or vice versa [12].

For example, "chemical element" is stated to be a second-order class but is also a subclass of "concrete object", which is stated to be a first-order class. As second-order classes are (correctly) stated to be disjoint from first-order classes, this produces a disjointness violation. The solution here is to remove "chemical element" from the subclasses of "concrete objects" as chemical elements, e.g., "Mercury" is a class with concrete object instances, it is not a concrete object itself.

Mixture

Wikidata has classes whose instances are mixtures of several components. In some cases the class is a subclass of the component classes. If the component classes are disjoint then a disjointness violation results. For example "BBP DANSAERT Saison x Lambic" is a subclass of both "saison" and "lambic", two classes of beer that are subclasses of two disjoint beer characteristic classes ("high fermentation beer" and "spontaneous fermentation beer").

There are two problems here that are typical of mixtures. First, a mixture class should have the component classes as parts, not superclasses. Second the disjointness comes from "beer" being the disjoint union of several characteristic classes with no allowance for beers that mix several different characteristics. An extra member of the disjoint union is needed to account for such mixtures, adding that extra member satisfies the exhaustive partition criteria of a disjointness statement.

The similar problem shows up in situations that are not physical mixtures, but where a combination of characteristics is somehow possible. The class "game" is the disjoint union of "game of chance", "game of skill", and "game combining chance and skill", explicitly allowing for combinations.

Confusing Items

For some classes their meaning is not ambiguous but how to categorize the class is unclear. The class may have multiple aspects that fit under disjoint superclasses. For example, "biological sequence" was described as several small chemical fragments (monomers) linked together into a polymer with biological utility. One aspect of instances of the class is their physicality, leading via "biomolecular structure" to the superclass "concrete object"; another is the arrangement of monomers, leading via "sequence" to the superclass "abstract entity". Together these result in a disjointness violation.

The solution for this sort of class is to determine which aspect is more fitting and remove the other, possibly creating a new class for this other aspect and link the two classes together. This kind of splitting of classes can be contentious and generally needs to be discussed within the Wikidata community. A safe way to do this split is to cross-check from other databases and/or knowledge graphs.

The disjointness violation on "biological sequence" was causing around fifty thousand subclass violations so we used it as a test case of how to eliminate violations before we did the bulk of our analysis. Because BioPortal's ontology [13] considers the equivalent class a subclass of "data" we marked "biological sequence" as a subclass of "data". Since data is subclass of "abstract entity" we decided to keep "sequence" as a superclass and to remove "biomolecular structure". We also cited Bioportal's ontology as a reference for new link to "data". Information in Wikidata is generally supposed to be supported by references so having solid references reduces potential pushback from the community.

Basic Mistakes

Some disjointness violations just appear to be the result of some misreading of either the label or description of a class. For example, "linguistic rights activist" is subclass of "social movement". While a class for linguistic rights activism would indeed be a subclass of "social movement", the person who is an activist cannot be considered an instance of a social movement. These sorts of basic mistakes are generally fixable by just removing the incorrect subclass statement.

Vandalism
Vandalism is a problem with every freely-editable shared resource. Wikidata has had problems with vandalism [14] and it is possible that a disjointness violation is the result of vandalism on an item. However, such occurrences are infrequent on Wikidata, as other shared resources, such as Wikipedia, are vandalized more often.

Many of the above issues come from a lack of examination of the relevant parts of the Wikidata ontology. For example, looking at the ancestors of a class in the ontology would both reveal the characteristics of a class (such as being natural) and which of several readings of a label or description is correct.

7 Suggestions for Improving Disjointness

Even though Wikidata has a significant number of disjointnesses because of the far reach of some of its disjointness statements, disjointness could play a larger positive part in Wikidata. Just adding more disjointness statements to Wikidata, especially at the middle levels of the ontology, without making any other changes would provide this, but changes to Wikidata tools and to Wikidata itself would help increase the use of disjointness in Wikidata.

One possibility is to just go through parts of the Wikidata ontology looking for cases where a class appears to be a disjoint union of other classes and adding a "disjoint union of" statement to the class and then determining whether the disjointness is correct based on examining the new disjointness violations. It is unclear, however, how effective this process would be if performed by users with limited domain knowledge, particularly if the violations can only be detected using QLever with the current week-long delay.

Better Use of Tools. One problem with disjointness in Wikidata is that disjointness violations are not shown to users when they are editing Wikidata unless they use a special plugin—www.wikidata.org/w/index.php?title=User%3ATomT0m %2Fclassification.js&action=raw&ctype=text%2Fjavascript. Making this useful tool a part of the core user interface would help in showing disjointness violations to users.

Another problem is that reporting on disjointness violations cannot be done in the official Wikidata query service because of its poor performance. Moving to a faster query service with performance on par with QLever would allow for real-time examination of disjointness violations.[9]

---

[9] As of August 2024 QLever only works on offline RDF dumps of Wikidata.

Better Disjointness Constructs. Disjointness in Wikidata is only possible as part of a disjoint union construct. This can require the creation of artificial classes to be the union, such as "award, award nominees, award recipients or award ceremony" (Q26877490). The disjoint union construct also requires the use of qualifiers and an artifical item to be the value of the statement. Further, the disjoint classes have to be specified inside the construct, separate from any other information about them.

Adding a construct that, for example, made all the instances or direct subclasses of a class pairwise disjoint would eliminate the need for a union class and make disjointness easier for users to state. As well, there are existing classes, such as some of the classes of classes in the biological domain whose instances are disjoint, eliminating the need to separately state the disjoint classes. This could be used, for example, to state that the different mammalian species are disjoint.

Crowdsourcing Fixes. To effectively reduce the number of disjointness violations, particularly instance violations that are not covered by our culprits, it will likely be useful to employ the Wikidata community as a whole. A tool that showed disjointness culprits in context and pointed out changes that would eliminate the issue would reduce the effort required to reduce disjointness violations and would consequently allow Wikidata to accomodate more disjointness statements that are regularly checked. If the tool had game-like rewards [15] that might help with its uptake in the community.

8 Summary and Future Work

Disjointness is a significant part of Wikidata. Many classes can be inferred to be disjoint based on the "disjoint union of" statements in Wikidata, particularly those at the highest levels of the Wikidata ontology. Unfortunately, there are many disjointness violations, where a class is a subclass of two disjoint classes or an item is an instance of two disjoint classes. Fortunately, a much smaller but still large number of culprits cause these violations.

Several empirically determined causes give rise to many of these violations. These causes appear to be mostly related to not taking into account information that helps define items when editing Wikidata, indicating that requiring confirmation for changes that create disjointness violations might induce users to investigate further, such as by looking at the superclasses of a class, fixing their erroneous edits before they adds to the number of disjointness violations.

We are continuing to fix disjointness violations. The current state of our work is described at www.wikidata.org/wiki/User:Egezort/Fixing_all_(yes_all)_Disjointness_Violations. We have tackled most of the disjointness statements that produced a small number of violations, leaving the larger ones for later, perhaps as a community effort.

The ultimate goal of this line of work is to have Wikidata have a large and useful set of disjointness statements with no disjointness violations. We plan to investigate a large sample of the culprits we have identified, fix the issues that resulted their violations, and see just how many other disjointness violations are resolved. We also plan to work on the suggestions above for improving disjointness in Wikidata in order to drive towards this goal.

Acknowledgments. Ege Atacan Doğan was partly supported by the ERASMUS Student Mobility for Traineeships program, hosted at the Free University of Bozen-Bolzano and supervised by Enrico Franconi.

References

1. Vrandečić, D., Krötzsch, M.: Wikidata: a free collaborative knowledgebase. Commun. ACM **57**(10), 78–85 (2014)
2. Shenoy, K., Ilievski, F., Garijo, D., Schwabe, D., Szekely, P.: A study of the quality of Wikidata. J. Web Semant. **72** (2022)
3. Patel-Schneider, P.F.: Barriers to using Wikidata as a knowledge base. In: WikidataCon (2019). www.wikidata.org/wiki/Wikidata:WikidataCon_2019, Berlin (2019). Accessed 23 July 2024
4. Abdulai, M., Lacroix, L.: Wikidata: ontology issues prioritization (2023). www.wikidata.org/wiki/Wikidata:Ontology_issues_prioritization. Accessed 17 July 2024
5. Pintscher, L.: Wikidata survey on ontology issues, potential solutions (2023). www.wikidata.org/wiki/Wikidata_talk:Ontology_issues_prioritization#Overview_of_potential_solutions. Accessed 17 July 2024
6. Almeida, J.P.A., Carvalho, V.A., Brasileiro, F., Fonseca, C.M., Guizzardi, G.: Multi-level conceptual modeling: theory and applications. In: Proceedings of the XI Seminar on Ontology Research in Brazil and II Doctoral and Masters Consortium on Ontologies (2018). https://ceur-ws.org/Vol-2228/
7. Welcome to Blazegraph. blazegraph.com (2013). Accessed 23 July 2024
8. Bast, H., Buchhold, B.: QLever: a query engine for efficient SPARQL+text search. In: CIKM 2017: ACM Conference on Information and Knowledge Management, Singapore (2017)
9. Völker, J., Vrandečić, D., Sure, Y., Hotho, A.: Learning disjointness. In: Franconi, E., Kifer, M., May, W. (eds.) Semant. Web: Res. Appl., pp. 175–189. Springer, Berlin, Heidelberg (2007)
10. Abdulai, M.: Wikidata: ontology issues prioritization (2023). www.wikidata.org/wiki/Wikidata:Ontology_issues_prioritization. Accessed 23 July 2024
11. Pintscher, L.: Ontology issues in Wikidata: everything in neat and tidy boxes? Not quite! In: WikidataCon 2023 commons. https://foundation.wikimedia.org/wiki/File:WikidataCon/_2023/_Ontology/_issues/_in/_Wikidata/_-/_Everything/_in/_neat/_and/_tidy/_boxes/_Not/_quite!.pdf (2023). Accessed 23 July 2024
12. Dadalto, A.A., Almeida, J.P.A., Fonseca, C.M., Guizzardi, G.: Type or individual? Evidence of large-scale conceptual disarray in Wikidata. In: 40th International Conference on Conceptual Modeling (ER 2021), pp. 367–377 (2021). https://doi.org/10.1007/978-3-030-89022-3_29
13. BioPortal: EDAM ontology (2024). https://bioportal.bioontology.org/ontologies/EDAM?p=classes&conceptid=data_2044. Accessed 28 Aug 2024

14. Heindorf, S., Potthast, M., Stein, B., Engels, G.: Vandalism detection in Wikidata. In: CIKM 2016: Proceedings of the 25th ACM International on Conference on Information and Knowledge Management, pp. 327–333 (2016)
15. Oceja, J., Sierra, A.O.: Gamifiying Wikipedia? In: 12th European Conference on Games Based Learning, Sophia Antipolis, France (2018)

Enhancing WebProtégé with Version Control Systems

Erhun Giray Tuncay(✉)[iD], Nenad Krdzavac[iD], and Felix Caspar Engel[iD]

TIB - Leibniz Information Center of Science and Technology, Hanover, Germany
{giray.tuncay,nenad.krdzavac,felix.engel}@tib.eu

Abstract. Collaborative ontology editors and modern version control systems (VCS, such as GitHub or GitLab) have a prominent role among the best practices of ontology development process. Integrating VCS functionality into collaborative ontology editors is essential for an effective and seamless ontology development workflow. Unfortunately, such an integration is not present in the current landscape of openly licensed software yet. In this study, we aim to bridge this gap between collaborative ontology development and VCS by adding the most common version control workflows into the widely used open-source editor WebProtégé. These common workflows are mostly based on steps such as add, commit, push, clone, checkout and branch commands that can be executed by the REST API or command line of a VCS for a particular repository. Our solution includes associating WebProtégé users with VCS users based on personal access tokens for authenticated utilization of VCS REST APIs, harmonization and abstraction of the non-standard VCS REST API calls into the most common VCS workflows and integrating these workflows into WebProtégé using service and servlet components as well as a convenient Graphical User Interface (GUI). The solution is compatible with the most popular versioning environment GitHub alongside its open-source equivalent GitLab and is evaluated by its users based on the System Usability Scale (SUS) approach in the scope of this study. The service is publicly available at http://service.tib.eu/wp4tib and the code base can be found at https://github.com/TIBHannover/webprotege.

Keywords: Knowledge Engineering · Ontology Development · Ontology Versioning · Collaborative Ontology Editors

1 Introduction

The National Research Data Infrastructure Germany (NFDI) [4] is working on the development of a comprehensive research data infrastructure. The NFDI4Ing Terminology Service [10] is one of the components of this infrastructure. It offers already extensive terminological resources (including 100 ontologies, 49565 terms, 13224 properties and 262205 individuals) [15] and tools for processing them (including browsing, keyword search and visualisation). However, the aim of the NFDI4Ing Terminology Service is to also provide an integrated, collaborative development environment for working on ontologies. Collaborative ontology editors and up-to-date Version Control Systems (VCS) are key components of best practice in the ontology development process. A tight coupling of both systems is essential for an effective and integrated ontology development workflow. Unfortunately, this coupling is not reflected in the current landscape of openly licensed software. The widely used WebProtégé is completely independent of any VCS support. For this reason, our objective in this study is to establish a close link between common VCS and WebProtégé. This requirement is not trivial for various reasons, as it presents us with the following challenges (CH):

- **CH01:** The commands and calls of common VCSs are not fully standardized. The essential Git commands and calls are tailored for GitLab and GitHub services differently (e.g. for the clone command and branch extraction with REST API).
- **CH02:** WebProtégé uses a customly implemented authentication mechanism that is not compatible to simultaneously authenticate with Git technologies such as GitHub or GitLab.
- **CH03:** WebProtégé records the creation history of ontologies. This concerns the user's interaction with the ontology, which is recorded semantically on a single linear chronology. This is in contrast to the Git-based approach, which stores changes in a tree structure without semantic support.

This article is structured as follows. Section 2 provides a comprehensive overview of the state of the art ontology development tools. Section 3 proposes a solution architecture for the above mentioned challenges. The implementation details of the publicly available minimum viable product are given in Sect. 4 along with demonstrative user interfaces. Later on, Sect. 5 presents and analyzes a user evaluation of the WebProtégé extension with Git features. The article concludes with an outlook and a summary of this study.

2 State of the Art and Related Work

Ontologies evolve over time along with the needs of their community. *Leenheer and Mens* [8] point out the necessity for a collaborative framework that is capable of versioning and merging such evolving ontologies. *Pittet et al.* [20] further categorize the list of semantic operations in the version history of an ontology as instance and schema operations that are further categorized into adding or deleting classes or properties. However, there has been limited success to make use of semantically aware VCS in Ontology Development and non-semantic VCS can alternatively be used for this purpose. GitHub and Gitlab are the most prominent non-semantic VCS that is used to improve collaboration in Ontology Development Tools. For this reason, we have evaluated their integration into selected collaborative ontology development tools based on the following fields on Table 1:

- General description of ontology tools. It includes basic information about selected ontology tools such as availability and developers.
- Level of integration of the chosen tools with the GitHub and GitLab.
- Access management. This criterion includes information on how selected ontology tools support the verification of user identities, and access rights to control versioning system repositories.
- Licensing models for ontology tools.

Table 1. Ontology tools and their integration with VCS

Tool name/ Supported Feature	WebProtégé [13]	TopBraid Composer Free Ed. [5]	Ontology Development Kit (ODK) [17]	Protégé [7]
Version Control Systems (VCS) Integration	No	Yes	Yes	No
Git Clone	No	Yes	Yes	No
Git Commit/Push	No	Yes	Yes	No
Delete branch	No	No	No	No
Listing existing branches	No	Yes	No	No
Access Management	Yes	Limited	Limited	No
License Model	BSD 2 clause	EULA	BSD 3 clause	BSD
Collaborative Development	Yes	Limited	Limited	Yes

Table 1 shows selected ontology development tools and a list of supported features. The desktop version of Protégé [18,23] is a project developed at Stanford University [7]. This tool is one of the most widely used ontology tools available online and is easy to install. It is an extensible tool with a plug-in mechanism that can be adapted to user needs [18]. Ontology versioning is supported in the desktop version [19], but not in the context of versioning files available in a control versioning systems such as GitLab or GitHub. The desktop version of the Protégé tool only allows users to load an ontology from a remote Git repository as a raw ontology file providing the ontology URI. Git operations like clone, pull, commit, push are not supported in the desktop version of the Protégé tool. Collaborative ontology development is supported in the desktop version of the Protégé tool via a plug-in [23]. The implementation is based on the client-server architecture. Any change made by the user on the Protégé client side is transmitted to the remote repository. These changes in ontology are available and immediately visible to all other clients [23]. The desktop version of the Protégé tool is available to users under the BSD license.

WebProtégé (ver. 3.0) is developed at the Stanford Center for Biomedical Informatics Research [13]. The main features are e.g. collaborative development, change tracking, project history and integration with third party applications. Registered users can comment on a project, submit issues in the form of threaded comments, tag entities with badges that can be searched. WebProtégé is also integrated with Slack[1]. All changes made to the ontology project are tracked and grouped into revisions, filtered and displayed in terms of entities. The tool does not support integration with GitHub or GitLab. WebProtégé is available under the BSD-2 license (see the license model in Table 1). Currently, WebProtégé is in the transition phase to a fine-grained microservice architecture.

TopBraid Composer is a commercial product developed by TopQuadrant [5]. It is available as a Maestro Edition which requires a license key, and as a Free Edition (TBC FE, for short) which does not require license key. For our purposes, we tested the TBC FE [5]. Git features integrated in the TBC FE are inherited from Eclipse IDE [22]. Users can import a Git repository containing ontology files into the TBC FE workspace. During the import process users can select or deselect one or more feature branches. In addition, users of the TBC FE have the ability to commit changes to a remote Git repository, merge, rebase, and switch between feature branches. Access management in the TBC FE is limited. For example, authorization is not supported by the TBC FE, although authentication is possible. It is optional to provide a username and password when importing a remote Git repository into the TBC FE workspace. However, using a personal access token key is not an option when importing an ontology from a remote Git repository. Collaborative ontology development is also limited in the TBC FE. Only after committing their changes to the remote Git repository users can share their changes with other users. Sharing an ontology project with specific people by specifying their username is not supported. TopQuandrant offers customers an end user license agreement (see license model row in Table 1).

The Ontology Development Kit (ODK) is a toolkit developed at the National Institutes of Health, United States. It provides a set of automatically executable workflows for managing an ontology lifecycle [17]. It uses Git functionality to manage the ontology lifecycle. The ontology stored in a repository can be pushed to a remote Git repository allowing for version control and limited collaborative ontology development among users. The ODK supports both push and pull requests when the ontology is hosted in a Git repository. The special task of deleting remote Git branches containing ontologies is not specified in [17]. Accessing a Git repository using a personal access token key is not supported in the ODK. The authors have published the ODK on GitHub under a BSD 3 clause license.

Among the listed ontology tools, only Protégé desktop version and its Could-based version do not have full integration with Git functionality. The advantage of the WebProtégé over to the other three shown in Table 1 is its expanded support for collaborative ontology development which is not provided by Git functionality. On the other hand, WebProtégé's support for access management

---

[1] https://slack.com/intl/en-gb/.

is better than the desktop version of the Protégé tool. Only TBC FE enables users to view current branches from the Git plugin utilized in the TBC FE tool.

3 VCS and WebProtégé Integration Architecture

3.1 System Overview

Our improved version of WebProtégé has a client-server architecture with a Git Commands Service as part of a Git module on the server side and and a Git REST API Service on the client side. In this context, the client can extract repository metadata by GitLab and GitHub REST APIs and then specify the execution parameters for the respective Git workflows on the Google Web Toolkit (GWT) based GUI. The execution parameters specified by the user on the client side are transferred to the Servlets for triggering the respective Git Workflows on the server side. Later on, Servlets employ the Git Commands Service that executes commands on the server Operating System in order to trigger the requested Git Workflows. The resulting changes in the project are stored in both the MongoDB database and the file directory of the server. The overall architecture of our VCS integrated WebProtégé is illustrated in Fig. 1.

3.2 Key Components

Our approach is focused on addressing the challenges of non-standard VCS commands and calls as well as incompatible authentication mechanisms between VCS and WebProtégé. The key components of our approach are mainly Git User Authentication in WebProtégé and Generic Git Functionality for GitHub and GitLab. We mainly addressed these challenges by using abstraction and personal access tokens in the context of our key components listed below:

A) Git User Authentication in WebProtégé: Our first challenge was the incompatibility of the authentication and authorization protocols of prominent VCS tools with WebProtégé as mentioned in **CH02**. Although prominent VCS such as GitHub and GitLab rely on some standards such as OAuth 2.0 [11] for this purpose, WebProtégé 5.0 does not support any standard. Consequently, it was not possible to merge a particular VCS user and a WebProtégé user into a single identity. The remaining possibility was mapping VCS users into WebProtégé users as part of the user metadata. In this context, manual login into VCS is far from being practical since our use cases require VCS workflows rather than single commands. Alternatively, personal access tokens proved to be the most practical authentication method into VCS while using WebProtégé due to their capability of automating the authenticated VCS operations both in command line and the REST API.

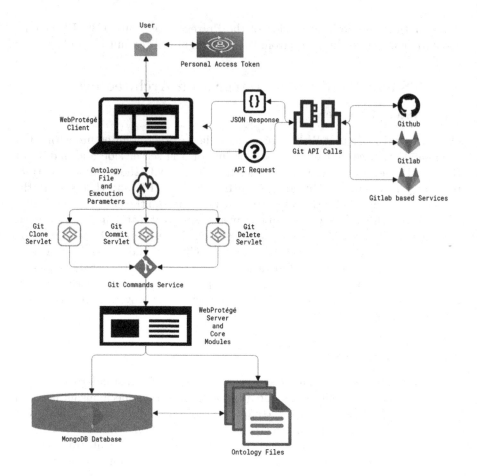

Fig. 1. WebProtégé and Git Integration Architecture

B) Generic Git Functionality: In order to achieve a Generic Git Functionality, we established a Git REST API Service on the client side and a Git module that encapsulates a Git Command Line Service on the server side. The purpose of using service patterns for both tasks was to create abstraction as mentioned in **CH01**. In this scope, it was only possible to utilize a REST API Service on the client side rather than a Command Line Service due to security concerns of the underlying technology GWT. Thus, we implemented the Git REST API Service to access GitHub and GitLab REST APIs for client side operations that mainly extract data (such as existing branches) from the repositories. It should be noted that the sequence of API calls for GitLab and GitHub were not identical as mentioned in CH01 and it was implemented for both technologies by abstraction.

For the server side, our main intention was to execute all the basic repository content modification calls or commands and furthermore establishing the common git workflows with specific parameters provided by the client based on this service. For this purpose, we initially evaluated the alternative methods of repository content modification and discovered that the prominent methods for implementing this feature were VCS API Calls and VCS commands on the command line. As a result of this evaluation, we identified that Git commands on the command line for both GitLab and GitHub were identical except Clone whereas their APIs did not follow a standard. Using repository modification commands rather than REST API also enabled ontology files being directly accessible by the WebProtégé server in its file system. Consequently, we implemented the service on the command line using the following basic Git commands and implemented GitHub and GitLab version of the clone command separately:

- checkout,
- clone,
- add,
- commit,
- push.

Next, we implemented the common Git workflows in Java Servlets using these basic commands in various combinations. Finally, the Servlets execute the Git workflows based on the execution parameters provided by the user. The most common workflows are listed below:

- **Extracting a Particular Ontology File**: This workflow employs clone and checkout commands respectively. It is used as an additional option for "Project Creation", "Merge Ontologies" and "Apply External Edits" features of WebProtégé.
- **Updating the File Tree of A Branch**: This workflow requires a branch choice from the list of branches extracted by the respective client side API calls. After that, it employs clone, add, commit and push commands respectively to change an existing branch or write into a branch that is created using an existing one. It is used for writing the Ontology under development to a particular branch of a repository in the desired format.
- **Delete an Unused Branch**: This workflow requires a branch choice from the list of branches extracted by the respective client side API calls. After that it employs clone and checkout commands to remove the unused branch.

4 Implementation Details and Demonstrative Interfaces

Our codebase is forked from WebProtégé version 5.0 that is the last monolithic version of WebProtégé. This version is capable of adding new modules with hooks unlike the latest released WebProtégé version 4.0. The improvements were developed in compliance with the existing technology stack that relies on GWT version 2.8 and a MongoDB database version 3.11. GWT automatically creates a JavaScript front end application with a JRE emulation library that allows a restricted and secure usage of the Java language. On the other hand, the

Java based server modules include customized Apache Lucene implementation based on the version 8.5 to enable full text indexing and searching capability on ontology entities. Java access to the server side Git commands on the Operating System Shell has been achieved with ProcessBuilder classes while Git REST API calls performed on the client side are implemented by RequestBuilder classes of GWT.

Figure 2 shows the alternative project creation methods with a File Chooser in (A) and Cloning from Repository in (B). Cloning from repository may require the relative path of the ontology in the repository as well as the branch name. If these values are not provided, cloning will be carried out with default branch name and possible ontology file names that partially match the repository name. An almost identical functionality also exists in the "Apply External Edits" and "Merge Ontologies" features accessible from the project menu.

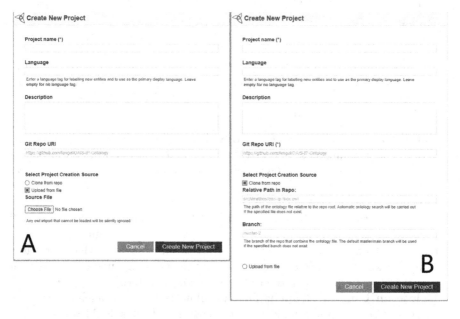

Fig. 2. Panel for creating a new project from Git repository (A) Create new project from local file (B) Create new project from Git repository

VCS functionality can effectively be used with a personal access token for the user and repository URI for the WebProtégé project. These variables can be modified as shown in Fig. 3. In the absence of a personal access token, it is still possible to clone from a public repository and display its existing branches given that the repository URI is present in Project Settings. It should also be noted that it is a common practice to fork from a public repository that does not provide write access to external users in order to commit changes and make a pull request to the original repository afterwards. The forked Repo URI should be used in WebProtégé for this purpose.

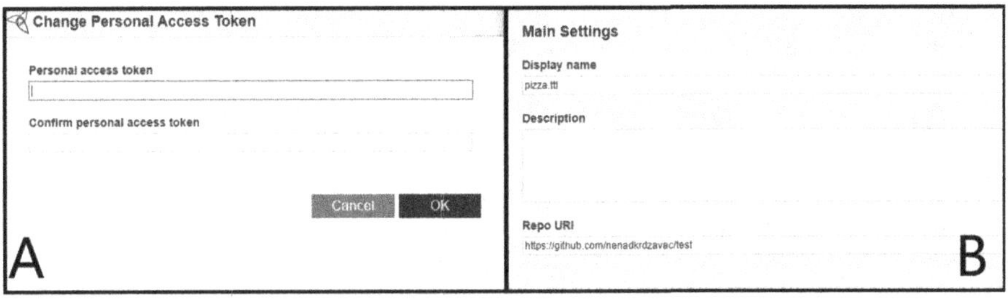

Fig. 3. A) Change access token key (B) Add Git repository URL to main setting.

Finally, the personal access token of the user must have write access to be able to push the changes committed to a local branch. Figure 4 shows a Commit & Push Dialogue Box that is popped up from the project menu in the Project Layout Page.

Fig. 4. Commit and push ontology project to Git repository from WebProtégé

5 The WebProtégé Service Usability Evaluation

This section intends to evaluate the usability of the VCS enhanced WebProtégé. We evaluated the usability based on the System Usability Scale (SUS) [16], which is a survey designed to assess cognitive usability. The survey consists of ten questions each of which can be responded by five response options ranging from strongly agreeing (rated as 5) to strongly disagreeing (rated as 1). Participants received instructions on how to efficiently use the VCS enhanced WebProtégé service as a prerequisite to the survey. Then, they were asked to provide their responses to the ten questions regarding the service's usability. Finally, they were asked to provide their free text opinion about the service.

Table 2. System Usability Scale (SUS) raw and final based on individual scores and their experience with using Git-enhanced WebProtégé

Question/ Participant ID	Q1	Q2	Q3	Q4	Q5	Q6	Q7	Q8	Q9	Q10	SUS raw score	SUS final score
1	5	1	1	1	5	1	5	2	5	2	34	85
2	4	3	3	2	4	1	5	2	4	1	31	77.5
3	5	1	4	1	3	1	3	2	4	1	33	82.5
4	5	1	5	2	4	2	4	1	5	2	35	87.5
5	2	1	3	2	4	2	4	2	3	3	26	65
6	4	1	5	1	5	1	4	1	5	2	37	92.5
7	5	1	5	1	5	1	4	2	4	2	36	90
8	3	2	5	3	5	1	5	1	4	3	32	80
9	5	5	5	5	5	5	5	5	5	5	20	50
10	3	2	4	2	1		4	2	3	3	21	52.5
11	5	1	5	3	4	2	5	1	5	2	35	87.5
12	4	1	3	3	4	2	3	1	4	3	28	70
13	1	3	3	2	3	3	4	3	3	5	18	45
14	5	4	4	2	4	2	5	2	4	2	30	75
Mean:											29.71	74.28

The survey results and user comments provided valuable insights and pointed out future directions regarding the convenience and practicality of the VCS enhanced WebProtégé service. Table 2 illustrates the survey results, in which fourteen software developers, ontologists and semantic web researchers participated. Table 2 presents the scores for each individual question in the SUS feedback, as well as the raw and final SUS scores for each participant, including the mean for all raw and final SUS scores. The calculation of the SUS raw score involves subtracting one from each odd numbered question value and subtracting each even numbered question value from five. The SUS final score for each participant is obtained by multiplying their SUS raw score by 2.5. As shown in the Table 2, a tenth participant did not answer the sixth question. In this case, in the SUS raw score calculation, we excluded the value of sixth question for the tenth participant.

Overall, the average of all the SUS final scores is equivalent to 74.28 which is marked as grade B (74.1–77.1) defined in the GitLab's proprietary SUS scale table [6]. This means that the overall experience of users is between good (C+, 71.1–72.5) and excellent (A, 80.8–4.0). Based on the SUS final score results shown in Table 2 five participants (first, fourth, sixth, seventh, and eleventh) have experience with using the Git-enhanced WebProtégé functionalities that is marked as *best Imaginable* by the SUS usability scale. However, there is still room for improvement of the Git-enhanced features of the WebProtégé service as highlighted by our thirteenth participant that evaluated our tool with a final SUS score of 45. Some of the reasons why this user would not like to use the WebProtégé service enhanced with Git functionalities are as follows (see scale values of questions in Table 2):

- The service is unnecessarily complex (Q2 value is equal to 3)
- The user found several features in the service not well integrated (Q5 value is equal to 3)
- The user observes inconsistencies in the use of the service (Q6 value is equal to 3).
- The user strongly agrees with the statement that there is a significant amount of material to learn before beginning to use the service (Q10 value is equal to 5)

Participants have additionally submitted their feedback on the VCS enhanced WebProtégé service in free text form. We have compiled these comments to analyze the future direction that our user community expects and their current satisfaction. The remarks were strongly positive and our tool was generally evaluated as exceptionally user-friendly, simple to use and a valuable tool for efficiently organizing and presenting data.

6 Discussion and Outlook

As it is mentioned in **CH03**, the versioning features of WebProtégé and VCS tools are not compatible with each other. The Revision History of an Ontology Project in WebProtégé is a linear list of all semantic operations performed on an Ontology and it is stored in a binary format in the file directory of the service. This infrastructure enables the user to revert revisions due to its semantic nature. On the other hand, the commit history of a VCS is more collaborative since it represents the linear history of each branch and the interactions between them. Furthermore, there is no restriction in the content of each commit to be semantically atomic in VCS tools. Another incompatibility issue is that there are multiple ontology files in each branch of most VCS repositories whereas only a single ontology is under development in a WebProtégé project. The future steps to accomplish the desired compatibility can be listed as follows:

- **Semantic Comparison for VCS**: The fundamental operation for version control is comparison or difference of a single file in a pair of its versions. VCS compare raw files as added and deleted lines. However, it make more sense to compute the difference of two ontology files based on added, modified or deleted semantic entities in ontologies. Tools such as ROBOT [14] and ContoDiff [12] can be used by VCS to extract a summary of semantic differences of two ontology versions similar to the implementation in NFDI4Ing Terminology Service Sandbox[2]. It would be much easier to map the WebProtégé revision history to semantically enhanced VCS that relies on such tools for ontology version comparison.

---

[2] https://terminology.nfdi4ing.de/ts/sandbox/ondet.

- **Ontology File Validation**: The ontology versions can be validated using tools such as Ontolo-CI [21] in a Github environment with continuous integration. Then, validated ontologies can be part of VCS releases which can be implemented by Git REST API.
- **Multiple Branches**: Each WebProtégé project should encapsulate multiple branches as inner projects that have their own respective revision history. This requires modifications in the underlying data structures. The second step after such encapsulation would be performing merge/pull requests across branches and automatic or manual conflict resolution strategies when there are conflicts between the branches. The second step can also be implemented using Git REST API.
- **Multiple Ontology Files in a Branch**: It would be useful to associate all ontology files in all branches of a repository with a single WebProtégé project.
- **Comments**: The structure of comments in VCS repositories and WebProtégé are also not compatible. Comments play an important role in collaboration and they are heavily used in WebProtégé and VCS tools. However, it is only possible to comment on entities in WebProtégé whereas it is possible to comment on pull requests, issues, commits or snippets in VCS tools. It may be possible to associate VCS snippets with WebProtégé provided that both of them support threading in comments as a future work. Plus, another improvement could be the ability to import and export WebProtégé comments on entities for the sake of mobility.

The architecture of WebProtégé has significant impact for future developments. The technology stack of the upcoming micro-service architecture is gradually converting from the current monolithic architecture. The current monolithic architecture of WebProtégé does not enable Single Sign On with NFDI4ING Terminology Service Sandbox or other external services. Due to this problem and other similar modularity problems on the horizon, we are considering to switch to the micro-service architecture. The original maintainers of WebProtégé has started this effort by converting the back end technology to Spring Boot for a robust REST API. In this architecture, Apache Pulsar [1] and Apache Zoo Keeper [2] services are used for asynchronous messaging, MinIO [3] service is used for blob storage of project snapshots and KeyCloak service is used for authentication. The KeyCloak [9] based solution of the new architecture will allow seamless integration for our purpose. However, it hasn't reached a stable release or version yet.

7 Summary

In this study, we introduced our improved version of WebProtégé that enables its users to extract particular Ontology Files, Update the File Tree of A Branch and Delete an Unused Branch in the VCS repositories of projects as part of various new features. These new features will benefit various stakeholders in the Ontology Development world such as Ontologists unfamiliar to VCS tools, Ontologists who prefer to fully automate an ontology release cycle or baselining process with Continuous Integration/Continuous Delivery (CI/CD) and Ontology Community members that do not use WebProtégé. In particular, Ontologists and Ontology communities will be able to transparently collaborate in publicly available repositories that enable comments and discussion about the features and the progress in a granular way.

One key component of our contribution was making use of personal access tokens to be able to identify VCS users and map their VCS authentication rights into WebProtégé users. Despite their difference to the traditional manual authentication methods, personal access tokens have gained a huge momentum among VCS users since several VCS providers are encouraging the usage of them for automated authentication to the repositories. Accordingly, using personal access tokens have eliminated the requirement to interactively ask for user credentials in server side Git command line operations. Furthermore, it enabled us to perform authenticated Git REST API queries with an hourly 5000 call limit on the client side. It should be noted that authenticated Git REST API calls can only be performed via tokens and the hourly limit for unauthenticated Git REST API calls is currently only 60 in GitHub and GitLab.

Another component of our contribution was providing abstraction across VCS implementations with service design patterns. In particular, a Git Commands Service was implemented on the server side due its more standardized commands and a Git REST API Service was implemented on the client side due to the security restrictions of the client side GWT technology. This approach enables us to extend the VCS functionalities with further VCS tools provided that they offer similar functionality with GitLab and GitHub. The upcoming releases of our service are intended to include abstraction of more VCS tools along with other common Git Workflows including pull requests, repository creation, file deletion, release creation, switch and roll back.

The VCS enhanced WebProtégé service, is tested by fourteen participants. Following the tests, the participants took a survey to indicate their satisfaction using the extended WebProtégé. The evaluation shows that users' experience with the Git-enhanced functionalities of WebProtégé is ranked between good and excellent. Nonetheless, there is a significant room for improvements in the service as we have mentioned above and in Sect. 6.

Acknowledgement. The authors acknowledge the financial support provided for this work by the
- German Research Foundation (DFG) - NFDI4ing, project number 442146713.
- Federal Ministry for Economic Affairs and Energy of Germany (BMWK) - CoyPu, project number 01MK21007[A-L].

References

1. Apache Pulsar—Apache Pulsar. https://pulsar.apache.org/. Accessed 07 Dec 2023
2. Apache ZooKeeper. https://zookeeper.apache.org/. Accessed 07 Dec 2023
3. MinIO—High Performance, Kubernetes Native Object Storage. https://min.io/. Accessed 07 Dec 2023
4. National research data infrastructure germany (nfdi). https://www.nfdi.de/?lang=en. Accessed 07 Dec 2023
5. Topbraid composer, free edition (2013). https://archive.topquadrant.com/topbraid-composer-install/. Accessed 06 Dec 2023
6. The gitlab handbook, system usability scale, interpreting sus scores (2023). https://handbook.gitlab.com/handbook/product/ux/performance-indicators/system-usability-scale/. Accessed 06 Dec 2023
7. Protege tool (2023). https://protege.stanford.edu/. Accessed 01 Dec 2023
8. De Leenheer, P., Mens, T.: Ontology Evolution, pp. 131–176. Springer US, Boston (2008). https://doi.org/10.1007/978-0-387-69900-4_5
9. Divyabharathi D. N., Cholli, N.G.: A review on identity and access management server (KeyCloak). Int. J. Secur. Priv. Perv. Comput. **12**(3), 46–53 (2020). http://services.igi-global.com/resolvedoi/resolve.aspx?doi=10.4018/IJSPPC.2020070104
10. Engel, F., Tuncay, G., Kraft, A.: NFDI4Ing Terminology Service (2021). https://publications.rwth-aachen.de/record/968598. number: RWTH-2023-08670
11. Hardt, D.: The OAuth 2.0 Authorization Framework. Technical Report. RFC6749, RFC Editor (2012). https://www.rfc-editor.org/info/rfc6749
12. Hartung, M., Groß, A., Rahm, E.: Conto–diff: generation of complex evolution mappings for life science ontologies. J. Biomed. Inf. **46**(1), 15–32 (2013). https://www.sciencedirect.com/science/article/pii/S1532046412000627
13. Horridge, M., Gonçalves, R.S., Nyulas, C.I., Tudorache, T., Musen, M.A.: Webprotégé 3.0-collaborative owl ontology engineering in the cloud. In: ISWC (P&D/Industry/BlueSky) (2018). https://ceur-ws.org/Vol-2180/paper-26.pdf
14. Jackson, R.C., Balhoff, J.P., Douglass, E., Harris, N.L., Mungall, C.J., Overton, J.A.: ROBOT: a tool for automating ontology workflows. BMC Bioinf. **20**(1), 407 (2019). https://doi.org/10.1186/s12859-019-3002-3
15. Kraft, A., Engel, F., Klinger, A.: Terminologies in RDM for engineering – a service approach: NFDI4Ing terminology service. In: Proceedings of the Conference on Research Data Infrastructure, vol. 1 (2023). https://www.tib-op.org/ojs/index.php/CoRDI/article/view/207
16. Lewis, J.R.: The system usability scale: past, present, and future. Int. J. Hum.–Comput. Interact. **34**(7), 577–590 (2018). https://www.tandfonline.com/doi/pdf/10.1080/10447318.2018.1455307
17. Matentzoglu, N., et al.: Ontology development kit: a toolkit for building, maintaining and standardizing biomedical ontologies. Database 2022, baac087 (2022). https://doi.org/10.1093/database/baac087
18. Musen, M.A.: The protégé project: a look back and a look forward. AI Matters **1**(4), 4–12 (2015)
19. Patel, A., Jain, S.: Ontology versioning framework for representing ontological concept as knowledge unit. ceur-ws. org (2021). https://ceur-ws.org/Vol-2786/Paper18.pdf
20. Pittet, P., Nicolle, C., Cruz, C.: Guidelines for a Dynamic Ontology - Integrating Tools of Evolution and Versioning in Ontology (2012). https://arxiv.org/abs/1208.1750

21. Publio, G.C., Gayo, J.E.L., Colunga, G.F., MenendÃz, P.: Ontolo-CI: continuous data validation With ShEx. In: SEMANTICS 2022 EU. Vienna, Austria (2022)
22. The Eclipse Foundation: Eclipse IDE. https://eclipseide.org/
23. Tudorache, T., Noy, N.F., Tu, S., Musen, M.A.: Supporting collaborative ontology development in protégé. In: Sheth, A., et al. (eds.) ISWC 2008. LNCS, vol. 5318, pp. 17–32. Springer, Heidelberg (2008). https://doi.org/10.1007/978-3-540-88564-1_2

Visual Presentation and Summarization of Linked Data Schemas

Lelde Lāce, Aiga Romāne-Ritmane, Mikus Grasmanis, Artūrs Sproģis, Jūlija Ovčiņņikova, Uldis Bojārs, and Kārlis Čerāns

Institute of Mathematics and Computer Science, University of Latvia, Riga, Latvia
karlis.cerans@lumii.lv

Abstract. A visually presented schema of a data set can help its user to gain understanding about its contents and use an appropriate vocabulary to build queries or develop applications accessing it. Still, most of Linked Open Data sets are provided without available visual schemas and for data sets of realistic size the schemas may be way too large to be visualized on a suitable diagramming canvas. In this paper we develop a concept of visualization-oriented data schema, involving property ascription point information and propose a method for creating visual presentations of the data schemas, including both the detailed schema visualization and visual schema summarization options. We evaluate the method on 24 prominent data sets from the Linked Open data cloud, where it can obtain legible visualizations of the full schema or at least its fragment.

Keywords: RDF · SPARQL · Linked data · Knowledge graph schema · Visual schema diagram

1 Introduction

Understanding the structure and contents of a Knowledge Graph is important for any user action over it, be that ad hoc data querying or building data consuming applications. The data structure presentation allows seeing if the data set contains the relevant information and in what form (e.g., using what vocabulary) it has been encoded.

A well-known means of data structure description is the data schema that is based on the data set entity (class and property) vocabularies and their connections.

An abstract data schema that involves a class and property mapping or its encoding in RDF data shape language, as SHACL [1] or ShEx [2], can be useful for machine processing, including data validation, user interface form generation or context-aware entity name suggestion e.g., during auto-completion of the text of a query over the data. For human perception a meaningful visual presentation of the data schema would well complement the textual schema presentation as it could invoke the user's visual perception capabilities in schema understanding.

Still, a visual presentation of a data schema in the paradigm of a graph of attributed nodes and edges faces a natural limitation of the size of the graph that can be expected to help the schema perception; often this size shall be less that the number of relevant entities making the data set structure.

© The Author(s), under exclusive license to Springer Nature Switzerland AG 2025
S. Tiwari et al. (Eds.): KGSWC 2024, LNCS 15459, pp. 290–305, 2025.
https://doi.org/10.1007/978-3-031-81221-7_20

To respond to this problem, we investigate the possibilities of presenting the schemas of knowledge graphs in the style of UML class diagrams (representing the classes as nodes and the properties as links or as attributes). We aim at legible presentations of possibly large schemas by developing methods that (i) recognize the relevance of a class as a property source or target and (ii) allow to merge nodes of classes with similar instance incoming and outgoing property characteristics.

We build upon related work that involves on-the-fly computing of schema diagram fragments in LD-VOWL [3] and LODSight [4] (with limitations regarding the schema and the data set size, as well as the details included in the schema), automated means for data schema extraction as (enriched) SHACL shapes (cf. [5]) and abstract class-to-property mappings (cf. [6]), as well as data schema visualization in various notations, involving VOWL [7], OWLGrEd [8], RDFShape [9] and ViziQuer [10].

The main contributions of this paper are:

- describing a mathematical framework for an extended data schema notation, that involves the relevance markers for property ascription at classes,
- developing a method for visual compact data schema representation, allowing to obtain legible summaries for data schemas with well over 100 classes, and
- presenting a library of schema summary or fragment visualizations for 24 prominent data sets from the Linked Open Data[1] cloud.

In what follows, Sect. 2 provides further Related work details and Sect. 3 presents the data schema and schema graph concepts. Section 4 then describes basic schema diagrams; Sect. 5 presents the schema summarization and Sect. 6 describes the implementation and evaluation. Finally, Sect. 7 concludes the paper.

The schema visualization methods have been integrated within the open-source ViziQuer tool[2] (where also visual SPARQL query creation is supported). The resources supporting the paper, including the data schema visualizations and the used software startup options are available from https://github.com/LUMII-Syslab/viziquer-tools-lod24. The schema visualization can be performed on ViziQuer Playground[3], as well.

2 Related Work

The visual presentation of relational database schemas is common in most of major database management systems as well as in a variety of custom database handling tools. Still, these tools, as the entire relational database management framework, work on the technical level of tables, columns, and links, therefore they can be considered just as a source of inspiration for the solutions in the knowledge graph and semantic data area.

In the knowledge graph and semantic technology realm there are visual notations and tools for presenting OWL ontologies, such as VOWL [7], OntoDia [11], and OWLGrEd [8] (cf. also [12] for an overview of ontology visualization methods). Visual ontology editors such as ChOWLk [13] and OWLGrEd allow visual ontology authoring, as well. There are also visualization tools for RDF data shape notations SHACL and ShEx,

---

[1] https://lod-cloud.net/.
[2] https://github.com/LUMII-Syslab/viziquer.
[3] https://viziquer.app.

including, e.g., RDFShape [9], Shacl Play! [14] and shacl2plantUML [15]. To obtain the visualization of the actual data schema using some of these tools, the data schema would need to be described in the respective notation first.

The methods for automated extraction of data shapes from a data set have been attracting research recently, as well, including [5], where a method of extracting an enriched SHACL description of the data set structure from the data dump in.n3 format has been provided. Still, the existing tools for SHACL extraction have not yet been evaluated together with the tools for SHACL visualization. It is also not clear how the SHACL specifications would encode the subclass relation essential for compact schema presentation in the UML class diagram form (e.g., so that a property does not get ascribed both to a super-class and a subclass).

The idea of automatically extracting the schema from an RDF data set has already been explored in [16], where the schema presentation in the form of a UML-style diagram is considered and compact schema presentations are discussed, as well. Notably, this work discusses the need to mine the subclass relation from the data set, although the implementation yielding mixed results. The options for creating the summary diagrams also seem limited to including one "most popular" or "most distinctive" data and object property for a class in the summary diagram, leaving open a desire for obtaining summary diagrams with richer contents.

A recent prominent RDF data summarization tool is RDFQuotient [17] that allows computing the data summary nodes and their relations just from the data set contents, paying attention also to the data resource class information. Although the tool would have an option of re-locating a property ascription from a subclass to a superclass, the possibilities of its finer-grained interaction with the abstraction possibilities and custom class node merging are less clear and the available examples of legible created data structure diagrams are quite size-limited.

Some of early end-user tools allowing visualization of existing Linked Data set schemas involve LD-VOWL [3] and LODSight [4]. These approaches attempt to create the visual data set structure on-the-fly, as the user starts to look at the upcoming diagram (which is quite admirable). Still, this imposes limitations on the size of the data schemas that can be analysed and on the details that can be included with each of the obtained data schemas.

The process of extracting a data schema from the data set by means of a series of SPARQL queries has been conceptually outlined in an earlier work by the authors [6], as well, where the schema had been made available for a visual query environment. An initial experiment of visualizing smaller-scale data set schemas (with up to 50 classes) using a limited schema compacting approach and an external diagram visualization module has been described in [10]. This work has been substantially expanded here in terms of mathematical precision, refinement of compacting methods, and expanding their reach to diagrams of schemas with well over 100 classes, simplified architecture (web-based schema diagram management), and presenting a library of 24 large data sets from the Linked Open Data cloud.

3 Data Schema Graphs

We start the description of the knowledge graph (KG) schema, and the schema graph concepts by introducing the used notation to describe the knowledge graphs themselves.

A *knowledge graph* $K = (R,L,C,P,rc,T)$ consists of a set of resources R (involving both external resources Re (corresponding to IRIs) and internal resources Ri (corresponding to blank nodes)), a set of literals L, a set of classes C, a set of predicates P (we allow P to overlap with Re), a resource class assignment $rc: R- > 2^{\wedge}C$ and a set of triples $T \subseteq R \times P \times (R \cup L)$; we write *(x a c)* for $c \in rc(x)$ and simply *(x,y,z)* for $(x,y,z) \in T$.

If not specified otherwise, we shall assume that the knowledge graph K is given throughout the rest of this section and shall refer to its components by the notation introduced in the previous paragraph.

Given the KG K, let its schema $S(K)$ be $(C,P,cp,pc,cpc,cc,\#_C,\#_P,\#_{cp}.\#_{pc}.\#_{cpc})$, where

- C is the set of K *classes*, P is the set of K *properties*, $\#_C(c) = \#\{x | (x\ a\ c)\}$ (# is the set size) and $\#_P(p) = \#\{(x,y) | (x,p,y)\}$ for $c \in C$ and $p \in P$,
- $cp \subseteq C \times P$ and $pc \subseteq C \times P$ are sets of *class-to-property* and *property-to-class links* and $cpc \subseteq C \times P \times C$ is the set of *class-property-class links* such that:

  - $\#_{cp}(c,p) = \#\{(x,y) | (x\ a\ c) \wedge (x,p,y)\}$ and $(c,p) \in cp \Leftrightarrow \#_{cp}(c,p) > 0$,
  - $\#_{pc}(p,c) = \#\{(x,y) | (x\ a\ c) \wedge (y,p,x)\}$ and $(p,c) \in pc \Leftrightarrow \#_{pc}(p,c) > 0$, and
  - $\#_{cpc}(c1,p,c2) = \#\{(x,y) | (x\ a\ c1) \wedge (x,p,y) \wedge (y\ a\ c2)\}$ and $(c1,p,c2) \in cpc \Leftrightarrow \#_{cpc}(c1,p,c2) > 0$.

- $cc \subseteq C \times C$ is the sublass relation (the set of *subclass-to-superclass pairs*), i.e. $(c1,c2) \in cc$ whenever $(x\ a\ c1) \Rightarrow (x\ a\ c2)$ for all x.

We write $(c_1,c_2) \in cc+$, if c_1 is a strict transitive subclass of c_2, and $c_1 \sim c_2$, if $c_1 = c_2$, $(c_1,c_2) \in cc+$ or $(c_2,c_1) \in cc+$ (i.e., c_1 is a (transitive) subclass or superclass of c_2).

For a data schema we define its *projection* to (C',P',cc'), where $C' \subseteq C$, $P' \subseteq P$ and $cc' \subseteq cc$ by restricting also the cp, pc and cpc components accordingly (the class restriction induces minimal necessary restriction on cc, as well). The schema projection concept provides a well-defined meaning to *fragments* of the full KG schema if the class and property sets are restricted; the restriction of subclass relation (besides the restriction due to the smaller class set) allows working with schema variants with limited information about the subclass relation.

There is a *natural presentation* of the *schema* in a form of an *attributed graph* with the schema classes as nodes and the cpc relation as property-labelled links among classes, while the cp and pc relations are encoded in node attributes designated to hold lists of properties outgoing from and incoming into the respective classes. The cc relation shall be encoded by dedicated subclass links among the nodes in the schema graph. The respective size information can be added to the graph using further decorations.

Drawing the full schema graph (in the presence of a non-empty subclass relation) may quickly indicate an overload of connections, e.g., due to ascribing properties both to the superclasses and subclasses. Figure 1 contains an illustration of a fragment of a simple Nobel Prizes data schema (cf. Fig. 2 for the full schema and source credit), including the full class and property connections (left) and only the "essential" ones (right).

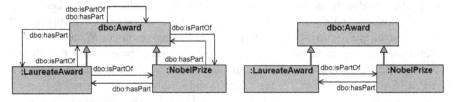

Fig. 1. Full (left) and essential (right) class and property connection example

To prepare for reduced schema graph presentation we first define the markers for determining the relevance of a class in the property source or target context.

Given a property p, a class set $A \subseteq C$ is a *principal source class set* for p, if

- for any class c and resource x, $(x\ a\ c) \wedge (x,p,y)$ implies $d \in A$ for some $d \sim c$ such that $(x\ a\ d)$ (*coverage condition*),
- if $c_1 \in A$ and $c_2 \in A$ such that $c_1 \neq c_2$, then $\neg(c_1 \sim c_2)$ (*minimality condition*: no property is ascribed both to a subclass and a superclass), and
- if $(x,p,y) \Rightarrow (x\ a\ c)$ (i.e., c is (ontological) domain for p) and $(c,d) \in cc+$, then $d \notin A$ (*specificity condition*: no property is ascribed to a superclass if it suffices to ascribe it to a subclass).

The principal source class set is the set of classes that characterizes the property presence at the schema class instances: if a property p is present at instances of some class c, then either p is ascribed to the class c itself or to some its superclass, or all instances of c that possess the property p are split among the subclasses of c that each has the property p ascribed to it. On the other hand, the minimality and specificity conditions guarantee that the property ascription is not too verbose (having unnecessary property ascription points) or too generic.

We note that in general the principal source class set for a property can be non-unique, e.g., if $c_i \subseteq c$ for $i = 1, 2, 3$, and the property p applies to instances of c_1 and c_2, but not to instances of c_3, then either the ascription of p to c, or the ascription of p to both c_1 and c_2 would allow to obtain a principal source class set.

We define a *principal target class set* B for p in a similar (dual) way:

- $(y\ a\ c) \wedge (x,p,y)$ implies $d \in B$ for some $d \sim c$ such that $(y\ a\ d)$ (coverage),
- if $c_1 \in B$ and $c_2 \in B$ such that $c_1 \neq c_2$, then $\neg(c_1 \sim c_2)$ (minimality), and
- if $(x,p,y) \Rightarrow (y\ a\ c)$, and $(c,d) \in cc+$ then $d \notin A$ (specificity).

Further on, the principal target and source class sets are similarly defined for a property in its "other end" class context.

For a property p and its source class c a class set B is a principal target class set, if:

- $(x\ a\ c) \wedge (x,p,y) \wedge (y\ a\ c')$ implies $d \in B$ for some $d \sim c'$ s.t. $(y\ a\ d)$ (coverage),
- if $c_1 \in B$ and $c_2 \in B$ such that $c_1 \neq c_2$, then $\neg(c_1 \sim c_2)$ (minimality), and
- if $((x\ a\ c) \wedge (x,p,y)) \Rightarrow (y\ a\ c')$ and $(c',d) \in cc^+$, then $d \notin A$ (specificity).

For a property p and its target class c a principal source class set is defined similarly.

The non-uniqueness observation applies to the principal target class sets for a property, and principal source and target class sets for a property in a context, as well.

Although the concepts of the principal source and target class sets for a property are defined on the level of the KG; one can compute some such principal sets just from the information of the KG schema (the frequency/count information # is essential here).

To compute for a property p a principal source class set, denoted by $PS(p)$, let $src(p) = c_1, c_2, ..., c_k$ be the sequence of all p source classes c (having $(c,p) \in cp$), ordered by their triple count $\#_{cp}(c_i, p)$ descending, then (if the triple counts are equal) by the class size $\#_c(c_i)$ ascending (a smaller class (e.g., a subclass) comes before a larger class). Consider then all classes $c_1, c_2, ..., c_k$ from $src(p)$ in the index ascension order and include c_i in $PS(p)$ if no $c' \sim c_i$ (i.e., no subclass or superclass of c_i) is already in $PS(p)$.

The computed set $PS(p)$ is a principal source class set for p:

- If $(x\ a\ c)$ and (x,p,y) then c is in $src(p)$ and either $c \in PS(p)$, or c is not included in $PS(p)$ because of some $c' \in PS(p)$ such that $c \sim c'$. This establishes the coverage condition.
- The minimality follows from not including any c into $PS(p)$, if there already is some $c' \in PS(p)$ such that $c \sim c'$.
- Regarding the specificity, let $(x,p,y) \Rightarrow (x\ a\ c)$ and $(c,d) \in cc+$. Let $n = \#_{cp}(c,p)$. By $(x,p,y) \Rightarrow (x\ a\ c)$ we have $\#_{cp}(c',p) \leq n$ for any c'.

  Since c is in $src(p)$, then either (i) $c \in PS(p)$, or (ii) $c_2 \in PS(p)$ for some c_0 such that $c_0 \neq c$, $c_0 \sim c$ and c_0 is before c in $src(p)$.

  Since $\#_{cp}(c_0, p) \leq n = \#_{cp}(c,p)$, the ordering of $src(p)$ entails $\#_{cp}(c_0, p) = n$ and $\#_C(c_0) \leq \#_C(c)$, what excludes $(c, c_0) \in cc+$, so $(c_0, c) \in cc+$.

  So, in either case, $(c^*, d) \in cc+$ for some $c^* \in PS(p)$ such that c^* is before d in $src(p)$ (c^* is either c, if $c \in PS(p)$, or c_0 otherwise); this implies $d \notin PS(p)$, qed.

Note. If the specificity condition would be dropped, a simple way of computing the (simplified) principal source and target class sets would be just to consider the maximal classes (according to the subclass relation) in the respective source/target class sets. In the example of Fig. 1, this would lead to all properties ascribed to *:Award*, so missing important details of, e.g., *:isPartOf* connecting just *:LaureateAward* to *:NobelPrize*.

Although the provided algorithm computes the principal source/target class sets for properties just from the data schema, we propose to work with enriched data schemas, where the principal class sets for properties are pre-computed. The presence of the data set itself in the process of the principal class set computation provides options for varying the algorithm to include more specific principal classes, e.g., when a property can be ascribed to several, but not all, subclasses of a given general class. Since the subclass relation computation can be resource-demanding, the schema enrichments built over a weaker version of the coverage condition that does not rely on the subclass relation can be considered, as well, if the data set is available during the enrichment process.

We also extend the schema information with cardinalities, that can be specified either for a property in general, or for a property in a class context. The property domain and range information (in the exclusive, ontological sense) can be ascribed to the properties, as well. Such information, if available, can be further on included in the schema diagram so making it more informative to its users (cf. Sect. 4)[4].

4 Basic Schema Diagrams

A schema diagram can be created from an extended data schema (with principality markers and, optionally, the cardinality and domain and range information, included) by depicting the schema classes as diagram nodes and showing the properties ascribed to a node either as node attributes or as links outgoing from a node. We shall have a property ascribed (in the attribute and/or the link form) to its principal source classes.

For each pair *(c,p)*, where *c* is in the ***principal source class set*** of *p* we shall draw an ***edge*** from the node corresponding to *c*, labelled by the property *p*, to each class *c'* in the ***principal target class set*** of *p **in the context*** of the source class *c*.

Note the asymmetry of the source and target ends of the class connections by the properties; this allows to obtain well-defined semantics of the created schema diagrams in the chosen schema visualization approach.

A property *p* ascribed to a class *c* is depicted as an ***attribute*** of the *c* node, if it ***cannot be established*** that all triples *(x,p,y)* for *(x a c)* are covered by some edge outgoing from the *c* node (i.e., for all *y*, such that *(x a c)* and *(x,p,y)*, we have *(y a c_1)* for some c_1 with a *p*-labelled edge from *c* into it); skipping the attribute form in the diagram can be done, e.g., if $\#_{cpc}(c,p,c') = \#_{cp}(c,p)$ for some *c'*.

Figure 2 contains a simple schema of the Nobel Prizes data set[5] (for better presentation an auto-introduced extension with an abstract superclass *dbo:City or dbo:Country* is allowed, that collects the attributes and links common to both subclasses). We draw the subclass-to-superclass relation by dedicated subclass edges in the schema diagram.

The frequency information associated with classes ($\#_c$), property availability at classes ($\#_{cp}$) and links ($\#_{cpc}$) are included in the diagram, as well. Further on, there are cardinalities (e.g., *[1]* and *[*]*) for properties in the context of the respective source class (in the case of an attribute presentation) and in the context of the source and target pair (in the case of a link labelled by a property), as well as domain (marked by *D*) and range (marked by *R*) indications telling if the property appearance place corresponds to the property domain or range (in the ontological, exclusive sense).

The attribute presentation of a property includes options for describing of the "other end" of the property links; in Fig. 2 such descriptions involve "IRI" (the property triple objects in the context can be IRIs) and "dgr" (some of the property triples correspond to links drawn in the diagram).

Note that the properties *dct:isPartOf* at *:LaureateAward* class and *:nobelPrize* at *:Laureate* are presented both in the link and the attribute forms since not all property triples in the data have objects that belong to *:NobelPrize*.

---

[4] Further schema constructs may involve, for instance, property adjacency relations (e.g., what properties can "follow" a given property, or what properties can have common subjects or common objects); this information can be important, e.g., if the schema is used to support autocompletion of queries over the data set. We do not focus on these aspects here, as they are less relevant to the current task of constructing schema presentation.

[5] The data were originally retrieved as a snapshot from https://data.nobelprize.org/sparql and are available on the paper's support page.

The mark ~ at the frequency of dbo:City or dbo:Country indicates that the size has been estimated (in this case – calculated as the sum of the subclass sizes, not taking into account the possibility of the subclass overlapping)[6].

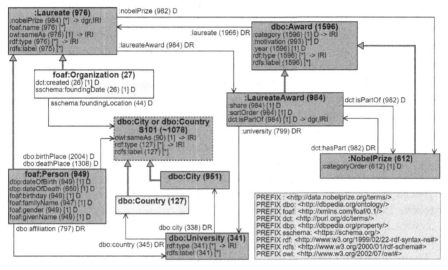

Fig. 2. A Nobel Prizes data schema

5 Advanced Schema Summarization

The described schema visualization method works well just for schemas of limited size, as visual diagrams with larger node sets and more involved interlinking patterns can quickly grow beyond easy comprehensibility. To enable the handling of larger schemas, we introduce the following schema summarization and visual tuning methods:

(i) *class node merging* (ascribe several classes to a single node; join their attribute and link end lists),
(ii) introducing *abstract super-classes*, and
(iii) *inlining links into attributes* (auto-inlining of circular links into dedicated "looping" attributes and "splitting" certain links into their presentation at the source node (by "outgoing" attributes) and the target node (by dedicated "incoming" attributes).

The parameters of the diagram creation can instruct not to do class node merging at all or to merge just the classes with identical attribute and incoming and outgoing property lists. Aside from these basic cases, the class node merging is performed based on the following principles:

---

[6] Fine-tuning the diagram by calculating the exact sizes of the aggregated items can, in principle, be done by consulting the data set itself during the schema diagram creation.

1. For each pair of classes (A,B) their **similarity measure** $s(A,B)$ and **difference measure** $d(A,B)$ are calculated, as follows:
   a. $s(A,B) = \sum sqrt(min(\#(p,A)/\#_C(A),1) * (min(\#(p,B)/\#_C(B),1))* w(p)$ over all properties p incoming into or outgoing out from both A and B. Here $\#(p,X)$ is $\#_{pc}(p,X)$ for p incoming into X and $\#_{cp}(X,p)$ for p outgoing from X (we consider all A and B properties here, not only those ascribed to them in the schema graph). $w(p)$ is the weight assigned to the property p in the similarity computation (usually $w(p) = 1$; some properties as *rdfs:label* or *owl:sameAs* can be excluded from similarity computation by setting $w(p) = 0$).
   b. $d(A,B) = \sum sqrt((min(\#(p,A)/\#_C(A),1))\wedge 2 * (\#_C(A)/(\#_C(A) + \#_C(B)))$ for all properties (incoming or outgoing) present for the class A and not present for the class B (the sum is obtained by considering both classes in both roles).

   The intuition is that properties belonging to both classes contribute to their similarity, while the properties belonging to one class and not to the other contribute to their difference. A larger property is expected to make a larger contribution.
2. Given a **size factor** $ex \geq 0$ (typically also $ex < 1$, sample values are 0, $1/100$, $1/10$, $1/5$ and $1/3$) let the ex-**weighted similarity and difference measures** sw and dw be as follows (*log* is decimal logarithm):
   a. $sw(A,B) = s(A,B)/pow((log(cA) + 1) * (log(cB) + 1),ex)$
   b. $dw(A,B) = d(A,B) * pow((log(cA) + 1) * (log(cB) + 1),ex)$

The intuition is that smaller classes with the same similarity/difference characteristics shall have higher weighted similarity and lower weighted difference and, so, might be used in merging faster.

The similarity and difference measures are generalized to groups of classes, as well. We consider a simple approximate implementation, where the size of the and property appearance in the group is estimated as the sum of the sizes over all group members (this is precise, if classes do not overlap; otherwise, an overestimation is possible).

A merged schema then can, in principle, be defined based on clustering of classes in accordance to some distance function $f(A,B)$, as e.g., $dw(A,B)/(sw(A,B) + \varepsilon)$ for a small ε (some normalization of the function f to make it Euclidean may be necessary).

Our approach, alternatively, gives the user explicit control over the conditions when the class nodes can be merged, so that the appearance of class names in a single node can be interpreted as a certain similarity level between these classes (this way we cannot have *a priori* guarantees on size of the obtained schema, though).

We introduce a **difference threshold** G (typical values are $0, 2, 5, 10$ and 20) and consider a pair of classes (A,B):

- **similar**, if both $dw(A,B) < s(A,B)$ and $dw(A,B) < G$,
- **neutral**, if $dw(A,B) < s(A,B)$ and not $dw(A,B) < G$, and
- **different**, if not $dw(A,B) < s(A,B)$.

We incrementally build (locally) maximal clusters of classes so that:

- all pairs of classes in a cluster are either similar or neutral, and
- for each class in a cluster at least 50% of the other classes are similar to it.

Should the size of the created clustered data set be above the desired target, another try with changed ex and G parameters can be done, or a fragment of the schema (in terms of classes and/or properties), e.g., the largest ones, can be considered. Several diagrams, based on different property sets, can be created, as well.

The *introduction of abstract super-classes*, if requested, is based just on the similarity of the classes (or groups of classes) and the properties that are present in at least two subclasses of the abstract superclass are brought up to the super-class level. This is a finer diagram abstraction mechanism if compared to the class node merging, as it allows the distinct properties (attribute and link ends) at each subclass to stay there and not be brought up to the superclass level. Still, it comes at a cost of introducing an extra node in the diagram (instead of merging several nodes into one), and its introduction needs to be justified mainly by simplifying the overall link structure in the diagram (note that the introduction of the *dbo:City* or *dbo:Country* abstract class in Fig. 2 allowed to merge the links both from *foaf:Person* and *foaf:Organization*, as well as the entire attribute list from both subclasses).

There are two types of conditions that induce the *inlining of a property link* into the nodes of its ends (at each end keeping the information of the other property ends):

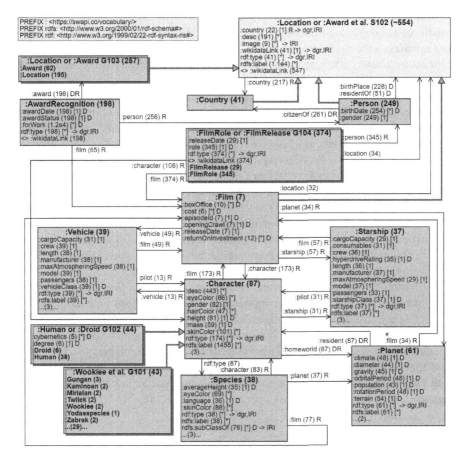

Fig. 3. StarWars data schema example

- maximum number of same-property edges in a diagram (e.g., 7), and
- a minimum number of triples for a property to be shown as line is set.

Figure 3 contains a presentation of an example schema of *StarWars* data set [18], obtained using the described schema summarization method (there are 51 classes in the data set, summarized into 13 nodes (the node *Wookie et.al.* is obtained as summarization of 36 classes); the presentation of the diagram in the tool allows seeing the class and property list elements that are not included in the visual presentation, as well).

6 Implementation and Evaluation

The visual schema presentation assumes the availability of an (enriched) data schema, as described in Sect. 3. To enable work with schemas of realistic size, we have the schemas pre-computed. We use the open-source OBIS Schema Extractor tool[7] that retrieves the schema from a SPARQL endpoint. The schemas are stored using the Data Shape Server (DSS) tool[8] (both schema storage and schema serving functionality included) and then are seamlessly accessed from the ViziQuer tool environment, where both the schema visualization and schema-based visual queries (cf. [19]) are available. The *links to the live examples* of the considered schemas on ViziQuer playground, as well as a *Docker-based environment* for setting up and running the examples locally are provided on the paper support resource.

The schema visualization is initiated by the 'Data Schema' button in the ViziQuer environment project (diagram list) view, after what the window with parameter setup is opened. The parameters to be tuned involve the list of classes and properties to be visualized (there are sliders available for choosing the largest classes and/or properties, as well as manual options for selecting classes and properties) and the diagram merging parameters: *merging strength* (difference threshold in Sect. 5), *size factor* (cf. Sect. 5) and link inlining parameters (*number of same-property lines* and *triple count threshold*). After the parameters have been set, there is an option (via the button 'Show merged classes') to see the counts of nodes and links, as well as the contents of nodes that are going to be created in the diagram. At this point the parameters can be adjusted to ensure that the created diagram is going to be of reasonable size. As a rule of thumb, it would not be recommended to draw diagrams with more than 100 nodes[9].

The button 'Create Schema diagram' creates the diagram and places a pointer to it in the project diagram view. When opening the diagram for the first time, it is automatically laid out (it may take a little time), after what it can be observed and manually tuned by moving the diagram nodes around. For smaller diagrams (around 20 nodes), the tuning is going to be rather easy, while for larger diagrams it gets more complicated. If the node count approaches 100, or even if it is about 40 – 60 with a complex line configuration, it can take a couple of hours for a professional to produce a reasonably well looking diagram (if the diagram tuning task appears too difficult, another schema diagram of

---

[7] https://github.com/LUMII-Syslab/OBIS-SchemaExtractor.

[8] https://github.com/LUMII-Syslab/data-shape-server.

[9] Node count above 150 can cause temporal freezing of the system in the current implementation.

smaller size (e.g., for a fragment of the schema, or obtained via stronger compacting parameters) can be created and worked with).

We note that if the property list at a node is too large to be shown in full in the node box, the full property list can be seen in the side panel, if selecting the respective node.

To evaluate the schema visualization method, we consider 24 prominent Linked Open Data sets registered in the Linked Open Data cloud[10]. For specificity, we consider the data sets meeting the following criteria:

- the data set is available via a SPARQL endpoint (only 145 out of 1584 data sets had a responsive SPARQL endpoint at the time of the experiment),
- the data set is not a part of multiple data set agglomeration on the same SPARQL endpoint[11] (so the data access information can be uniformly retrieved from the LOD cloud data file *sparql* section that does not list named graphs),
- the endpoint triple count, list of classes (with instance count), and the list of properties (with triple count) are accessible via direct aggregated SPARQL queries, and the schema extractor produces a valid data schema[12],
- the (full) class count is between 15 and 2000, and the property count is below 10000 (the smaller schemas are easy and are well-handled, e.g., in [10]; we do not pretend to show diagrams of too heterogenous data sets either).

We order the obtained list in descending order by the actual triple count. For the evaluation, we take the top 21 of the obtained data sets, add a custom *AcademySampo* practical data set, and two example data sets (with a reasonably large triple count) from the *foodie cloud* aggregation, resulting in 24 data sets for which we create the diagram visualizations (either the full diagrams, or the fragments with the largest classes). The diagram visualization experience is summarized in Table 1. The columns *Classes* and *Properties* may contain two numbers each, indicating the full class/property count and the relevant class/property count (excluding, e.g., the *OpenLink Virtuoso* system classes and properties). In the column Lines the numbers are given both for drawing lines of all sizes (excluding just the same-property lines, if more than 7), and for drawing lines corresponding to at least 100 triples). The node and line counts, if not stated otherwise, are given for the maximum strength merging (strong merging (20) and size factor 0); if the diagram size and appearance permits, larger and semantically more nuanced diagrams can be obtained by lowering the merging strength and raising the size factor.

The considered data schemas are available for experimenting on the ViziQuer playground (e.g., creating schema diagrams using various parameters) or by running the ViziQuer tools software locally. The created schema diagrams are also available both in the image form and within a project that can be loaded into the visual tool.

The performed experiment allows to make the following observations:

---

[10] https://lod-cloud.net/lod-data.json.
[11] This excludes, e.g., the data sets from https://www.foodie-cloud.org/sparql (two data sets are brought back into the experiment manually), http://publications.europa.eu/webapi/rdf/sparql, https://linked.opendata.cz/sparql and http://opendatacommunities.org/sparql.
[12] In the case of the SPARQL endpoints considered here, the extractor had been able to produce a valid schema in all but one cases when the class and property lists were possible to obtain.

Table 1. SPARQL endpoint schema presentation experiment

Endpoint URL	Triples	Classes	Properties	Nodes	Lines	Notes
https://libris.kb.se/sparql	2695020210	540	939	225	467(0) 417(100)	A fragment with 40 largest classes can be shown as a diagram with 22 nodes (larger diagrams are more difficult due to high line count)
https://sparql.nextprot.org/	2130123293	165	238	39	66(0) 65(100)	OK
http://affymetrix.bio2rdf.org/sparql	1377023559	661	1328	132	307(0) 271(100)	A fragment with 100 largest classes can be shown as a diagram with 45 nodes
https://ruian.linked.opendata.cz/sparql	870638775	85	200	43	53(0) 26(100)	OK
http://data.bnf.fr/sparql	651506623	38 (26)	886	19	39(0)	Basic merging, OK
http://kaiko.getalp.org/sparql	522998164	140 (128)	332 (245)	31	34(0) 28(100)	OK (size factor 1/5 or 1/3 recommended)
http://cr.eionet.europa.eu/sparql	482077457	272	2001	220	350(0) 308(100)	A fragment with 50 largest classes can be shown as a diagram with 46 nodes
http://dati.isprambiente.it/sparql	385222839	135 (122)	383	86	135(0) 101(100)	Legible diagram (large); for presentation purposes fragments can be considered. OK
http://dati.camera.it/sparql	322885735	104 (92)	367 (283)	61	119(0) 113(100)	OK
http://data.allie.dbcls.jp/sparql	287461727	55 (43)	201	28	37(0)	OK
http://datos.bne.es/sparql	258140051	28 (16)	329	15	32(0)	Basic merging, OK (externally fetched labels)[a]
http://rdf.disgenet.org/sparql/	99381703	122 (110)	665	46	78(0) 48(100)	OK
https://taxref.mnhn.fr/sparql	82745498	1925 (1913)	697 (608)	58	96(0) 73(100)	OK. (Note: very large namespace list)
http://opendata.aragon.es/sparql	70049160	218 (206)	1355 (1259)	88	119(0) 103(100)	OK
Muziekweb[b]	37114240	30	59	16	20(0) 20(100)	OK
http://premon.fbk.eu/sparql	32611819	123 (95)	234 (146)	46	42(0) 36(100)	OK
http://geo.linkeddata.es/sparql	29884998	360	212	41	26(0) 12(100)	OK (textual form of the class names maintained, as in the data set)

(*continued*)

Table 1. (*continued*)

Endpoint URL	Triples	Classes	Properties	Nodes	Lines	Notes
http://en.openei.org/sparql	27317782	1612 (1600)	5163	306	209(0) 81(100)	A fragment with 100 largest classes can be shown as a diagram with 37 nodes
http://datos.bcn.ck/sparql	52057935	542	357	93	128(0) 115(100)	Large, yet legible diagram. Fragments recommended for the first impression. OK
http://id.eaufrance.fr/sparql	17743557	88 (76)	420 (318)	56	58(0) 46(100)	OK
http://ldf.fi/warsa/sparql	14385118	90	310	57	110(0) 96(100)	OK
http://ldf.fi/yoma/sparql	6627922	267	123	34	60(0) 53(100)	OK
poi.rdf[c]	410867958	290	46	5 41[d]	1(0)	OK (subclass lines not counted)
catchrecords Norway[e]	192867166	18	49	17	17(0) 17(100)	Basic merging, OK

a The labels were not present in the dataset and not all entities had labels provided in a structured form at the definition page https://datos.bne.es/def/index-en.html.
b https://data.muziekweb.nl/MuziekwebOrganization/Muziekweb/sparql/Muziekweb.
c https://www.foodie-cloud.org/sparql, Named graph: http://www.sdi4apps.eu/poi.rdf.
d Merge equivalent classes only.
e https://www.foodie-cloud.org/sparql, graph: http://w3id.org/foodie/open/catchrecord/norway/.

- For all considered schemas either full or fragment-based well-legible presentations can be created (sample diagrams are available in the diagram library).
- For endpoints with up to 200–250 classes legible summaries of the full data schema can be reasonably expected if strong summarization is applied and if the properties corresponding to smaller triple counts are inlined. Still, the actual visual complexity of the diagrams may vary from one data endpoint to another
- It might be worthwhile to attempt summarization of full data set schemas of larger size, as well (there are successful legible summaries of full schemas for endpoints with 290, 360, 542 and even 1913 classes in the diagram library).
- The light-weight schema merging techniques (e.g., merging of equivalent classes or basic merging with higher penalties for large class merging) can be recommended also for schemas of smaller size, as this would enable both faster schema summary creation and easier comprehension.
- For the first impression, maximum strength summarization can be chosen without introducing higher summarization penalties for larger classes; in the case of large line counts consider property inlining. The fragment-based presentation or considering just a fraction of larger classes can also be beneficial to obtain the first impression.

7 Conclusions

We have developed a mathematically precise concept of an abstract knowledge graph schema based on the knowledge graph data classes, properties and their relations (involving the class-to-property and subclass relations) and including the principal property ascription points that allow presenting the schema diagram while avoiding the noninformative property-to-class attachments (e.g., avoiding the display of a property both at a superclass and at a subclass).

To handle visual presentation of larger schemas that would not fit on a typical diagramming canvas, possible methods for the schema node grouping and property inlining have been presented. An experiment was conducted in creating visual schema diagrams for realistic Linked Open Data sets (with the class limit of 2000 and property limit of 10000), where for 20 out of 24 data sets the summaries of the full data set schema has been legibly visualized, while for the rest 4 data sets the schema fragment visualization has been successfully performed.

The library of the visual schema presentations for the selected data sets has been created and is offered to the community to support the understanding of the structure of the these data sets by their users; the schema diagrams are offered both in the visual image form, as created by the authors of this paper, and in a live form within the ViziQuer tool, where the interested parties can do further tuning, or create alternative versions of diagrams using different visualization parameters.

The anticipated further work on visual schema presentation would involve expanding the library of data schema visualizations, systematic comparison of different schema summarization strategies, and involving potential end users in the evaluation.

Acknowledgments. This work has been supported by the Latvian Science Council Grant lzp-2021/1-0389 "Visual Queries in Distributed Knowledge Graphs".

References

1. Shapes Constraint Language (SHACL). W3C Recommendation (2017). https://www.w3.org/TR/shacl/
2. Shape Expressions Language 2.1. Final Community Group Report (2019). https://shex.io/shex-semantics/
3. Weise, M., Lohmann, S., Haag, F.: LD-VOWL: extracting and visualizing schema information for linked data. In: 2nd International Workshop on Visualization and Interaction for Ontologies and Linked Data, pp. 120–127 (2016)
4. Dudáš, M., Svátek, V., Mynarz, J.: Dataset summary visualization with LODSight. In: The Semantic Web: ESWC 2015 Satellite Events. LNCS, vol. 9341 (2015)
5. Rabbani, K., Lissandrini, M., Hose, K.: Extraction of validating shapes from very large knowledge graphs. Proc. Very Large Databases **16**(5), 1023–1032 (2023)
6. Čerāns, K., Ovčiņņikova, J., Bojārs, U., Grasmanis, M., Lāce, L., Romāne, A.: Schema-backed visual queries over europeana and other linked data resources. In: Verborgh, R., et al. (eds.) The Semantic Web: ESWC 2021 Satellite Events: Virtual Event, June 6–10, 2021, Revised Selected Papers, pp. 82–87. Springer International Publishing, Cham (2021). https://doi.org/10.1007/978-3-030-80418-3_15

7. Lohmann, S., Negru, S., Haag, F., Ertl, T.: Visualizing Ontologies with VOWL. Semant. Web **7**(4), 399–419 (2016)
8. Bārzdiņš, J., Bārzdiņš, G., Čerāns, K., Liepiņš, R., Sproģis, A.: UML style graphical notation and editor for OWL 2. In: Forbrig, P., Günther, H. (eds.) BIR 2010. LNBIP, vol. 64, pp. 102–114. Springer, Heidelberg (2010). https://doi.org/10.1007/978-3-642-16101-8_9
9. Labra Gayo, J.E., Fernández-Álvarez, D., Garcıa-González, H.: RDFShape: an RDF playground based on shapes. In: CEUR Workshop Proceedings, vol. 2180 (2018)
10. Lāce, L., Romāne, A., Fedotova, J., Grasmanis, M., Čerāns, K.: A method and library for visual data schemas. To appear in Proceedings of ESWC'2024 Satellite Events. LNCS. Springer (2024)
11. Mouromtsev, D., Pavlov, D., Emelyanov, Y., Morozov, A., Razdyakonov, D., Galkin, M.: The simple, web-based tool for visualization and sharing of semantic data and ontologies. In: ISWC P&D 2015, vol. 1486. CEUR (2015). http://ceur-ws.org/Vol-1486/paper_77.pdf
12. Dudáš, M., Lohmann, S., Svátek, V., Pavlov, D.: Ontology visualization methods and tools: a survey of the state of the art. Knowl. Eng. Rev. **33** (2018)
13. Chávez-Feria, S., García-Castro, R., Poveda-Villalón, M.: Chowlk: from UML-based ontology conceptualizations to OWL. In: Groth, P., et al. (eds.) ESWC 2022, vol. 13261, pp. 338–352. Springer, Cham (2022). https://doi.org/10.1007/978-3-031-06981-9_20
14. Draw UML from SHACL. https://shacl-play.sparna.fr/play/draw. Accessed 24 Oct 2024
15. Shacl2plantuml. https://github.com/rosecky/shacl2plantuml. Accessed 24 Oct 2024
16. Li, H., Zhang, X.: Visualizing RDF data profile with UML diagram. In: Li, J., Qi, G., Zhao, D., Nejdl, W., Zheng, H.-T. (eds.) Semantic Web and Web Science, pp. 273–285. Springer, New York (2013). https://doi.org/10.1007/978-1-4614-6880-6_24
17. Goasdoué, F., Guzewicz, P., Manolescu, I.: RDF graph summarization for first-sight structure discovery. VLDB J. **29**(5), 1191–1218 (2020)
18. Star Wars, Example Dataset. https://platform.ontotext.com/semantic-objects/datasets/star-wars.html. Accessed 05 July 2024
19. Čerāns, K., et al.: ViziQuer: a web-based tool for visual diagrammatic queries over RDF data. In: Gangemi, A., et al. (eds.) ESWC 2018. LNCS, vol. 11155, pp. 158–163. Springer, Cham (2018). https://doi.org/10.1007/978-3-319-98192-5_30

A Proposed Ontology Evaluation Tool to Assist Ontology Engineers in Selecting Ontologies During the Reuse Phase

Lina Nachabe[(✉)] and Nushrat Jahan

Mines Saint-Étienne, Univ. Clermont Auvergne, INP Clermont Auvergne, CNRS, UMR 6158 LIMOS, 42023 Saint-Étienne, France
lina.nachabe@emse.fr

Abstract. European data spaces are increasingly promoting interoperability among systems by the use of shared vocabularies and standardized data models. Ontologies, which provide structured vocabularies to uniformly interpret data, are crucial for achieving this interoperability. One essential step in ontology development is the reuse of existing ontologies. However, in the last decade, many efforts have been spent to develop ontologies that, in some cases, cover the same domain. Thus, selecting the most suitable one for reuse is becoming a challenging task for ontology engineers. To address this, we propose a tool that assists ontology engineers in evaluating and selecting ontologies based on various criteria, including structural features, lexical coverage, fairness, and maturity. This tool offers a comprehensive scoring system to guide decision-making during the reuse phase, ensuring efficient ontology integration. It will be validated through feedback from domain experts and applied to develop a unified semantic data model for energy data spaces.

Keywords: Ontology Reuse · Ontology Evaluation · FAIR Principles · Ontology Selection · Semantic Web

1 Introduction

The European Union's initiative on European Data Spaces seeks to foster innovation and stimulate economic growth by enabling secure and interoperable data sharing across critical sectors, including health, energy, and mobility [9]. Semantic interoperability is primordial for the integration and reuse of data across diverse systems. Ontologies, serving as semantic models, are essential for achieving this interoperability, as they offer a unified framework for data access, discovery, and comprehension.

A critical phase in the ontology development cycle is the ontology reuse phase, where engineers decide which existing ontologies to reuse. This practice conserves resources and prevents unnecessary duplication [15,26]. Over the past decade, there has been a surge in the number of ontologies, often with multiple ontologies designed for the same purpose or to cover the same domain [7]. This proliferation has complicated the selection process. To address this, we propose developing *OntoReuse* tool to assist ontology engineers in choosing the most appropriate ontology for reuse during their development process [27].

Evaluating ontologies for reuse involves various aspects, including the structural evaluation (e.g., concept structure and relations, inheritance), the lexical coverage (used terms and descriptions), and the usage aspects (adoption in other projects). Other factors such readability, adaptability, and documentation are also considered [27]. The maturity of the ontology, its versioning, availability, community and social references, and relation to standards are also crucial [13]. These factors are considered essential to ensure the FAIRness of the ontology [1].

Although different metrics have been proposed to evaluate ontologies, however none of the existing solutions combine these metrics in one tool. Thus, we are proposing a comprehensive tool that integrates diverse evaluation metrics into a unified scoring system. This tool will aid ontology engineers in the reuse phase of the ontology development cycle calculating a score for each metric in order to facilitate decision-making, ensuring that the most suitable ontologies are selected for reuse. Ontology engineers will be able to compare the metrics calculated for different ontologies, and choose which ontology to rely on during the reuse phase.

The proposed tool will be validated by collecting feedback from ontology engineers. For validation purposes, the tool is tested on ontologies related to energy domain found in Lov4IoT repository[1].

The rest of this paper is structured as follows: Sect. 2 discusses the opportunities and challenges associated with ontology reuse. Section 3 reviews the existing ontology evaluation metrics and tools, providing a comprehensive overview of the current state of the art. Finally, Sect. 4 presents the OntoReuse tool, detailing its implementation, features, and testing results.

2 Ontology Reuse: Opportunities and Challenges

Ontologies, as formal specifications of shared concepts, address interoperability issues by providing a common vocabulary and defining the relationships between terms, thereby enhancing communication and facilitating data integration across systems [14].

Ontology engineering, a discipline within computer science, information science, and systems engineering, focuses on developing methods for creating ontologies. This involves representing, naming, and defining categories, properties, and relationships among concepts, data, and entities using formal representations like OWL/RDF [20]. It encompasses steps such as ontology requirement specifications, implementation, publication, and maintenance, with an emphasis on reusing existing ontologies during implementation phase.

---

[1] http://www.lov4iot.appspot.com/?p=lov4iot-energy.

Ontology reuse is an essential phase in ontology development process. It involves selecting and incorporating previously developed ontologies to avoid duplication and benefit from collective expertise [18]. There are two main approaches: soft reuse and hard reuse. Soft reuse references specific elements from other ontologies by their URIs, allowing selective and flexible reuse without importing the entire ontology [22,23]. Hard reuse imports entire ontologies using owl:imports, incorporating them as single entities without modification, ensuring full integration into the new structure [22,23].

Ontology reuse promotes knowledge sharing, ensures consistent and accurate data across different systems, and supports standardization. It enhances community collaboration and innovation [5], keeps knowledge representations up-to-date and supports the evolving Semantic Web, fostering a more interconnected and intelligent web ecosystem [6]. While ontology reuse offers significant benefits, it is accompanied by various challenges that must be addressed to ensure successful implementation. These main challenges are:

- Quality assessment: absence of comprehensive ontology evaluation frameworks and automatic measurement techniques for evaluating and ensuring the selection of suitable ontologies [11].
- Selection mechanisms: selection criteria is subjective, as stakeholders have varying preferences. Automated evaluation for large-scale applications requires a thorough understanding of ontology evaluation methods and metrics [16].
- Semantic interoperability and alignment: managing semantic heterogeneity, concept misalignment, and domain specificity are critical challenges in ontology reuse [21].
- Evolution and Maintenance:managing changes, updates, and versioning of ontologies for reuse, while ensuring consistency and backward compatibility [21]

Addressing these obstacles is crucial for maximizing the efficiency and effectiveness of ontology reuse in diverse applications. Ontology evaluation can solve the challenge of quality assessment and the selection mechanisms. It is the process of assessing the quality and effectiveness of ontologies based on specific criteria, such as accuracy, completeness, consistency, usability, and relevance, to ensure it is suitable for its intended purpose [29]. Many existing works conduct evaluations based on different axes such as lexical, structural, and performance criteria. These works are depicted in Sect. 3.

3 Existing Ontology Evaluation Metrics and Tools

Various evaluation metrics as well as different evaluation tools have been proposed to evaluate the ontologies. While Sect. 3.1 discusses relevant existing metrics, Sect. 3.2 depicts some existing tools that uses these metrics to evaluate an ontology.

3.1 Ontology Evaluation Metrics

As depicted in Fig. 1, we have grouped these metrics into four primary axes: Lexical, which deals with the terms and concepts covered by the ontology; Structural, which examines the organization and formal properties of the ontology; FAIRness, which evaluates the Findability, Accessibility, Interoperability, and Reusability of the ontology; Maturity that assesses the development stage and readiness of the ontology.

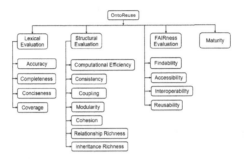

Fig. 1. Evaluation Axes and Metrics

Lexical Evaluation. Lexical evaluation refers to the quality and effectiveness of the vocabulary and terminology used within an ontology. This evaluation focuses on how well the terms and labels in the ontology cover the domain. Key metrics include:

- *Accuracy*: Ensures that the ontology accurately represents domain knowledge without errors or inconsistencies, aligns with domain experts' knowledge, and maintains correct and precise information [1,4,16,28].
- *Completeness*: Assesses whether the ontology covers all relevant concepts and relationships within the domain, adequately addressing the domain's scope and answering relevant questions [1,4,16,19,24,28].
- *Conciseness*: Evaluates the efficiency of the ontology in representing information without unnecessary duplication or complexity, ensuring clear representation of concepts and relationships [4,16,24,28].
- *Coverage*: Measures how well the ontology represents the modeled domain [16].

Structural Evaluation. Structural evaluation focuses on assessing the formal properties and aspects of the ontology's structure, ensuring it is logically coherent, consistent, and well-organized. Key metrics include:

- *Computational Efficiency*: Evaluates the performance of computational processes involving the ontology, such as reasoning and querying, and assesses the speed at which tools like reasoning engines interact with the ontology [4,16].
- *Consistency*: Ensures that the ontology does not contain contradictions or inconsistencies in its definitions and relationships, evaluating the coherence and absence of contradictions within the ontology schema [1,2,4,16,24,28].
- *Coupling*: Assesses the number of classes from imported ontologies referenced in the ontology [16,19].
- *Modularity*: Evaluates the degree to which the ontology schema is modular and well-organized [28].
- *Cohesion*: Measures how closely related entities within an ontology are. High cohesion indicates strong connections between classes. Key metrics include the Number of Root Classes (NoR), the Number of Leaf Classes (NoL), and the Average Depth of Inheritance Tree of Leaf Nodes (ADIT-LN), which averages the depth of paths from roots to leaf nodes. These metrics collectively assess the structural organization of the ontology [8,16,19].
- *Relationship Richness (RR)*: Reflects the diversity of relations within the ontology, indicating richness beyond simple class-subclass relationships [28].
- *Inheritance Richness*: Describes the distribution of information across the ontology's inheritance tree, helping to categorize knowledge effectively. It distinguishes between horizontal ontologies (few levels with many subclasses) and vertical ontologies (many levels with few subclasses) [19,28].

FAIRness Evaluation. FAIRness evaluation focuses on how well the ontology adheres to the principles of Findability, Accessibility, Interoperability, and Reusability. Key metrics include:

- *Findability*: Assesses how easily ontologies and semantic resources can be found, evaluating aspects like persistent identifiers, metadata quality, and searchability. It checks ontology URI persistence, resolvability, unique version IRI, presence of minimum descriptive metadata, and availability of ontology prefix and namespace in external registries [1,12].
- *Accessibility*: Evaluates how easily ontologies and semantic resources can be accessed and retrieved, considering factors like open access, documentation availability, and download options. It assesses proper content negotiation and open URI protocol [1,12].
- *Interoperability*: Focuses on the ability of ontologies and semantic resources to work together seamlessly, examining the use of standard formats, vocabularies, and alignment with existing ontologies. It identifies references to pre-existing vocabularies in metadata annotations, classes, properties, or data properties [1,2,12].
- *Reusability*: Assesses how well ontologies can be reused for different purposes, considering licensing, versioning, and documentation. It verifies the availability of human-readable documentation, provenance metadata, licensing information, detailed vocabulary metadata, and proper descriptions of ontology

terms with labels and definitions [1,12,28]. It includes the adaptability of the ontology evaluating its flexibility and ability to be extended without removing existing axioms [4,16].

Maturity Evaluation. assesses the development stage and stability of the ontology, including aspects like version history, community adoption, and ongoing maintenance. The maturity can be evaluated using the Technology Readiness Level (TRL) scale. Applying the TRL framework to ontologies involves determining the stages of their conceptualization, validation, and implementation. For instance, at TRL 1, the basic principles of the ontology are observed, where initial concepts and potential use cases are identified. As the ontology matures to TRL 3, an experimental proof of concept is developed, demonstrating the feasibility of its structure and vocabulary. At TRL 4 and TRL 5, the ontology is validated in a controlled laboratory setting and then in a relevant environment, such as a specific domain or application area. Moving further up the TRL scale, TRL 6 and TRL 7 involve demonstrating the ontology's functionality in increasingly realistic and operational environments. Finally, at TRL 8 and TRL 9, the ontology is fully developed, qualified, and proven in an operational environment, indicating its readiness for integration into real-world systems and applications [10].

In the following paragraph, we will explore the existing ontology evaluation tools that align with these metrics and provide insights into their functionalities and applications.

3.2 Existing Ontology Evaluation Tools

In this section we review relevant existing tools for ontology evaluation based using metrics cited in Sect. 3.1.

Lexical Evaluation Tools

- OntoVal: Covers various lexical aspects such as Accuracy, Completeness, Conciseness and Clarity. It assesses how well an ontology represents domain knowledge and handles the precision and comprehensiveness of terms and relationships [3].
- OntoKeeper: Focuses on Accuracy, Consistency, and Clarity, providing insights into the quality and effectiveness of the ontology's vocabulary and terminology [2].

Structural Evaluation Tools

- OOPS! (OntOlogy Pitfall Scanner): Provides automated detection of potential structural issues such as inconsistencies, completeness gaps, and conciseness problems. It helps developers address common pitfalls and improve the ontology's formal properties and organization [24].

- OQuaRE: Assesses the cohesion of ontologies by evaluating the relationships and organization of classes, such as the Number of Root Classes (NoR), Number of Leaf Classes (NoL), and Average Depth of Inheritance Tree of Leaf Nodes (ADIT-LN) [8].

FAIRness Evaluation Tools

 - O'FAIRe: Offers metadata-based automated assessment of the FAIRness of ontologies. It evaluates Findability, Accessibility, Interoperability, and Reusability by examining metadata descriptions, persistent identifiers, and searchability [1].
 - FOOPS!: Provides a comprehensive FAIRness assessment by conducting 24 different checks across the FAIR dimensions. It evaluates aspects such as ontology URI persistence, metadata quality, content negotiation, and interoperability and provides explanations for failures and actionable suggestions for improvement [12].

Currently, there are no specific tools dedicated to evaluating the maturity of ontologies using the Technology Readiness Level (TRL) framework. However, integrating maturity evaluation as a crucial axis in ontology assessment is essential because it reflects the degree of trust and reliability of an ontology. Consequently, existing tools offer various functionalities for evaluating ontologies, however none of these tools combine the metrics mentioned in Sect. 3.1.

4 OntoReuse Proposed Tool

To address the gaps in current ontology evaluation tools, we propose a novel solution called OntoReuse. This tool integrates the metrics outlined in Sect. 3.1 to compute a comprehensive score for an ontology across four critical axes: Lexical, Structural, FAIRness, and Maturity. In the subsequent sections, we will first detail the methodology used to calculate these metrics (Sect. 4.1), followed by a description of the implementation phases of OntoReuse (Sect. 4.2).

4.1 Methodology for Metrics Scoring

In this section, the methodology for metric calculation for each axis is depicted.

Lexical Scoring. Lexical Evaluation assesses the quality of an ontology's lexical terms, ensuring it includes all necessary elements while avoiding redundancy. To evaluate this, we calculate a lexical score that measures how well the ontology aligns with the target domain. For a given ontology and domain, we compute *Ontology Relevance (OR)* using Eq. 1 and *Domain Coverage (DC)* using Eq. 2 [30]. These metrics reflect the overlap between the ontology's concepts and those extracted from the domain's text corpus. The metrics are defined as [30]: - **O**: Total number of concepts in the candidate ontology. - **D**: Number of related terms in the domain. - **S**: Number of shared concepts between the candidate ontology and the domain concepts.

$$\text{Domain Coverage} = \frac{S}{D} \qquad (1)$$

$$\text{Ontology Relevance} = \frac{S}{O} \qquad (2)$$

Structural Scoring. This evaluation focuses on metrics related to the structural elements of the ontology to determine its effectiveness in representing complex relationships and hierarchies. Key metrics include relationship richness (RR), inheritance depth (ID), cohesion, computational efficiency and consistency.

– *Computational efficiency:* To assess it, we examine the ontology's performance in various tasks, determining tool compatibility and task speed. Key metrics include the time to load (T_{Loading}) the ontology, to perform reasoning ($T_{\text{Reasoning}}$), and to execute queries ($T_{\text{QueryExecution}}$). We are using 3 basic queries depicted in Table 1 [25].

Table 1. SPARQL Queries Used for Ontology Evaluation

Query Number	SPARQL Query
Q1	SELECT ?class WHERE {
	?class a owl:Class .
	}
Q2	SELECT ?property WHERE {
	?property a owl:ObjectProperty .
	}
Q3	SELECT ?subject ?predicate ?object WHERE {
	?subject ?predicate ?object .
	} LIMIT 100

– *Consistency:* Ensures that an ontology is free from contradictions. We used the Pellet reasoner to evaluate the ontology's consistency, assigning a score of 1 if it is consistent and 0 if it is not [25].
– *Relationship Richness (RR):* RR is calculated as the ratio of the total number of relationships (P) to the sum of subclasses (SC) and relationships as depicted in 3 [28]. An RR value near zero signifies that most relationships are class-subclass relationships, while an RR value close to one hundred indicates that most relationships are beyond class-subclass.

$$RR = \frac{|P|}{|SC| + |P|} * 100 \qquad (3)$$

- **Inheritance Richness (IR)**: IR is defined as the average number of subclasses per class. The number of subclasses (C1) for a class C_i is given by $H^C(C_1, C_i)$. as calculated in 4 [19,28]. A low IRs indicates a vertical structure, suggesting detailed, specialized knowledge. Conversely, a high IRs indicates a horizontal structure, reflecting a broad range of general knowledge.

$$IR_s = \frac{\sum_{C_i \in C} H^C(C_1, C_i)}{|C|} * 100 \qquad (4)$$

- **Cohesion**: it includes the Number of Root Classes (NoR); the Number of Leaf Classes (NoL); and the Average Depth of Inheritance Tree of Leaf Nodes (ADIT-LN) [16]. These metrics collectively assess the ontology's structural organization and modular relatedness.

FAIRness Evaluation Scoring. For FAIRness scoring, we reused the !FOOPS tool [12] because it comprehensively covers all the necessary metrics. In addition, we introduced the concept of content negotiation, which reflects the reuse potential of an ontology. Ontologies that are accessible in multiple formats are more likely to be reused. It assesses an ontology's base URL's capability to provide data in various formats, such as ".ttl" (Turtle), ".rdf" (RDF/XML), ".owl" (OWL/XML), ".jsonld" (JSON-LD), ".n3" (Notation3), and ".nt" (N-Triples). The scoring ranges from 1 to 6, incrementing by one for each supported format.

Maturity Scoring. The maturity level of the ontology is currently assigned manually by the user according to [10] as detailed in Metrics Calculation Details

To improve this process, future efforts will focus on dynamically determining the maturity level based on the quality of the documentation, the number of related projects utilizing the ontology and other factors.

After exploring the methodology of metrics evaluation, the following paragraph will delve into the implementation of the tool.

4.2 OntoReuse Implementation

The tool is implemented using Python and has been deployed on GitHub[2] to gather community feedback and encourage contributions, aiming to refine the existing code and incorporate additional relevant metrics. It is a web page solution hosted on the local machine.

Implementation Phases

1. Ontology Loading: The user has the option to either provide the URL of the ontology or upload a local version directly from his/her device. Once the ontology is loaded, an extended knowledge graph is automatically computed by incorporating all its imported ontologies, ensuring a comprehensive representation of the entire ontology ecosystem.

---

[2] https://github.com/Nushrat-Jahan/OntoReuse.

2. Lexical score: The user inputs specific terms or expressions that they want the ontology to cover. The tool leverages the Datamuse API[3] to generate lexical related terms for each input, then calculates the DC 1 and OR 2 metrics. Only terms with a similarity score above 0.5 are retained.
3. Structural score: SPARQL queries are used to calculate the structural metrics as explained in Sect. 4.1. It also checks the consistency of the ontology using the `owlready2` library and measures computational efficiency through query execution times.
4. FAIRness score: FOOPS! API is called to assess the FAIRness of the ontology. Following this, the tool analyzes the ontology's base URL webpage to identify the availability of different formats, such as .ttl, .rdf, .owl, .jsonld, .n3, and .nt. Based on the number of supported formats found, the tool calculates the content negotiation score.
5. Maturity score: is set manually ranging from 1 to 9 as depicted in [17].
6. Display the result in form of a table.

Testing. Figure 2 depicts the OntoResue tool main page. Table 2 depicts the results for the SEAS Photovoltaic ontology[4] giving as input text for lexical testing *solar energy*. Only 3.5% shared concepts are found with a domain coverage of 9.375%. This means that additional terms related to the domain are needed to be modeled. With 65.12% IR, it shows that a significant portion of the ontology's classes are specialized, suggesting a rich hierarchical structure. The ontology loads and reasoning time are almost instantaneously, demonstrating efficiency in terms of size and structure.

In general, ontology engineers will rely on these metrics when selecting an ontology for reuse. If domain coverage is a priority, they may choose an ontology that encompasses a broader range of terms and concepts. However, this often comes at the cost of increased reasoning and loading times. Therefore, if the goal is to embed the ontology in resource-constrained devices, these performance aspects become critical. Moreover, maturity is a key consideration when aiming to use a well-established ontology, particularly one that has been standardized and implemented in real-world applications. It is important to note that the structural calculations are based on importing the entire ontology (hard reuse). The tool was tested on 20 ontologies from the LOV4IOT repository within the energy domain, including ontologies such as SAREF, SEAS, INTECONNECT, and OPENADR. Results indicate that while the tool provides a lexical coverage accuracy of 60%, it does not always generate accurate related terms. This accuracy largely depends on the input term and the performance of the DataMuse API. Further research is recommended to improve the generation of related terms.

The structural evaluation component of the tool allows for a better understanding of the richness in relationships and properties within the ontologies. To gather user feedback, a survey was distributed to ontology engineers, with

---

[3] https://www.datamuse.com/api/.
[4] SEAS Photovoltaic Ontology v1.0.

Fig. 2. User Interface of OntoReuse for Ontology Analysis

15 responses received from 40 potential participants. The respondents included researchers and ontology engineers from both enterprise and academic backgrounds. The survey results showed that 82% of respondents found the tool easy to use, 90% felt it was well-documented, but only 53% believed that the related terms were generated accurately.

Additionally, respondents appreciated the tool's ease of installation and its coverage of different metrics, and indicated that it can be useful in both the reuse and evaluation phases of ontology development. However, they suggested that the tool should be hosted on the web rather than run locally and recommended the inclusion of additional metrics, such as modularity and the accuracy of term definitions, especially if these terms are derived from standardized sources. Finally, some respondents suggest to add a script for automatic maturity score calculation.

4.3 Conclusion

Ontology reuse is crucial for resolving interoperability issues across diverse domains, facilitating effective communication and data integration among disparate systems. By reusing ontologies, organizations can promote knowledge sharing, standardization, and more efficient ontology development by leveraging existing, well-established ontologies. To assist ontology engineers in choosing which ontology to reuse, ontology evaluation tools are needed. Although different tools for ontology evaluation exist, such as O'FAIRe, OOPS!, and FOOPS!, however, they are not comprehensively covering all necessary evaluation metrics. Thus, this paper introduces the OntoReuse tool, which addresses these gaps by integrating a wide range of evaluation criteria into a single, unified framework. The framework includes lexical, structural, fairness, and maturity evaluations, providing a comprehensive approach to ontology assessment. OntoReuse supports key metrics such as accuracy, completeness, consistency, relationship richness, inheritance richness, computational efficiency, findability, accessibility, interoperability, reusability, and content negotiation. This thorough evaluation enables users to make informed decisions when selecting ontologies for reuse, ensuring they choose the most appropriate ones for their needs. The tool has been tested on 20 ontologies and evaluated by 15 ontology engineers. While these initial tests are promising, the tool still requires more extensive user feed-

Table 2. Results for PhotoVoltaic Ontology using OntoReuse

Metric	Value
Lexical Evaluation	
Domain Coverage (S/D)	9.375%
Ontology Relevance (S/O)	3.5%
Structural Evaluation	
Relationship Richness	47.46%
Inheritance Richness	65.12%
Sum of the number of subclasses	56
Average number of subclasses per class	0.6511627906976745
Inheritance Depth	4
Number of object properties	130
Number of datatype properties	1
Total number of relationships (properties)	131
Number of Roots (NoR)	2
Number of Leaves (NoL)	61
Average Depth of Inheritance Tree of Leaf Nodes (ADIT-LN)	65.12%
Consistency	1
Time to load ontology	0.0001 s
Time to perform reasoning	0.0001 s
Time to execute query 1	0.2147 s
Time to execute query 2	0.0063 s
Time to execute query 3	0.0138 s
FAIRness Evaluation	
ontology_URI	https://ci.mines-stetienne.fr/seas/PhotovoltaicOntology-1.0.ttl
overall_score	0.083
content_negotiation_score	3
Maturity Level	
TRL_score	6

back and validation to confirm its effectiveness and utility in various real-world scenarios. Enhancements are needed in the lexical evaluation component and automatic maturity metric calculation. Additional metrics such as modularity can be incorporated to further improve the tool's functionality.

Acknowledgement. This work is funded by the European Union's Horizon Europe innovation action program under grant agreement no [101069287] (OMEGA-X). https://omega-x.eu/.

References

1. Amdouni, E., Bouazzouni, S., Jonquet, C.: O'faire makes you an offer: metadata-based automatic fairness assessment for ontologies and semantic resources. Int. J. Metadata Semant. Ontol. **16**(1), 16–46 (2022)
2. Amith, M., et al.: Architecture and usability of ontokeeper, an ontology evaluation tool. BMC Med. Inf. Decis. Mak. **19**, 1–18 (2019)
3. Avila, C.V.S., Maia, G., Franco, W., Rolim, T.V., da Rocha Franco, A.D.O., Vidal, V.M.P.: Ontoval: a tool for ontology evaluation by domain specialists. In: ER Forum/Posters/Demos, pp. 143–147 (2019)
4. Brank, J., Grobelnik, M., Mladenic, D.: A survey of ontology evaluation techniques. In: Proceedings of the Conference on Data Mining and Data Warehouses (SiKDD 2005), pp. 166–170. Citeseer (2005)
5. Cantador, I., Fernández, M., Castells, P.: Improving ontology recommendation and reuse in webcore by collaborative assessments (2007)
6. Cantador, I., Fernández, M., Castells, P.: Webcore: a web application for collaborative ontology reuse and evaluation. Available at researchgate. net. Accessed 10 Nov (2020)
7. Carriero, V.A., et al.: The landscape of ontology reuse approaches. In: Applications and Practices in Ontology Design, Extraction, and Reasoning, pp. 21–38. IOS Press (2020)
8. Duque-Ramos, A., Fernández-Breis, J.T., Stevens, R., Aussenac-Gilles, N.: Oquare: a square-based approach for evaluating the quality of ontologies. J. Res. Pract. Inf. Technol. **43**(2), 159–176 (2011)
9. European Commission: Data Spaces Panel Report (2020). https://data.europa.eu/sites/default/files/report/Data_Spaces_Panel_Report_EN.pdf. Accessed 8 Mar 2024
10. European Commission: Guiding notes to use the trl self-assessment tool. Horizon 2020 (2020)
11. Fernández, M., Cantador, I., Castells, P.: Core: a tool for collaborative ontology reuse and evaluation (2006)
12. Garijo, D., Corcho, O., Poveda-Villalón, M.: Foops!: an ontology pitfall scanner for the fair principles. In: ISWC (Posters/Demos/Industry) (2021)
13. Gouriet, M., et al.: The energy data space: the path to a European approach for energy. In: Designing Data Spaces: The Ecosystem Approach to Competitive Advantage, pp. 535–575. Springer, Cham (2022). https://doi.org/10.1007/978-3-030-93975-5_33
14. Guarino, N., Oberle, D., Staab, S.: What is an ontology? In: Handbook on Ontologies, pp. 1–17 (2009)
15. Halper, M., Soldatova, L.N., Brochhausen, M., Sabiu Maikore, F., Ochs, C., Perl, Y.: Guidelines for the reuse of ontology content. Appl. Ontol. **18**(1), 5–29 (2023)
16. Hlomani, H., Stacey, D.: Approaches, methods, metrics, measures, and subjectivity in ontology evaluation: a survey. Semant. Web J. **1**(5), 1–11 (2014)

17. Jahan, N.: OntoReuse: Metrics Calculation Details (2024). https://github.com/Nushrat-Jahan/OntoReuse/blob/main/Metrics%20Calculation%20Details.md. Accessed 03 Sept 2024
18. Katsumi, M., Grüninger, M.: What is ontology reuse? In: FOIS, pp. 9–22 (2016)
19. Khan, Z.C.: Evaluation metrics in ontology modules. In: Description Logics (2016)
20. Kotis, K.I., Vouros, G.A., Spiliotopoulos, D.: Ontology engineering methodologies for the evolution of living and reused ontologies: status, trends, findings and recommendations. Knowl. Eng. Rev. **35**, e4 (2020)
21. Maedche, A., Staab, S.: Measuring similarity between ontologies. In: Gómez-Pérez, A., Benjamins, V.R. (eds.) EKAW 2002. LNCS (LNAI), vol. 2473, pp. 251–263. Springer, Heidelberg (2002). https://doi.org/10.1007/3-540-45810-7_24
22. Pinto, H.S., Martins, J.: Reusing ontologies. In: AAAI 2000 Spring Symposium on Bringing Knowledge to Business Processes, vol. 2, p. 7. AAAI, Karlsruhe (2000)
23. Poveda-Villalón, M., Fernández-Izquierdo, A., Fernández-López, M., García-Castro, R.: Lot: an industrial oriented ontology engineering framework. Eng. Appl. Artif. Intell. **111**, 104755 (2022)
24. Poveda-Villalón, M., Suárez-Figueroa, M.C., Gómez-Pérez, A.: Validating ontologies with OOPS! In: ten Teije, A., et al. (eds.) EKAW 2012. LNCS (LNAI), vol. 7603, pp. 267–281. Springer, Heidelberg (2012). https://doi.org/10.1007/978-3-642-33876-2_24
25. Raad, J., Cruz, C.: A survey on ontology evaluation methods. In: International Conference on Knowledge Engineering and Ontology Development, vol. 2, pp. 179–186. SciTePress (2015)
26. Sowiński, P., Wasielewska-Michniewska, K., Ganzha, M., Paprzycki, M., Bădică, C.: Ontology reuse: the real test of ontological design. In: New Trends in Intelligent Software Methodologies, Tools and Techniques, pp. 631–645. IOS Press (2022)
27. Talebpour, M., Sykora, M., Jackson, T.: The evaluation of ontologies for quality, suitability for reuse, and the significant role of social factors. In: Fred, A., Salgado, A., Aveiro, D., Dietz, J., Bernardino, J., Filipe, J. (eds.) IC3K 2018. CCIS, vol. 1222, pp. 161–177. Springer, Cham (2020). https://doi.org/10.1007/978-3-030-49559-6_8
28. Tartir, S., Arpinar, I.B., Moore, M., Sheth, A.P., Aleman-Meza, B.: Ontoqa: metric-based ontology quality analysis (2005)
29. Trokanas, N., Cecelja, F.: Ontology evaluation for reuse in the domain of process systems engineering. Comput. Chem. Eng. **85**, 177–187 (2016)
30. Zaitoun, A., Sagi, T., Hose, K.: Automated ontology evaluation: evaluating coverage and correctness using a domain corpus. In: Companion Proceedings of the ACM Web Conference 2023, pp. 1127–1137 (2023)

Manufacturing Commonsense Knowledge

Muhammad Raza Naqvi[1(✉)], Arkopaul Sarkar[1], Farhad Ameri[2],
Linda Elmhadhbi[3], and Mohamed Hedi Karray[1]

[1] Laboratoire Génie de Production (LGP), Université de Technologie Tarbes Occitanie Pyrénées (UTTOP), 47 Av. d'Azereix, 65000 Tarbes, France
razisyed4@gmail.com, snaqvi@enit.fr
[2] School of Manufacturing Systems and Networks, Ira A. Fulton Schools of Engineering, Arizona State University, W, Mesa, AZ 85212, USA
[3] INSA Lyon, Université Lumière Lyon 2, Université Claude Bernard Lyon 1, Université Jean Monnet Saint-Etienne, DISP UR4570, 69621 Villeurbanne, France

Abstract. This paper underscores the critical role of integrating common knowledge across the manufacturing industry by introducing and formalizing a new concept called Manufacturing Commonsense Knowledge (MCSK). Although commonsense knowledge is crucial for enhancing AI-driven operational intelligence and decision-making, its structured application within the broader manufacturing sector has been insufficient. To bridge this gap, we present a structured methodology for translating MCSK into first-order logic (FOL), employing standard ontological frameworks such as the Basic Formal Ontology (BFO), Industrial Ontologies Foundry (IOF), Relations Ontology (RO), and Machine Services Description Language (MSDL). This translation process is pivotal, whether the underlying AI systems employ symbolic, sub-symbolic, or hybrid approaches, as it transforms intuitive MCSK into organized semantic rules. Our findings demonstrate that structured MCSK patterns enhance knowledge representation's clarity and utility and significantly improve the explanatory capabilities of AI decision-making processes across the industry. The broader impacts of our research extend to enhancing machine interoperability, predictive analytics, and advanced manufacturing practices, thus paving the way for a more informed and efficient industrial future.

Keywords: Common-Sense knowledge · Manufacturing Common-Sense knowledge · Knowledge engineering

1 Introduction

In the quest to create AI systems, particularly those that combine symbolic and subsymbolic methods to mimic human reasoning capabilities, embedding Commonsense Knowledge (CSK) remains a formidable challenge [1]. CSK, intuitive to humans and foundational to routine decision-making processes, relies on an understanding of the world shaped by human interactions [2].

However, for AI systems—especially those integrating symbolic logic and subsymbolic learning—mastering this natural grasp of social and physical environments poses a significant challenge. Embedding AI systems with CSK aims to bridge the gap between machine and human interaction, enhancing AI systems' ability to perform and interact within real-world environments in a manner aligned with human reasoning.

This is particularly relevant for AI systems that integrate symbolic logic, such as MCSK, with sub-symbolic approaches [3]. This challenge is especially pronounced in the manufacturing industry, where operations are dynamic and diverse, and decision-making processes are highly complex [4]. Our work focuses on the manufacturing domain, specifically on formalizing MCSK patterns into first-order logic and converting these into semantic rules using a standard vocabulary.

These rules are designed to be embedded within hybrid AI systems that combine symbolic reasoning with machine learning techniques, facilitating advanced reasoning and decision-making processes. This approach is particularly practical in manufacturing environments with complex and dynamic decision-making processes.

Reminder of the paper is organized as Sect. 2; We overview CSK, different patterns of CSK, CSK benchmarks, and the importance of formalizing the MCSK Sect. 3 discusses the definition of MCSK and its various patterns with examples; and later in Sect. 4 explain the formalization of these patterns into FOL and semantic rules using standard vocabularies using BFO, IOF, RO, and MSDL, and last in Sect. 5 is about the conclusion and future directions.

2 Literature Review

Common sense is the basic practical knowledge and reasoning most people use in daily life [5,6]. According to DARPA [7], it is "the basic ability to perceive, understand, and judge things that are shared by ('common to') nearly all people and can reasonably be expected of nearly all people without need for debate." They argue that every human may not notice the large number of common sense assumptions they make that constitute everything they say and do [7].

This is based on how the physical world is called intuitive physics; intuitive psychology means a basic comprehension of human behaviors and motives and a general understanding of the subject matters that most people possess [8]. Since the arrival of AI, one of its main goals has been to collect and represent CSK so that machines can understand and reason with it, which is still a challenge to date; this has also led to work on designing, collecting, and constructing CSK as Cyc [9], ConceptNet [10], and WordNet [11] are some renowned examples among many others [12].

Literature about CSK patterns shows that it always comes with certain patterns such as lexical, similarity, distinctness, taxonomic, part-whole, spatial, creation, utility, desire/goal, quality, comparative, temporal, relational, as detailed in the correspondence of each relation and dimensions by different sources by

Ilievski et al. [13]. In terms of CSK benchmarks about certain areas, there are some benchmarks such as social, physical, visuals, and numbers, etc. [14–17].

Despite the well-established frameworks for CSK, there remains a distinct lack of similar benchmarks in the context of manufacturing [18,19]. This gap highlights the unique challenges and the critical need for MCSK [20]. MCSK aims to address these challenges by tailoring CSK to the specific needs and conditions of the manufacturing industry.

3 Manufacturing Commonsense Knowledge

The National Institute of Standards and Technology (NIST)[1] defines manufacturing knowledge (MK) as a phrase that includes knowledge about machine and process capabilities, material properties and understanding unintended consequences of design in manufacturing [21]. MK encompasses knowledge about manufacturing processes, assembly, quality, materials handling, and operation planning [22,23].

MCSK builds upon this foundation by integrating background knowledge and practical experience, such as that derived from Machinist's handbooks, to create a specialized form of CSK that supports decision-making and operational intelligence in manufacturing environments.

However, because of the intricacy and unpredictability of the data, directly extracting MCSK from text sources presents considerable difficulties. Initially, we concentrated on finding and organizing patterns associated with manufacturing processes and operations. By taking this approach, we can systematically collect relevant knowledge and facilitate the development of MCSK, which in turn makes it more accessible and actionable for practical applications.

There are four pillars of manufacturing engineering according to Michigan State University[2], which include;

1. Materials and manufacturing processes
2. Product, tooling, and assembly engineering
3. Manufacturing systems and operations
4. Manufacturing competitiveness

MCSK patterns focused on covering all these points. We identify 7 CSK patterns relevant to manufacturing and correspond to the four pillars of manufacturing: Requirement, Precedence, Causation, Similarity, Distinctness, Part hood, and performance.

When we refer to MCSK patterns, "we specifically mean there are recognizable and repetitive patterns associated with phrases related to the manufacturing context. Each pattern represents a way of understanding that typically emerges from the Natural language used to discuss manufacturing processes, plans, and scenarios."

The following section will discuss the MCSK patterns we identify and how to formalize them.

---

[1] https://www.nist.gov/manufacturing.
[2] https://www.mtu.edu/mmet/graduate/manufacturing-engineering/four-pillars/.

3.1 MCSK Patterns

Based on general common sense patterns and dimensions already available in the literature [13]. We put it into the manufacturing context to understand how it is relevant and then sub-categorize each CSK pattern to relate to manufacturing by using the example of a wooden chair.

Requirement. Requirement in MCSK is the understanding that specific tasks or processes in manufacturing require specific machine, equipment, or conditions to be completed successfully. This knowledge is considered 'common sense' because it aligns with the generally accepted understanding that certain tasks necessitate certain prerequisites.

- Material Requirement: A wooden chair requires wood.
- Process Requirement: Cutting and assembling are required to manufacture a wooden chair.
- Machine Requirement: A saw is required to cut wood.
- Skill/Capability Requirement: The capability to cut and assemble wood properly is required.
- Environment/Location Requirement: A workshop is required for cutting and assembling the wooden chair.

Precedence. Many manufacturing processes must follow a specific order or sequence of steps. This is often called the "process flow" or "workflow." This is what we refer to as "precedence."

- Process Precedence: Designing the wooden chair comes before cutting the wood.
- Object Precedence in Workflow: The wood must be present before it can be cut.
- Existence Precedence: The wooden chair design must exist before beginning the manufacturing process.

Causation. Causation in manufacturing refers to the cause-and-effect relationships that exist between different steps in a manufacturing process.
-Material Causation: Using low-quality wood causes the wooden chair to be less sturdy.

Distinctness. The manufacturing processes are distinct due to the materials used, the steps involved, and the products created.
-Material Distinctness: A book is made with paper; a wooden chair is made with wood.

Similarity. Many manufacturing processes share similarities due to the nature of the tasks involved. For example, making a wooden chair and a wooden bed involves the same materials, e.g., wood, and tools, e.g., a cutter, etc.

Part-Hood. Part-hood in manufacturing refers to the components or materials that make up a product or the group or category that a product belongs to. This concept of part-hood is fundamental to understanding how different products are made and how they relate to each other.

- Process Part-hood: Cutting and joining are parts of the wooden chair-making process.
- Object Part-hood: A wooden chair consists of parts like legs, a backrest, and a seat.
- Grouping: A wooden chair can be grouped with other types of furniture.

Performance. It's important to note that assessing performance and competitiveness often involves specific knowledge and context that might go beyond 'common' knowledge. It can require detailed information about the materials, tools, and processes used, the design's specifics, the market's characteristics, and so on. Some aspects of performance competitiveness are 'common knowledge.' For example, it's generally understood that using higher-quality materials can improve a product's durability, a more efficient manufacturing process can lower costs, or a unique and appealing design can make a product more competitive in the marketplace.

4 Formalizing MCSK Patterns

Formalizing these MCSK Patterns within a first-order logic framework and subsequently converting them into semantic rules using standard Vocabulary, we employ four standard vocabularies: BFO[3], IOF[4], RO[5], and MSDL[6].

4.1 Process Requirement

Capturing the relationship between material product and manufacturing process requirements.

$$\text{IOF:Materialproduct}(p) \rightarrow \exists x (\text{IOF:Manufacturingprocess}(x) \land \text{IOF:isOutputOf }(p,x)) \tag{1}$$

---

[3] https://obofoundry.org/ontology/bfo.html.
[4] https://github.com/iofoundry/Core.
[5] https://github.com/oborel/obo-relations.
[6] https://labs.engineering.asu.edu/semantics/ontology-download/msdl-ontology/.

Equation 1 asserts that if p is identified as a type of material product, then there must exist at least one manufacturing process x such that p is the output of x. The property IOF:isOutputOf explicitly connects p and x, indicating that every instance of IOF:Materialproduct is necessarily produced by some IOF:Manufacturingprocess. This relationship captures the dependency of the material product on the manufacturing process.

4.2 Material Requirement

Every product is dependent on at least one material. We suppose P to be the set of all Materialproduct, where $\forall p \in P$ we have:

$$\text{IOF:Materialproduct}(p) \rightarrow \exists t (\text{IOF:Materialcomponent}(t) \land \text{RO:memberPartOf}(t, p)) \tag{2}$$

Equation 2 asserts that for any entity p identified as an IOF:Materialproduct, there exists at least one IOF:Materialcomponent t, such that t is a constituent part of p, as defined by the relation RO:memberPartOf.

This relationship uses RO:memberPartOf to express a specific kind of material constitution, where the material component t is not merely spatially adjacent to the material product p, but is an integral part of its structure.

In mereological terms, particularly within the framework of the BFO, the relation RO:memberPartOf is understood as a specific type of part-hood that reflects the dependency of the material product p on its material components t. For instance, if p represents a table (the material product) and t represents wood (the material component), then the RO:memberPartOf relation implies that the table is constituted by the wood, underscoring that the wood is materially integrated into the table, not just co-located.

4.3 Machine Requirement

$$\text{IOF:Manufacturingprocess}(x) \rightarrow \exists e (\text{MSDL:Productionequipment}(e) \land \text{BFO:participatesInAtSomeTime}(e,x)) \tag{3}$$

Equation 3 logical formulation stipulates that for any given manufacturing process x, there must exist at least one piece of production equipment e such that e is involved in executing the process x at some point in time. This relationship is supported by the domain and range constraints inherent in the ontologies used. Specifically, if e, the production equipment, is classified as an 'independent continuant'—excluding 'spatial regions'—it is eligible to participate in a manufacturing process x as per the constraints set by the BFO. The property BFO:participatesInAtSomeTime ensures that this participation is contextualized temporally, indicating that e actively engages in x during a specific time frame. This explanation aligns with the domain-specific constraints and the role qualifications defined within the BFO.

4.4 Capability Requirement

To capture the relationship between a process and the capability required for that process, we define $\mathcal{X}$ as the set of all manufacturing processes where $\forall x \in \mathcal{X}$ we have:

$$\text{IOF:Manufacturingprocess}(x) \rightarrow \exists e, c \big(\text{IOF:ProductionEquipment}(e)$$
$$\wedge \text{MSDL:Capability}(c) \wedge \text{BFO:participatesInAtSomeTime}(e, x) \quad (4)$$
$$\wedge \text{BFO:inheres}(c, e) \wedge \text{BFO:realizes}(x, c)\big)$$

Equation 4 asserts that for each manufacturing process x, there must exist corresponding production equipment e and a capability c. The equipment e is involved in the process x at some point in time, as described by BFO:participatesInAtSomeTime(e, x). Moreover, the capability c, which is essential for the process, inheres in the equipment e as denoted by BFO:inheres(c, e), and is realized through the unfolding of the process x as indicated by BFO:realizes(x, c). This relationship highlights that the realization of a capability within a process is dependent not only on the process itself but also on the equipment through which this capability is expressed and utilized.

4.5 Location Requirement

To express the relationship between a manufacturing process and its required location or environment, we define the following:

$$\text{IOF:Manufacturingprocess}(x) \rightarrow \exists s (\text{BFO:Site}(s) \wedge \text{BFO:occursIn}(x, s)) \quad (5)$$

In Eq. 5 express that if x is an instance of IOF:Manufacturingprocess, there must exist a site s classified under BFO:Site, such that the process x occurs within site s. This relation, defined by BFO:occursIn, establishes that every manufacturing process is inherently linked to a specific site or environment that accommodates or facilitates its operations.

4.6 Process Precedence

Given that there's a manufacturing Process $x \in \mathcal{X}$, then there must exist another manufacturing Process $y \in \mathcal{X}$ that occurs before x.

$$\text{IOF:ManufacturingProcess}(x) \rightarrow \exists y (\text{IOF:ManufacturingProcess}(y)$$
$$\wedge \text{BFO:precedes}(y, x)) \quad (6)$$

In Eq. 6 for any manufacturing process x, there exists another manufacturing process y that precedes x. The relation BFO:precedes(y, x) formalizes this temporal order, specifying that process y must occur before process x can commence. This establishes a precedence relationship where "precedence" indicates the sequence in which manufacturing processes should unfold. In this context, it ensures that every process x is contingent upon the completion of some preceding process y, thus structuring the temporal dynamics within the manufacturing operations.

$$\text{IOF:ManufacturingProcess}(x) \rightarrow (\neg \exists y (\text{IOF:ManufacturingProcess}(y) \\ \wedge \text{BFO:precedes}(y, x))) \vee \exists y (\text{IOF:ManufacturingProcess}(y) \\ \wedge \text{BFO:precedes}(y, x)) \quad (7)$$

Equation 7 logical expression accommodates the possibility that a manufacturing process x may not have a preceding process.

It states that for each IOF:ManufacturingProcess x, it is either the case that no other manufacturing process y precedes x (expressed as $\neg \exists y (\text{IOF:ManufacturingProcess}(y) \wedge \text{BFO:precedes}(y, x)))$, or there exists at least one manufacturing process y that does precede x (expressed as $\exists y (\text{IOF:ManufacturingProcess}(y) \wedge \text{BFO:precedes}(y, x)))$. This formulation respects the fact that some processes might be initial steps in a production chain, thus having no predecessors, while others are part of a sequential chain of processes.

4.7 Object Precedence

To model the notion that certain material components are prerequisites for a given manufacturing process, we present the following equation:

$$\text{IOF:Manufacturingprocess}(x) \rightarrow \exists y (\text{IOF:MaterialComponent}(y) \\ \wedge \text{IOF:isInputOf}(y, x)) \quad (8)$$

Equation 8 formalizes the requirement that for any IOF:Manufacturingprocess x, there must exist a material component y that acts as an essential input for the process. The relation IOF:isInputOf(y, x) specifies that y, defined as an IOF:MaterialComponent, is necessary for the execution of process x. This is the inverse of the relation "IOF:hasInput", where y being an input of x implies that y is a requisite component for the occurrence of process x. This formalization ensures that every manufacturing process is associated with specific required material components, emphasizing the dependency of processes on these components for their initiation and execution.

4.8 Existence Precedence

Given two events or entities x and y, if x precedes y in existence, then the existence or occurrence of y is contingent upon the prior existence or occurrence of x. This notion of precedence captures the temporal and dependency relationships that are crucial in manufacturing and other processes where order and timing are essential for the system's functionality.

$$\text{IOF:ManufacturingProcess}(y) \land \text{MSDL:PlanSpecification}(l) \land \text{IAO:prescribes}(l, y)$$
$$\rightarrow \exists t, t' \left(\text{BFO:existsAt}(l, t) land \text{BFO:existsAt}(y, t') \land \text{BFO:precedes}(t, t')\right) \quad (9)$$

In Eq. 9 pattern establishes that for a manufacturing process y and a plan specification l where l prescribes y, there must exist times t and t' representing when each respectively exists. Specifically, the plan specification l exists at time t and the manufacturing process y exists at time t'. Crucially, the existence of t precedes t', ensuring that the planning phase is completed before the initiation of the manufacturing process. This temporal ordering confirms that planning must occur before any manufacturing action, aligning with standard operational protocols.

4.9 Material Causation

Equation 10 express the notion that the specific quality of raw material directly influences the final product; we use the relation 'RO:CausallyInfluencedBy', which specifies a causal influence between material entities:

$$\text{IOF:RawMaterial}(r) \rightarrow \exists q, p(\text{BFO:Quality}(q) \land \text{BFO:qualityOf}(q, r)$$
$$\land \text{IOF:MaterialProduct}(p) \land \text{RO:CausallyInfluencedBy}(q, p)) \quad (10)$$

This pattern establishes that if a raw material r possesses a specific quality q, then this quality causally influences the final product p.

The relation RO:CausallyInfluencedBy(q, p) is used to indicate that the quality q of the raw material r has a direct causal impact on the characteristics of the final product p. Both the domain and range of RO:CausallyInfluencedBy are confined to material entities, underscoring that the influence is strictly material in nature. This formulation highlights the significance of material qualities in shaping the attributes of manufactured products, thereby emphasizing the material dependency of the product's properties on the raw material's characteristics.

4.10 Material Distinctiveness

Equation 11 express that two raw materials a and b are distinct if they have different qualities.

$$(a \neq b) \implies \exists q(\text{BFO:quality}(q) \land ((\text{RO:qualityOf}(q, a)$$
$$\land \neg \text{RO:qualityOf}(q, b)) \lor (\neg \text{RO:qualityOf}(q, a) \land \text{RO:qualityOf}(q, b)))) \quad (11)$$

This pattern asserts that for any two raw materials a and b, if they are considered distinct, then there must exist at least one quality q that helps differentiate them. This quality q, as identified under the BFO:quality, is possessed uniquely by one material and not by the other. Specifically, the logical expression states that if q is a quality of a and not of b, or it is a quality of b and not of a, then a and b are distinct. This formulation captures the concept that materials often possess multiple qualities, and if they have at least one differing quality, they are distinct. The expression supports the principle that distinctiveness in materials is fundamentally linked to their qualitative differences.

4.11 Manufacturing Process Parthood

In manufacturing, one process x can be part of another process y, and this part-whole relationship is represented by the relation hasOccurrentPart. Simultaneously, in Eq. 13, if x is part of y, they share the same temporal parts, represented by the relation hasTemporalPart.

$$\forall x, y (\text{IOF:ManufacturingProcess}(x) \land \text{BFO:hasOccurrentPart}(x, y) \rightarrow \text{IOF:ManufacturingProcess}(y)) \quad (12)$$

In Eq.12, logical pattern dictates that within the domain of industrial manufacturing, if a process x has an occurrent part denoted by y, using the relation BFO:hasOccurrentPart, then y must also be recognized as a IOF:ManufacturingProcess. This ensures consistency in the classification and understanding of manufacturing activities, emphasizing that sub-processes inherent within a larger process share operational and temporal characteristics. This relation asserts the importance of viewing manufacturing processes as potentially composite, where parts of processes are themselves complete processes that fit within a larger operational framework.

$$\forall x, y (\text{IOF:ManufacturingProcess}(x) \land \text{BFO:hasTemporalPart}(x, y) \rightarrow \text{IOF:ManufacturingProcess}(y)) \quad (13)$$

In Eq.13, if a manufacturing process x, defined within the IOF:ManufacturingProcess, possesses a temporal part y as defined by the relation BFO:hasTemporalPart, then y must also be classified as an IOF:ManufacturingProcess. This principle is founded on the notion that any distinct time segment or interval of a primary manufacturing process retains the characteristics of a manufacturing process. It highlights that sub-processes or phases within a larger manufacturing process, although temporally bounded, embody the essential properties of manufacturing processes, ensuring consistency in classification and understanding across different time scales.

5 Application of MCSK Patterns

This section details how MCSK patterns is utilized to formalize Semantic Rule Specialization, showcasing the practical application of these MCSK patterns.

Definition 1: Rule Template

A Rule Template, RT, is a logical expression containing placeholders representing general classes or relationships. This placeholders are meant to be replaced with specific class/instances from MCSK template-based statements to create a concrete Rule.

Example:

$$process(x) \rightarrow \exists y(process(y) \land \text{precedes}(x,y)) \qquad (14)$$

Definition 2: MCSK. An MACS is an NL statement that provides MCSK with information about a particular case in manufacturing, such as the chasis assembly process. The MACS serves as the source from which specific information related to manufacturing to replace placeholders in the Rule Template.

Example: Sanding Machine needs to be used in the Chair Assembly Process"

Example: Chair Assembly Process needs to be performed for making a Wooden Chair "

Definition 3: Concrete Rule

A Concrete Rule, CR, is derived from a Rule Template by replacing its placeholders with specific classes or instances extracted from a MACS.

Function: SpecializeRule

Formally defined, the function SpecializeRule is:

$$\text{SpecializeRule} : RT \times MACS \rightarrow CR \qquad (15)$$

Process:

1. **Input**: A Rule Template, RT, and manufacturing CSK, $MACS$.
2. **Extraction**: Identify and extract specific classes or instances from the $MACS$.
3. **Substitution**: Systematically replace the placeholders in RT with the extracted classes/instances.
4. **Output**: Return a Concrete Rule, CR.

Rule 1. Given the Rule Template, RT based on Standard vocabulary classes and property relations from BFO and IOF according to Process requirement MACS pattern, Which is RT herein:

$$IOF : MaterialProduct(x) \rightarrow \exists y(IOF : ManufactringProcess(y) \\ \land \text{IOF:isOutputof}(x,y)) \qquad (16)$$

Based on the given RT, the following MACS statement will be transformed into a specialized rule.

MACS: "Chair Assembly Process needs to be performed for making a Wooden Chair"

Applying **SpecializeRule**:

$$WoodenChair(x) \rightarrow \exists y(ChairAssemblyProcess(y) \land \text{IOF:isOutputof }(x,y)) \qquad (17)$$

Rule 2. Given the Rule Template, *RT*:

$$IOF : ManufactringProcess(x) \rightarrow \exists y(MSDL : Productionequipment(y)$$
$$\wedge \text{BFO:participatesInAtSomeTime(y,x))} \quad (18)$$

MACS: "Sanding Machine needs to be used in Chair Assembly Process" Applying `SpecializeRule`:

$$ChairAssemblyProcess(x) \rightarrow \exists y(SandingMachine(y) \\ \wedge \text{BFO:participatesInAtSomeTime(y,x))} \quad (19)$$

The `SpecializeRule` function efficiently tailors broad rule templates to generate specific, concrete rules based on MCSK patterns suited for particular manufacturing contexts. These examples illustrate how the general rule template is adapted to express detailed manufacturing requirements, ensuring precise adherence to manufacturing processes' procedural and tool-based necessities.

6 Future Work and Conclusion

Our research has been dedicated to formalizing manufacturing commonsense knowledge (MCSK) using first-order logic under the guidance of established frameworks like the BFO, IOF, MSDL, and RO. This approach has significantly enhanced AI systems' reasoning and explanation capabilities in these settings. A critical area of our research will focus on how to transform and adapt these MCSK patterns into various schema languages such as SWRL (Semantic Web Rule Language), SPARQL (SPARQL Protocol and RDF Query Language), and Datalog rules. This transformation is essential for enabling MCSK to be utilized effectively for reasoning within neuro-symbolic AI systems.

This endeavor promises to bridge the gap between traditional knowledge representation and the dynamic capabilities of contemporary AI technologies, thereby expanding the potential for AI to operate with enhanced efficiency and accuracy in complex manufacturing environments.

Disclaimer

The views expressed in this publication are the responsibility of the authors and do not necessarily reflect the views of the European Commission nor of the European Innovation Council and SMEs Executive Agency. The European Commission or the European Innovation Council and SMEs Executive Agency are not liable for any consequence stemming from the reuse of this publication.

Acknowledgments. This work is performed within the CHAIKMAT project funded by the French National Research Agency (ANR) under grant agreement "ANR-21-CE10-0004-01".

References

1. Zhang, M.: A survey: the advancements and societal effects of artificial intelligence and machine learning across various domains. In: The 13th International scientific and practical conference "Information and innovative technologies in the development of society"(April 02-05, 2024) Athens, Greece. International Science Group, vol. 321 p. 236 (2024)
2. Koporcic, N., Nietola, M., Nicholson, J.D.: Imp: it's time to get emotional! understanding the role of negative emotions in dynamic decision-making processes. J. Bus. Ind. Mark. **35**, 2151–2163 (2020)
3. Rodosthenous, C., Michael, L.: A hybrid approach to commonsense knowledge acquisition. In: STAIRS 2016, pp. 111–122. IOS Press (2016)
4. Li, C., Chen, Y., Shang, Y.: A review of industrial big data for decision making in intelligent manufacturing. Eng. Sci. Technol. Int. J. **29**, 101021 (2022)
5. Forguson, L.: Common Sense, Routledge London (1989)
6. Paine, T.: Common sense: 1776, Infomotions, Incorporated (1776)
7. Tujillo, J.: The DARPA Machine Common Sense (MCS) Program: A Phenomenological Diagnosis of its Interpretational Challenges, pp. 1–2. Incorporated, Infomotions (2018)
8. Epstein, S.: Demystifying intuition: what it is, what it does, and how it does it. Psychol. Inq. **21**(4), 295–312 (2010)
9. Lenat, D.B.: Cyc: a large-scale investment in knowledge infrastructure. Commun. ACM **38**, 33–38 (1995)
10. Speer, R., Chin, J., Havasi, C.: Conceptnet 5.5: an open multilingual graph of general knowledge. In: Proceedings of the AAAI Conference on Artificial Intelligence, vol. 31 (2017)
11. Fellbaum, C.: Wordnet. In: Theory and Applications of Ontology: Computer Applications, pp. 231–243. Springer (2010)
12. Zang, L.-J., Cao, C., Cao, Y.-N., Wu, Y.-M., Cao, C.-G.: A survey of commonsense knowledge acquisition. J. Comput. Sci. Technol. **28**, 689–719 (2013)
13. Ilievski, F., Oltramari, A., Ma, K., Zhang, B., McGuinness, D.L., Szekely, P.: Dimensions of commonsense knowledge. Knowl.-Based Syst. **229**, 107347 (2021)
14. Sap, M., Rashkin, H., Chen, D., LeBras, R., Choi, Y.: Socialiqa: commonsense reasoning about social interactions, arXiv preprint arXiv:1904.09728 (2019)
15. Zellers, R., Bisk, Y., Farhadi, A., Choi, Y.: From recognition to cognition: visual commonsense reasoning. In: Proceedings of the IEEE/CVF Conference on Computer Vision and Pattern Recognition, pp. 6720–6731 (2019)
16. Bisk, Y., Zellers, R., Gao, J., Choi, Y., et al.: PIQA: reasoning about physical commonsense in natural language. In: Proceedings of the AAAI Conference on Artificial Intelligence, vol. 34, pp. 7432–7439 (2020)
17. Qin, L., Gupta, A., Upadhyay, S., He, L., Choi, Y., Faruqui, M.: Timedial: temporal commonsense reasoning in dialog, arXiv preprint arXiv:2106.04571 (2021)
18. Sarkar, A., Naqvi, M.R., Elmhadhbi, L., Sormaz, D., Archimede, B., Karray, M.H.: CHAIK- MAT 4.0 - commonsense knowledge and hybrid artificial intelligence for trusted flexible manufacturing. In: Kim, KY., Monplaisir, L., Rickli, J. (eds) FAIM 2022. LNME, pp. 455–465. Springer, Cham (2023) https://doi.org/10.1007/978-3-031-17629-6_47
19. Naqvi, M.R., Elmhadhbi, L., Sarkar, A., Archimede, B., Karray, M.H.: Survey on ontology- based explainable AI in manufacturing. J. Intell. Manuf. (2024). https://doi.org/10.1007/s10845-023-02304-z

20. He, L., Jiang, P.: Manufacturing knowledge graph: a connectivism to answer production problems query with knowledge reuse. IEEE Access **7**, 101231–101244 (2019)
21. Hedberg, T.D., Jr., Hartman, N.W., Rosche, P., Fischer, K.: Identified research directions for using manufacturing knowledge earlier in the product life cycle. Int. J. Prod. Res. **55**, 819–827 (2017)
22. Leo Kumar, S.: Knowledge-based expert system in manufacturing planning: state-of-the-art review. Int. J. Prod. Res. **57**, 4766–4790 (2019)
23. Gillespie, R.: Manufacturing Knowledge: A History of the Hawthorne Experiments. Cambridge University Press, Cambridge (1993)

Construction and Canonicalization of Economic Knowledge Graphs with LLMs

Hanieh Khorashadizadeh[1(✉)], Nandana Mihindukulasooriya[2], Nilufar Ranji[1], Morteza Ezzabady[3], Frédéric Ieng[1,2,3], Jinghua Groppe[1], Farah Benamara[3], and Sven Groppe[1]

[1] University of Lübeck, Lübeck, Germany
{hanieh.khorashadizadeh,jinghua.groppe,sven.groppe}@uni-luebeck.de,
nilufar.ranji@student.uni-luebeck.de
[2] IBM Research, New York, USA
nandana@ibm.com
[3] Université de Toulouse, Toulouse, France
{morteza.ezzabady,farah.benamara}@irit.fr

Abstract. Ontology-based knowledge graphs, such as YAGO and Wikidata, rely on pre-defined schemas to organize and connect information. While effective, these systems are inherently domain-specific, requiring tailored ontologies that are costly, time-consuming, and demand expert knowledge to develop. To address these limitations, Open Information Extraction (OpenIE) offers a complementary approach by extracting structured information directly from unstructured text without needing a predefined schema. However, OpenIE results in a vast number of relations, often leading to redundancy and inconsistencies. To overcome this, we propose a novel approach that leverages Large Language Models (LLMs) for constructing a knowledge graph and for canonicalizing relations within it. Our method includes generating question-answer pairs from text, extracting triples from these pairs, and applying a two-step canonicalization process to ensure consistency and reduce redundancy. This paper presents our approach in detail, exploring related work, the construction of the knowledge graph, the canonicalization process, and the evaluation of our methods.

Keywords: Knowledge Graph · Large Language Model · Canonicalization · Open Information Extraction

1 Introduction

Ontology-based knowledge graphs (KGs) are those with a pre-defined ontology. These knowledge graphs rely on a fixed schema or ontology that defines the entities, relationships, and rules for organizing and connecting information. Examples of them are YAGO [1] and Wikidata [2]. There are some limitations associated with ontology-based KGs. One is that they are domain-specific and must have a tailor-made ontology for that domain. Also, developing an ontology for a new domain is costly, time-consuming, and requires expert knowledge. Therefore, Open Information Extraction (OpenIE) is needed as a complement to ontology-based knowledge graphs to cover their limitations.

Open Information Extraction (OpenIE) [3] is a method for extracting structured information in the form of (noun phrase, relation phrase, noun phrase) from unstructured text without requiring a pre-defined schema or set of relations. This process leads to a vast knowledge graph with an enormous number of relations that might imply the same. To avoid redundancy, these relations must be clustered together. For example, consider the following two triples: (inflation, impacts, housing prices) and (Lockdown, influences, food industry). The relations 'impacts' and 'influences' have equivalent meanings, and using both in a knowledge graph might lead to incomplete results whenever queries do not consider all synonyms in the set of relations of the knowledge base. So instead of these two relations, we would use a single term like 'influences'. This process is called canonicalization. Another example is the below triples, "price pressures, cause, inflation" and "money creation, bring about, inflation", the two verbs cause and bring about can be clustered together, and we might use the term 'bring about' or 'cause' for both. In this work, we propose an LLM-based approach to construct a KG and an LLM-based approach to canonicalize the relations. Another challenge of OpenIE is complex and long sentences [4]. Documents and reports often contain lengthy and complex sentences with multiple clauses, making it difficult to parse and extract relations accurately. Long or complex sentences can be challenging for direct triple extraction. We effectively break down the complexity into manageable parts by converting them into Question Answer pairs. Each question can target a specific aspect of the sentence, simplifying the extraction process.

Our main contributions are

- question-answer generation from each paragraph,
- extracting triples from each QA pair,
- and our two-step canonicalization method.

The paper is organized as follows: Sect. 2 explores related work, and Sect. 3 investigates the steps in which the KG is constructed. In Sect. 4, we examine the canonicalization steps. Section 5 entails canonicalization evaluation. Finally, we conclude in Sect. 6.

2 Related Work

Producing open knowledge graphs leads to enormous verb phrases. Verb phrases might mean the same thing or have an identical concept. To reduce these redundancies, we must cluster them and use a common term for that cluster [5]. The approach in [5] necessitates that the head and tail entities are pre-clustered. They used the AMIE algorithm [6] as the rule-mining approach to find equivalent verb phrases. The authors in [7] proposed CESI that canonicalizes Open-KBs with the help of side information and learned embeddings. Their clustering method is based on the distance in the clustering space. Canonicalizing using Variational Auto-encoders was introduced in [8]. The authors in [9] proposed a joint OKB canonicalization-linking method to deal with the problem. OKB canonicalization makes noun phrases and verb phrases into their canonicalized form. The other one is OKB linking, which links noun phrases and verb phrases to their equivalent term in an ontology-based knowledge base. The authors in [10] created a dataset for evaluating relation canonicalization, which can serve as a ground truth annotated benchmark. They selected NELL KG [11] that contains canonical relations. They first selected representative relations and extracted the source sentences from NELL, paraphrased the sentences, and extracted the triples from the paraphrased sentences. Recent work has also explored using LLMs in various aspects of KG construction [12]. For instance, LLMs have demonstrated significant advancements in entity and relation extraction, thereby contributing to identifying semantic relationships and constructing comprehensive KGs. Additionally, foundation models such as ChatGPT have been used in in-context learning to generate knowledge graphs from text [13]. This approach utilizes pre-trained models with minimal retraining, reducing computational resources and manual effort required. Another approach demonstrates the use of prompt-based models, like GPT-4, for generating high-quality relation extraction datasets, linking the knowledge graph to Wikidata for enriched, interoperable datasets in domains such as COVID and health [14]. The authors in [15] proposed a three-step LLM-based framework for knowledge graph construction, including Open Information Extraction, Schema Definition, and Schema canonicalization. In their third step, schema canonicalization, they canonicalize the open KG made in the previous step by creating embeddings for schema parts with a sentence transformer. Unlike all the other approaches mentioned above, our approach avoids over-generalization and provides more precise canonical terms.

3 Knowledge Graph Construction

The COVID-19 pandemic brought about significant disruptions across all sectors of the global economy. Artificial Intelligence [16] supported research and modeling for disease progression and economic analysis during this period. To navigate these challenges, governments, organizations, and research institutions produced numerous reports and analyses to better understand and respond to the rapidly changing economic conditions. These reports contain valuable insights into how different industries, markets, and economies were affected and the strategies implemented to mitigate the impact of the pandemic. Recognizing the need to systematically organize and extract knowledge from this vast body of information, we decided to build a Knowledge Graph (KG) from economic reports produced during the COVID-19 pandemic. Figure 1 depicts the pipeline for KG construction. We explain the pipeline details in the following paragraphs:

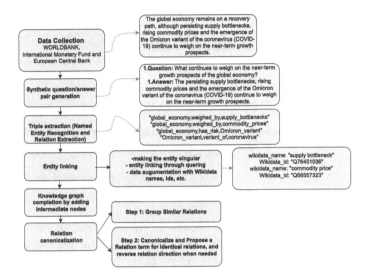

Fig. 1. Knowledge Graph Construction Pipeline. The pink ones are the steps, and the green ones are an example of each step (Color figure online)

3.1 Data Collection

The data for KG construction is from several sources including WORLDBANK[1], International Monetary Fund[2] and European Central Bank[3]. We pre-processed each document, omitting references and short texts. Overall, the data included 62 journals, reports, and economic bulletins.

3.2 Question-Answer Pair Generation

Questions often provide a clear structure and context for extracting information. Answers are expected to contain specific, relevant information, making it easier to extract triples. Hence first generating question-answer pairs can help build better structured knowledge graphs from structured QA datasets. We chunked the preprocessed file into different paragraphs and then made various questions and answers from each paragraph with the help of LLM. We utilized gpt-4 in this paper. Figure 2 shows the question-answer generation prompt. The extraction process can be complex, as you need to identify entities, relationships, and context within unstructured text.

3.3 Triple Extraction from Each QA Pair

Triple extraction is performed on each question-answer (QA) pair, where the Large Language Model (LLM) is prompted to identify and extract as many

---

[1] https://data.worldbank.org (visited on 31.8.2024).
[2] https://www.bookstore.imf.org (visited on 31.8.2024).
[3] https://www.oecd.org/coronavirus/en/policy-responses (visited on 31.8.2024).

> You are a helpful assistant that creates questions with their related answers from the following paragraph. Please make as many questions and answers as you can. Each question should only contain one question word. Please number questions and answers. Also state question and answer terms, for example: 1:Question: , 1:Answers:

Fig. 2. Question-Answer Generation Prompt

triples as possible from the provided data. Each QA pair is analyzed to identify subject-predicate-object structures that can be represented as triples within the knowledge graph.

3.4 Entity Linking

Querying Wikidata for each entity allows us to verify the accuracy and consistency of the data. By cross-referencing with Wikidata, we can ensure that the entities in our knowledge graph are correctly identified and aligned with authoritative sources.

Entity Singularization and Wikidata Query: Making entities singular and querying Wikidata again is a strategy often employed to improve the accuracy of entity linking and matching. Many entities, especially nouns, are typically stored in their singular form in databases like Wikidata. For instance, "Supply bottlenecks" might be stored as "Supply bottleneck" in Wikidata. By converting plural entities to their singular form before querying, you increase the chances of finding an exact match.

Add Wikidata ID and Wikidata Name as Entities Properties: The Wikidata ID and the Wikidata name are added as properties of each entity. This inclusion ensures that each entity is uniquely identifiable and linked to its corresponding entry in Wikidata, enhancing the reliability of the knowledge graph.

3.5 Knowledge Graph Completion by Adding Intermediate Nodes

Adding intermediate nodes in a knowledge graph is often necessary for several reasons, such as when the relationship between two entities cannot be adequately captured by a direct link. Intermediate nodes allow for the representation of more complex relationships, where additional context or steps are needed to fully understand the connection between entities. For instance, if we have entities "Company A" and "Product B," and the relationship is that "Company A produces Product B in Country C," an intermediate node representing "Production in Country C" can clarify the nature of this relationship. [17] addressed the challenge of numerical relation extraction, where one entity in the relation is numeric. However, they highlight a limitation in their approach: it does not account for the time scope of the relation. This is crucial because numerical relations, such as population or GDP, are typically only valid for a specific period. In our KG, most relations are time-dependent as they deal with economic data. To address this challenge, we added intermediate nodes. Figure 3 depicts an example of blank node addition.

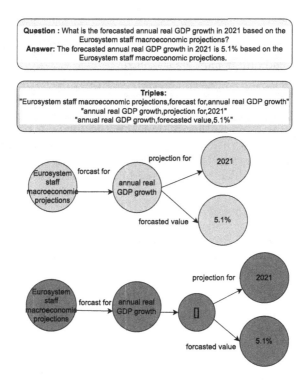

Fig. 3. Adding Blank Nodes. A blank node is added as an empty bracket. If triples involve events or actions that take place over time or under specific circumstances, adding an intermediate node can help represent time, location, or other contextual information.

3.6 Canonicalization of Knowledge Graph

In the knowledge graph, the presence of numerous relations can lead to redundancy and inconsistency if not properly managed. To address this, it is crucial to canonicalize the relations. Canonicalization involves the process of standardizing relations by grouping similar or synonymous ones together and proposing a unified relational term that accurately reflects the underlying semantics.

This process not only enhances the coherence and usability of the knowledge graph but also improves query efficiency and data integration across different domains. However, a key challenge in canonicalization is avoiding overgeneralization, which can result in the loss of important distinctions between relations. For example, while relations such as 'corresponds to,' 'relates to,' and 'associated with' share similarities, they may carry subtle differences that are contextually significant. Therefore, these relations could be consolidated into a single standardized term, but careful consideration must be given to preserving the specific nuances where they are critical. We also have to consider not to inverse the meaning of the triple while canonicalizing. For example, if we have the relations 'impacted by' and 'has impact on', and propose the canonicalized phrase 'impact on', then the triple of 'impacted by' needs a reverse. It means that the subject and object must be swapped.

In our proposed approach, canonicalization is done in two prompting steps (see Fig. 4): In the first step of canonicalization, all the relations are given to the LLM, asking to group relations with the same semantics. In the second step, each single group is given to the LLM, asking to propose a relation for those with the same semantics. The LLM also outputs if a reversal is needed or not. For instance, 'affected by risk' and 'are risks to' relations both need a reversal of the subject and the object when canonicalized to 'risk of'. Figure 5 shows a sample of canonicalization which is the output of the second step.

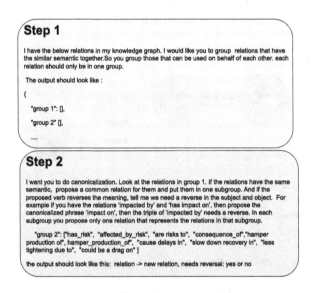

Fig. 4. Canonicalization Steps

4 Evaluating the Canonicalization

Twelve paragraphs of one source document (ECB report on December 2021) are selected for canonicalization evaluation. Overall, 149 triples are made from them. The original triple and the canonicalized triple are given to the LLM to judge whether the meaning is identical. Two manual annotators do the evaluation[4].

---
[4] The evaluation details for the other approaches and our approach are available at GitHub at this link: https://github.com/haniehkh18/Canonicalization-.

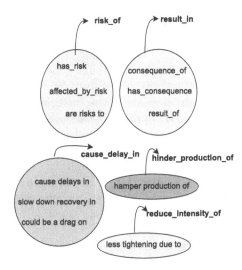

Fig. 5. Canonicalization Example- Second Step

We apply the evaluation criteria used in [7]. Hence, we focus on micro, macro, and pairwise precisions (see Table 1). *Micro precision* is calculated as the number of correct predictions (i.e., relations with the same semantic) divided by the total number of relations. *Macro precision* evaluates clusters individually, treating a cluster as correct if all phrases within it belong to the same meaning. It's a more stringent measure because it requires purity within the cluster. A cluster is considered "pure" if all the phrases within it share the same meaning. It is calculated as the proportion of clusters that are pure (i.e., all phrases belong to the same meaning) out of the total number of clusters. *Pairwise precision* is calculated by dividing the ratio of correct pairs to the total pairs. Comparisons are made with two other approaches, one of which is CESI [7], widely regarded as a pioneering clustering-based method for KG canonicalization. The other one is EDC [15], a recent canonicalization method based on LLM.

The result shows that our proposed approach outperforms previous approaches. What we noticed is that on CESI [7] over-generalization happens. For example, for the relations, "pertains to, intends to reinvest, contributes to, improves to," the CESI approach proposes to substitute the relation contributes to.

Table 1. Evaluation

Method	Precision (Micro)	Precision (Macro)	Precision (Pairwise)
CESI [7]	0.66	0.8	0.36
EDC [15]	0.79	0.76	0.67
Ours	0.99	1	1

5 Summary and Conclusions

This work emphasizes the importance of canonicalization in enhancing the effectiveness of knowledge graphs generated through Open Information Extraction (OpenIE). While OpenIE extracts vast amounts of information, it often produces redundant or synonymous relations, leading to inefficiencies.

We introduced an LLM-based approach to canonicalization that standardizes these relations by grouping equivalent terms under a unified label. This process reduces redundancy, improves semantic coherence, and ensures more accurate and efficient querying within the knowledge graph.

Focusing on canonicalization is crucial for building robust knowledge graphs, and future work could explore more advanced techniques to further refine this approach.

Acknowledgments. This work is funded by the German Research Foundation under the project number 490998901.

References

1. Suchanek, F.M., Kasneci, G., Weikum, G.: YAGO: a core of semantic knowledge. In: Proceedings of the 16th International Conference on World Wide Web, pp. 697–706 (2007)
2. Vrandečić, D., Krötzsch, M.: Wikidata: a free collaborative knowledgebase. Commun. ACM **57**(10), 78–85 (2014)
3. Anthony Fader, Stephen Soderland, and Oren Etzioni. Identifying relations for open information extraction. In *Proceedings of the 2011 conference on empirical methods in natural language processing*, pages 1535–1545, 2011
4. Kamp, S., Fayazi, M., Benameur-El, Z., Yu, S., Dreslinski, R.: Open information extraction: a review of baseline techniques, approaches, and applications. arXiv preprint arXiv:2310.11644 (2023)
5. Galárraga, L., Heitz, G., Murphy, K., Suchanek, F.M.: Canonicalizing open knowledge bases. In: Proceedings of the 23rd ACM International Conference on Conference on Information and Knowledge Management, pp. 1679–1688 (2014)
6. Galárraga, L.A., Teflioudi, C., Hose, K., Suchanek, F.: AMIE: association rule mining under incomplete evidence in ontological knowledge bases. In: Proceedings of the 22nd International Conference on World Wide Web, pp. 413–422 (2013)
7. Vashishth, S., Jain, P., Talukdar, P.: CESI: canonicalizing open knowledge bases using embeddings and side information. In: Proceedings of the 2018 World Wide Web Conference, pp. 1317–1327 (2018)
8. Dash, S., Rossiello, G., Mihindukulasooriya, N., Bagchi, S., Gliozzo, A.: Open knowledge graphs canonicalization using variational autoencoders. In: Proceedings of the 2021 Conference on Empirical Methods in Natural Language Processing (EMNLP 2021), pp. 10379–10394 (2021)
9. Liu, Y., Shen, W., Wang, Y., Wang, J., Yang, Z., Yuan, X.: Joint open knowledge base canonicalization and linking. In: Proceedings of the 2021 International Conference on Management of Data, pp. 2253–2261 (2021)

10. Lomaeva, M., Jain, N.: Relation canonicalization in open knowledge graphs: a quantitative analysis. In: European Semantic Web Conference, pp. 21–25. Springer (2022)
11. Mitchell, T., et al.: Never-ending learning. Commun. ACM **61**(5), 103–115 (2018)
12. Khorashadizadeh, H., et al.: Research trends for the interplay between large language models and knowledge graphs. arXiv preprint arXiv:2406.08223 (2024)
13. Khorashadizadeh, H., Mihindukulasooriya, N., Tiwari, S., Groppe, J., Groppe, S.: Exploring in-context learning capabilities of foundation models for generating knowledge graphs from text. In: Proceedings of the 2nd International Workshop on Knowledge Graph Generation From Text (Text2KG) in Conjunction with the Extended Semantic Web Conference (ESWC 2023), Hersonissos, Greece (2023)
14. Ezzabady, M., Ieng, F., Khorashadizadeh, H., Benamara, F., Groppe, S., Sahri, S.: Towards generating high-quality knowledge graphs by leveraging large language models. In: The 29th Annual International Conference on Natural Language & Information Systems (NLDB 2024), Turin, Italy (2024)
15. Zhang, B., Soh, H.: Extract, define, canonicalize: an LLM-based framework for knowledge graph construction. arXiv preprint arXiv:2404.03868 (2024)
16. Gruenwald, L., Jain, S., Groppe, S. (eds.): Leveraging Artificial Intelligence in Global Epidemics. Elsevier, Amsterdam (2021)
17. Madaan, A., Mittal, A., Ramakrishnan, G., Sarawagi, S., et al.: Numerical relation extraction with minimal supervision. In: Proceedings of the AAAI Conference on Artificial Intelligence, vol. 30 (2016)

Author Index

A
Abbas, Asim 62
Ahola, Annastiina 1
Ambre, Shruti 225
Ameri, Farhad 320
Amini, Reihaneh 17
Amini, Reza 17
Arevalo, Kiara Marnitt Ascencion 225

B
Barba-González, Cristóbal 199
Belarbi, Noureddine 153
Benamara, Farah 334
Bhattacharyya, Pramit 78
Bojārs, Uldis 290
Boman, Patrik 1
Bukhari, Syed Ahmad Chan 62

C
Čerāns, Kārlis 290
Chadalawada, Archit 139
Chadli, Abdelhafid 153
Chauhan, Samarth 78
Chen, Jiaoyan 168
Chy, Tareq Md Rabiul Hossain 183
Cruz, Christophe 32

D
Deepak, Gerard 139
Dibowski, Henrik 47
Doğan, Ege Atacan 259
Dorsch, Rene 109, 225

E
Elmhadhbi, Linda 320
Engel, Felix Caspar 275
Ezzabady, Morteza 334

F
Freund, Michael 93, 109

G
Genest, Olivier 183
Ghanem, Hussam 32
Grasmanis, Mikus 290
Groppe, Jinghua 334
Groppe, Sven 334
Gyrard, Amélie 183

H
Harth, Andreas 93, 109
Hasegawa, Rie 212
Hitzler, Pascal 17
Hooshafza, Sepideh 123
Hyvönen, Eero 1

I
Ichise, Ryutaro 212
Ieng, Frédéric 334

J
Jahan, Nushrat 306
Jiménez-Ruiz, Ernesto 168

K
Karray, Mohamed Hedi 320
Khalid, Mutahira 62
Khorashadizadeh, Hanieh 334
Krdzavac, Nenad 275
Kumari, Sushama 62
Kung, Antonio 183

L
Lāce, Lelde 290
Lamboro, Henon 183
Leskinen, Petri 1
Little, Mark 123

M
Mihindukulasooriya, Nandana 334
Mutharaju, Raghava 78

© The Editor(s) (if applicable) and The Author(s), under exclusive license to Springer Nature Switzerland AG 2025
S. Tiwari et al. (Eds.): KGSWC 2024, LNCS 15459, pp. 345–346, 2025.
https://doi.org/10.1007/978-3-031-81221-7

N

Nachabe, Lina 306
Naqvi, Muhammad Raza 320
Navas-Delgado, Ismael 199
Norouzi, Sanaz Saki 17
Nuzhat, Faiza 242

O

Ouared, Abdelkader 153
Ovčiņņikova, Jūlija 290

P

Patel-Schneider, Peter F. 259

R

Rabrait, Cécile 183
Rahman, Raihana 62
Randles, Alex 123
Ranji, Nilufar 334
Rantala, Heikki 1
Rodríguez-Revello, Jorge 199
Romāne-Ritmane, Aiga 290
Rybinski, Maciej 199

S

Sarkar, Arkopaul 320
Schmid, Sebastian 93, 109
Sebilleau, Dune 183
Seddik, Kebbal 153
Shivashankar, Kanchan 242
Sproģis, Artūrs 290
Steinmetz, Nadine 242
Stephens, Gaye 123

T

Teymurova, Sevinj 168
Tuncay, Erhun Giray 275

W

Wajahat, Iram 62
Wehr, Thomas 109
Weyde, Tillman 168

Y

Yaman, Beyza 123

Printed in the United States
by Baker & Taylor Publisher Services